Hybridomas and Cellular Immortality

Hybridomas and Cellular Immortality

Edited by

Baldwin H. Tom

University of Texas
Medical School at Houston
Houston, Texas

and

James P. Allison

University of Texas
System Cancer Center
Smithville, Texas

Plenum Press • New York and London

Library of Congress Cataloging in Publication Data

Main entry under title:

Hybridomas and cellular immortality.

Proceedings of a national symposium held Nov. 5–6, 1981, in Houston, Tex.
Includes bibliographical references and indexes.
1. Hybridoma—Congresses. 2. Antibodies, Monoclonal—Diagnostic use—Congresses. 3. Antibodies, Monoclonal—Therapeutic use—Congresses. I. Tom, Baldwin H. II. Allison, James P. [DNLM: 1. Antibodies, Monoclonal—Immunology—Congresses. 2. Hybridomas—Immunology—Congresses. 3. Lymphocytes—Immunology—Congresses. QH 425 H992 1981]
QR185.8.H93H93 1983 574.2′93 83-13847

ISBN 978-1-4615-9354-6 ISBN 978-1-4615-9352-2 (eBook)
DOI 10.1007/978-1-4615-9352-2

Proceedings of a national symposium on Hybridomas and Cellular Immortality, held November 5–6, 1981, in Houston, Texas

Softcover reprint of the hardcover 1st edition 1983

A Division of Plenum Publishing Corporation
233 Spring Street, New York, N.Y. 10013

ACKNOWLEDGEMENTS

The Organizers and the Office of Continuing Education, on behalf of The University of Texas Health Science Center at Houston, the Medical School, the Dean's Office, the Division of Immunology and Organ Transplantation, Department of Surgery, The Department of Pharmacology (especially Professor R. Juliano) and the Science Park Research Division, Smithville, gratefully acknowledge the following organizations for support:

Co-Sponsors

Ortho Pharmaceutical Corporation
Ortho Diagnostic Systems Incorporated
American Cancer Society, Texas Division, Inc.

Contributors

Abbott Laboratories
American Cyanamid Company
Becton Dickinson
Beckman Instruments Incorporated
Bethesda Research Laboratories, Inc.
Boehringer Mannheim Biochemicals
Bellco Biological Glassware
Bethesda Research Laboratories, Inc.
Cedarlane Laboratories Limited
Centocor
Costar
Coulter Electronics
Elsevier North-Holland, Inc.
Gamma Biologicals, Inc.
Hybritech Incorporated
Jamari Gallery
Kallestad Laboratories, Inc.
Merck Sharp & Dohme
Nikon
Nuclear Medical Labs, Inc.
Olac 1976 Ltd.
Plenum Publishing Corporation
Quadroma, Inc.
W. B. Saunders Company
Syntex Research
Totalab
The Upjohn Company
Zymed

PREFACE

The ability to "immortalize" immunologically-useful cells by hybridization with a unique cancer cell has revolutionized serological studies and has revealed new potential applications in all fields of biological sciences. This volume presents the studies from a highly successful national symposium on Hybridomas and Cellular Immortality held November 1981 in Houston, Texas. The individual chapters exhibit the diversity of topics discussed during the meeting. These include emphasis on the origin of antibody diversity, B and T lymphocyte differentiation, applications of monoclonal antibodies in studies of histocompatibility, tumor, and viral antigens, plus the use of somatic cell hybridizations for studying T cell products. Three papers focus on the emerging methodologies of in vitro primary immunizations for both humoral and cell-mediated immunities, relevant for coupling with hybridoma technology. There is a useful mix of general (methods) and specific (applications) chapters. A unique aspect of the book is the presentation of both recent research findings with concise descriptions of the state of the art methodologies. It is anticipated that this work will be of interest to a wide audience of practioners in biomedical research. Hopefully, the information contained will foster new and imaginative ideas in hybridoma applications.

Baldwin H. Tom, Ph.D.
James P. Allison, Ph.D.

CONTENTS

PART I. INTRODUCTION TO HYBRIDOMAS

PART II. ANTIBODY DIVERSITY AND LYMPHOCYTE DIFFERENTIATION

3 Hematopoietic Tumors: Normal or Abnormal Models of Leukocyte Differentiation?

Noel L. Warner, Lewis L. Lanier, Edwin Walker, Carlton Stewart, Gary Wood, and Charles Balch

4 Dissection of Human Immunoregulatory T Lymphocytes: Implications for Understanding Clinical Disease

Ellis L. Reinherz, Chikao Morimoto, Stefan Meuer and Stuart F. Schlossman

PART IV. MONOCLONAL ANTIBODIES IN DIAGNOSIS AND THERAPY

V. NEW AND RELEVANT METHODOLOGY

PART VI. POSTER ABSTRACTS

PART I

INTRODUCTION TO HYBRIDOMAS

1

SOMATIC CELL HYBRIDS AND HYBRIDOMAS

Baldwin H. Tom

University of Texas Medical School at Houston, 6431
Fannin, MSMB 6.252, Houston, Texas 77030

This minireview seeks to note some of the highlights in somatic cell hybrid research that led to the subsequent development of the unique, monoclonal antibody secreting, hybrid cell, the hybridoma. When Kohler and Milstein (1975) published their historical paper on producing monoclonal antibodies, their study represented the birth of a new technology, resulting from the marriage between cell biology and immunology.

1. SOMATIC CELL HYBRIDS

The observed presence of multinucleated cells were first made by pathologists on tissue sections (Table I), including tissue lesions of viral etiology. Interest in multinucleated cells derived from suggestions by early cell biologists that cell division began through the multiplication of nuclei within a single cell, followed by segmentation of the giant cell into daughter cells. As tissue culture techniques developed, investigators found that similar multinucleated cells were present in the in vitro cultures (Lewis, 1927). Subsequently, efforts to deliberately induce these cells were initiated. Since viruses were suspected to be associated with multinucleate cell production, both DNA and RNA viruses were tested for their ability to fuse cells (Ringertz and Savage, 1976). These fusions resulted in the generation of multinucleated cells in culture, but the hybrid cells were intermingled with parental cells. The report by Barski and colleagues (1960) demonstrating the feasibility for isolating multinucleated cells marked the beginning of a decade of intense somatic cell hybrid research. Barski et al (1960) reported that intraspecies mouse hybrid cells could be produced by culturing the cells together and allowing for fusion to occur through a random spontaneous process. Obviously this approach was very tedious and indeed required up to three months to obtain isolated hybrid cells. However, in 1962, Okada and colleagues (1962) utilized a RNA containing Sendai (previously named hemagglutinating virus, Japan, HVJ) virus to induce fusion between two cells. This turned out to be a very efficient technique that was modified in 1965 by use of ultraviolet radiation inactivated

TABLE I

Multinucleated cells, somatic cell hybrids, hybridomas.

I.	1838-1858:	Multinucleated cells observed in vertebrate tissue and lesions (in Harris, 1970; in Ringertz and Savage, 1976)
	1873-1931:	Multinucleated cells found in lesions of viral etiology (in Harris, 1970; in Ringertz and Savage, 1976)
II.	1912-1927:	Multinucleated cells in *in vitro* cultures (in Harris, 1970; in Ringertz and Savage, 1976)
III.	1954-1958:	*In Vitro* induction of hybrids with viruses (Enders and Peebles, 1954; Okada, 1958)
IV.	1960 :	Intraspecies mouse hybrids (Barski, Sorieul, and Cornefert, 1960)
	1962 :	Sendai virus fusion (Okada and Tadokoro, 1962)
	1964 :	HAT selection medium (Littlefield, 1964)
	1965 :	U.V. -inactivated Sendai fusion (Harris and Watkins, 1965; Okada and Murayama, 1965); Interspecies hybrids (mouse-man) (Ephrussi and Weiss, 1965); Viable hybrids
	1967 :	Chromosome mapping (Weiss and Green, 1967)
	1970 :	*In vivo* induced cell hybrids (in Sinkovics, 1981)
	1970 :	Lysolecithin fusion (Lucy, 1970)
	1975 :	Polyethylene glycol (PEG) fusion (Pontecorvo, 1975)
	1976 :	PEG + DMSO fusion (Norwood, Zeigler, and Martin, 1976)
V.	1975 :	HYBRIDOMAS (Kohler and Milstein, 1975)
	1980 :	Human-human hybridomas (Olsson and Kaplan, 1980; Croce, Linnenbach, Hall, Steplewski, and Koprowski, 1980)

Sendai viruses. The latter method for cell fusion was the very method subsequently used by Kohler and Milstein (1975) to produce the first monoclonal antibody secreting hybrid cells.

The difficulty in studying hybrid cells is illustrated in Figs. 1. and 2. When two cells A and B are fused, they not only fuse between each other to produce

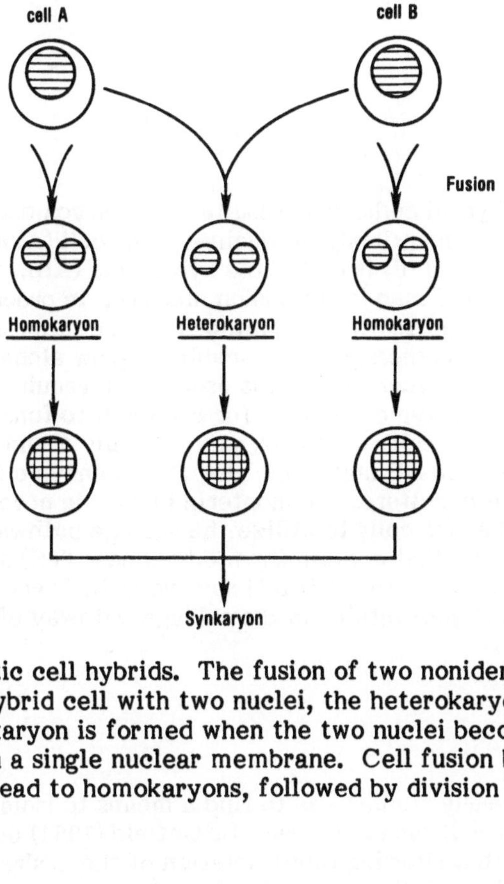

Figure 1. Somatic cell hybrids. The fusion of two nonidentical cells first results in a hybrid cell with two nuclei, the heterokaryon. After cell division, a synkaryon is formed when the two nuclei become encompassed within a single nuclear membrane. Cell fusion between identical cells lead to homokaryons, followed by division into synkaryons.

AB heterokaryons, but also with themselves to form AA or BB homokaryons (Fig. 1). Following a cycle of division, the multinucleated cell becomes a uninucleated cell, the synkaryon. The probability of forming a viable synkaryon is greatest when only two cells, thus two nuclei are fused. Multinucleated cells containing three or more nuclei do not usually survive. Altogether, five species of cells are represented following a hybridization (Fig. 2).

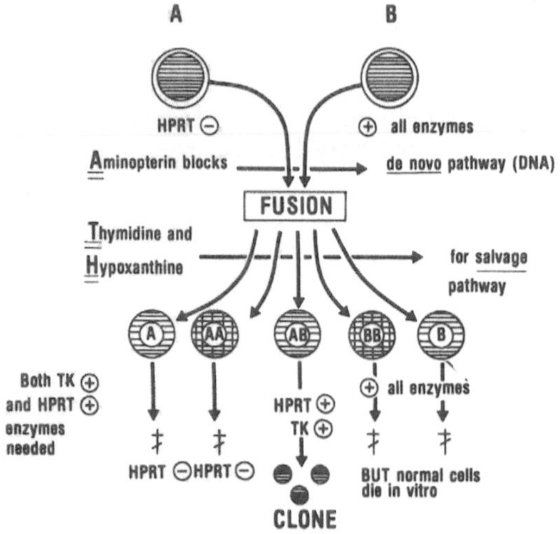

Figure 2. Selection of hybrid cells. The resultant synkaryon derived from the fusion between an antibody-producing tumor cell (A) and a normal immune spleen cell (B) is called the hybridoma (AB). Four other cell types can be found in the fusion mixture. In order for the desired synkaryon (AB) to be isolated from the other cells, both parental fusing partners must be unable to grow alone. If a selective medium (such as HAT) is used which requires a full complement of enzymes in order for each cell to function, then enzyme-deficient cells would not survive (cells A and AA). Normal spleen cells (B and BB), although not enzyme deficient, do not survive in culture. Aminopterin blocks de novo DNA synthesis and shifts cells to utilize the salvage pathway. Thus hybrid cells that contain thymidine kinase (TK) and hypoxanthine phosphoribosyltransferase (HPRT) survive in HAT and utilize thymidine and hypoxanthine in the salvage pathway of DNA synthesis.

The problem in the early studies was to find a means to isolate the AB hybrid cells from the other four cell types. In 1964, Littlefield (1964) developed a selecti medium and a fusion system allowing rapid isolation of the desired A x B hybrid cells. Aminopterin, a dihydrofolic acid analog (Fig. 3) was used to block de novo DNA synthesis. The result of the block was to force the cells to reutilize thymidin and hypoxanthine (guanine) in the salvage pathway for DNA synthesis. However, both enzymes, thymidine kinase (TK) and hypoxanthine (guanine) phosphoribosyltransferase (HPRT), are required in initiating the salvage pathway. Consequently, if a parental or a resultant hybrid cell lacked either enzyme, it would not survive in aminopterin-containing medium. Accordingly, an important consideration in determining the cell fusion partners is in the selection of appropriate enzyme-

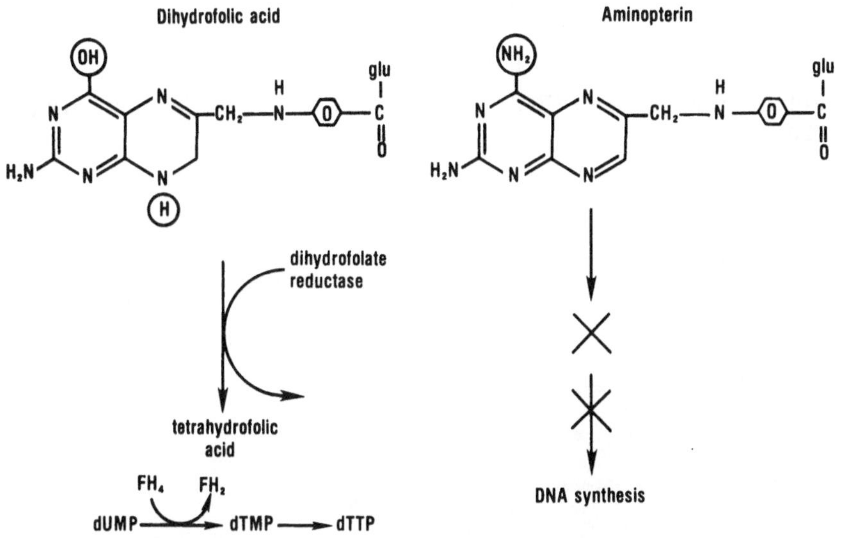

Figure 3. Aminopterin: Analog of dihydrofolic acid.

deficient cell mutants. In this selective medium thymidine and hypoxanthine are added as the substrates for TK and HPRT enzymes. This medium is known as the HAT medium, taken from the first letters of hypoxanthine, aminopterin, and thymidine, respectively. Note that if one of the cell fusion partners is a normal cell, this cell will not survive prolonged in vitro culture and does not constitute a selection problem.

It is important to note that prior to 1975, somatic cell hybrid studies found their greatest importance in the production of interspecies hybrids (Ephrussi and Weiss, 1965), such as between mouse and man, mouse and rat, etc. A key finding in these studies was the random elimination of the chromosomes of one species over the other; in the case of mouse and rat, mouse chromosomes were lost. In the mouse-human system, human chromosomes were deleted. This random segregation (loss or retention) of chromosomes results in an array of hybrid cells each bearing a different chromosome and presumably exhibiting different surface markers This latter situation was thus exploited to provide a most important tool in chromosome mapping studies (Weiss and Green, 1967). The identification of a human product on a hybrid cell bearing a single residual human chromosome among the murine chromosomes, thus allows one to assign the synthetic message for the human marker to the remaining human chromosome. This approach has been employed to assign hundreds of genes to given chromosomes, e.g., chromosome 6 and 17, respectively, bear the genes for synthesis of human and mouse major histocompatibility antigens (in Ringertz and Savage, 1976).

Although Sendai virus was successful in cell fusions, other, chemically defined, fusing agents were sought which would provide a higher hybrid cell yield. Thus

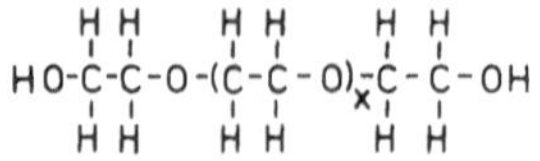

Figure 4. Polyethylene glycol (PEG). PEG, 1020 MW, x=19; PEG, 6022 MW, x= 123.

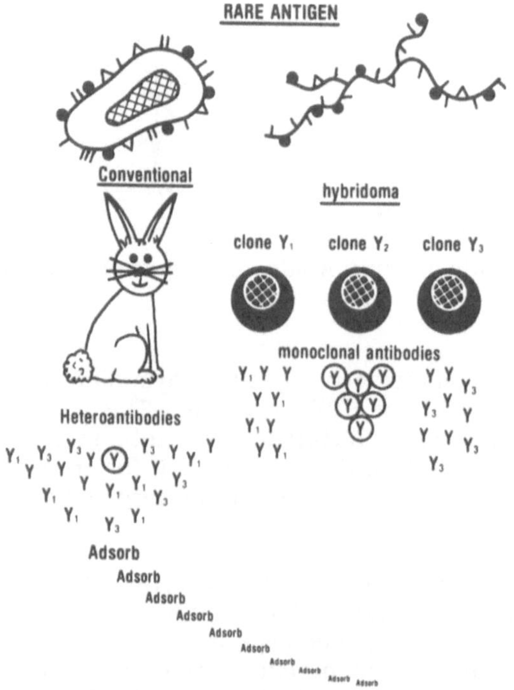

Figure 5. Search for rare antigen. Conventional approach with polyclonal expression of heteroantibodies requires extensive adsorptions. Hybridoma technology eliminates the adsorption requirement. Thus antibody to the rare antigen "Y" is readily obtained.

lysolecithin (lysophosphatidylcholine) was used as an alternative fusogenic agent in place of Sendai virus in the early 1970s (Lucy, 1970). This phospholipid was tested because early studies with Sendai virus revealed that phospholipase A enzymes would block fusion. Although lysophosphatidylcholine was extremely effective in producing cell fusion, the resultant hybrids did not survive well, a result of the toxicity of the agent. At the same time that Kohler and Milstein published their paper, another fusogenic agent, polyethylene glycol (PEG), was reported by

Pontecorvo (1975) to efficiently induce cell fusions. This chemically-defined polymeric compound does not have the toxicity of lysophosphatidylcholine and has proven to be almost as efficient as Sendai virus. Furthermore, PEG (Fig. 4) can be obtained with different molecular weights (usually between 1500-6000 MW) and is presently the agent of choice for fusions for hybridomas.

2. HYBRIDOMAS

The uniqueness of the hybridoma technique developed by Kohler and Milstein (1975) is illustrated in Fig. 5 and resides in the production of monoclonal antibodies specific for the rarest determinant without need for adsorptions, a necessity characteristic of conventional polyclonal antibodies. Kohler and Milstein (1975) and subsequent investigators used rodent myeloma cells as the fusing partners with immune spleen cells for producing the antibody-secreting hybrids. In 1980, Olsson and Kaplan (1980) and Croce et al (1980) reported the first production of human myeloma, hybrid cell-derived monoclonal antibodies.

A series of studies in the late 1960's by Sinkovics and associates (1981) adds an interesting footnote to this review. They demonstrated that a murine lymphoma cell, when coincubated with immune spleen cells in vitro, or coinoculated intraperitoneally into mice led to tetraploid cells capable of producing simultaneously, immunoglobulins and leukemia viruses derived from the lymphoma. The immunoglobulins neutralized the leukemia viruses. They suggested that such fusions might occur in the natural process of lymphoproliferative diseases. Thus, a malignant cell would become immunoresistant and escape host rejection by covering its neoantigens with "autoantibodies". The latter observations may constitute the first (inadvertant) production of a hybridoma. Further readings in the historical development of hybridoma technology and the applications for monoclonal antibodies are found in a series of publications (Herzenberg, et al., 1978; Melcher, et al., 1978; Kennett, et al., 1980; Kohler, 1980; Hammerling, et al., 1981; and McMichael and Fabre, 1982.). Indeed, hybridoma technology extends to virtually all areas of biomedical sciences, with only one's imagination the limiting factor.

3. REFERENCES

Barski, G., and Belehradek, J., Jr. 1977. Cytoplasmic Membranes in somatic cell interaction and hybridization. In "Mammalian cell Membranes" G.A. Jamieson and D.M. Robinson, eds., vol. 5, pp. 284-305. Butterworth, London.

Barski, G., Sorieul, S., and Cornefert, F. 1960. Production dans des cultures in vitro de deux souches cellulaires en association, de cellules de cellules de caractere "hybride". C. R. Hebd. Seances Acad. Sci. 251:1825-1827.

Croce, C.M., Linnenbach, A., Hall, W., Steplewski, Z., and Koprowski, H. 1980. Production of human hybridomas secreting antibodies to measles virus. Nature 288:488-489.

Enders, J.F., and Peebles, T.C. 1954. Propagation in tissue cultures of cytopathogenic agents from patients with measles. Proc. Soc. Exp. Biol. Med. 86:277-286.

Ephrussi, B. 1972. Hybridization of somatic cells. Princeton Univ. Press, New Jersey, pp. 174.

Ephrussi, B., and Weiss, M.C. 1965. Interspecific hybridization of somatic cells. Proc. Natl. Acad. Sci. U.S.A. 53:1040-1042.

Hammerling, G.J., Hammerling, U., and Kearney, J.F., eds. 1981. Monoclonal antibodies and T-cell hybridomas. Elsevier/North-Holland Biomedical Press, Amst 587 pp.

Harris, H. 1970. "Cell Fusion, The Dunham Lectures."Oxford Univ. Press, London and New York.

Harris, H., and Watkins, J.F. 1965. Hybrid cells derived from mouse and man: Artificial heterokaryons of mammalian cells from different species. Nature London 205:640-646.

Herzenberg, L.A., Herzenberg, L.A., Milstein, C. 1978. Cell hybrids of myelomas with antibody forming cells and T-lymphomas with T cells. In Weir, D.M. ed, Handbook of Experimental Immunology, Ch. 25, 3rd. ed., F.A. Davis.

Kennett, R.H., McKearn, T.J., and Bechtol, K.B. eds. 1980. Monoclonal antibodies. Plenum Press, New York. 423 pp.

Kohler, G.1980. Hybridoma techniques. Cold Spring Harbor Laboratory, New York, 67 pp.

Kohler, G. and Milstein, C. 1975. Continuous cultures of fused cells secreting antibody of predefined specificity. Nature London 256:495-497.

Lewis, W.H. 1927. The formation of giant cells in tissue culture and their similarity to those in tuberculous lesions. Am. Rev. Tuberc. 15, 616628.

Littlefield, J.W. 1964. Selection of hybrids from matings of fibroblasts in vitro and their presumed recombinants. Science 145:709-710.

Lucy, J.A. 1970. The fusion of biological membranes. Nature London 227:815-817.

Melcher, F., Potter, M., Warner, N.L. ed. 1978. Lymphocyte hybridomas. "Current topics in microbiology and immunology". Springer- Verlag.

McMichael, A.J. and Fabre, J.W. eds. 1982. Monoclonal antibodies in clinical medic Academic Press, New York., 664 pp.

Norwood, T.H., Zeigler, C.J., Martin, G.M. 1976. Dimethyl Sulfoxide enhances polyethylene glycol-mediated somatic cell fusion. Som. Cell Genet. 2:263.

Okada, Y. 1958. The fusion of Ehrlich's tumor cells caused by HVJ virus in vitro. Biken's J 1:103-110.

Okada, Y., and Murayama, F. 1965. Multinucleated giant cell formation by fusion between cells of two different strains. Exp. Cell Res. 40:154-158.

Okada, Y., and Tadokoro, J. 1962. Analysis of giant polynuclear cell formation caused by HVJ virus from Ehrlich's tumor cells. II. Quantitative analysis of giant polynuclear cell formation. Exp. Cell Res. 26:108-118.

Olsson, L., and Kaplan, H.S. 1980. Human-human hybridomas producing monoclonal antibodies of predefined antigenic specificity. Proc. Natl. Acad. Sci. U.S.A. 77(9):5429-5431.

Pontecorvo, G. 1975. Production of mammalian somatic cell hybrids by means of polyethylene glycol treatment Som. Cell Genet. 1:397.

Ringertz, N.R. and Savage, R.E. 1976. Cell Hybrids. Academic Press, New York, 366 pp.

Sinkovics, J.G. 1981. Early history of specific antibody-producing lymphocyte hybridomas. Cancer Res., 41:1246- 1247.

Weiss, M.C., and Green, H. 1967. Human-mouse hybrid cell lines containing partial complements of human chromosomes and functioning human genes. Proc. Natl. Acad. Sci. U.S.A. 58:1104-1111.

PART II

ANTIBODY DIVERSITY AND LYMPHOCYTE DIFFERENTIATION

2

THE NATURE AND ORIGIN OF ANTIBODY DIVERSITY

Paul D. Gottlieb

Department of Microbiology, University of Texas
at Austin, Austin, Texas 78712

1. INTRODUCTION

This article is intended as a brief summary of what is known about the nature of the amino acid sequence diversity of antibody V regions, and the mechanisms by which that diversity is encoded in the germ line of a species and expressed by antibody-producing cells. It is this diversity that is responsible for the very existence of an antibody system which can specifically recognize the wide variety of antigens that exist. A more detailed review with more extensive literature citations has been published previously (Gottlieb, 1980).

2. V REGION SEQUENCE DIVERSITY AT THE PROTEIN LEVEL

2.1. Nature of the sequence diversity

That antibody (Ab)[1] chains have variable (V) and constant (C) regions was suspected in the early 1960's from peptide maps of human myeloma light (L) chains. These showed that individual L chains had a number of peptides in common, but each had additional peptides which

[1]Abbreviations: Ab, antibody; V, variable region; C, constant region; L, light chain; H, heavy chain; κ, kappa chain; λ, lamba chain; V_L, V_H, V_κ, V_λ, variable region of the light chain heavy chain, etc.; C_L, C_κ, C_λ, C_μ, C_δ, C_γ, C_ε, C_α, constant region of the light chain kappa chain, etc.; Ig, immunoglobulin; HV, hypervariable region; CDR, complementarity determining residue; J, joining segment; D, diversity segment; Ag, antigen.

were unique. Amino acid sequence analysis of myeloma L chains demonstrated that the unique variations resided in the amino-terminal half of the L chains, whereas the carboxy-terminal half was essentially identical from individual to individual (Hilschmann and Craig, 1965). Heavy (H) chains also proved to have an amino-terminal V region and a carboxy-terminal C region, the structure of the latter being characteristic of each class of H chain. Since that time, amino acid sequence analysis —largely of human mouse myeloma proteins, mouse hybridoma proteins, and homogeneous antibodies of hyperimmune rabbits (see Kabat et al., 1979)— has been performed to characterize the extent and nature of V region sequence diversity.

Comparison of amino acid sequences of human (or mouse) V_{κ} regions shows that they can be arranged into groups, where V regions within groups show much greater similarity with each other than with those in different groups (Weigert and Riblet, 1976). The greatest differences within groups are concentrated in segments called hypervariable (HV) regions (Wu and Kabat, 1970), where the differences among proteins are extraordinary. The HV regions are separated by the so-called frame-work regions which differ among groups and which actually define the groups. The HV regions are so diverse that it is generally impossible to recognize the group to which a protein belongs by its HV regions. In the mouse, subgroups could be defined within groups, based largely upon consistent differences in framework regions (Weigert et al., 1978). Examination of V_H regions of man and mouse demonstrated that these also had framework and HV regions and could be arranged into groups.

Analysis of the three dimensional structures of several immunoglobulin (Ig) molecules by X-ray crystallography demonstrated that the walls of the antigen binding site were largely composed of amino acid residues which lie in the HV portions of V_H and V_L (Amzel, et al., 1974). Hence the HV regions have also been called complementarity-determining residues (CDR). The first and third HV regions of V_L and the first, second, and fourth HV regions of V_H contribute most of the Ag-binding site. The second V_L HV region is sometimes in the Ag-binding site, but the third V_H HV region is unlikely ever to lie in the site and the function of its high variability, if any, remains obscure.

2.2 Origin of the sequence diversity

It is apparent from the above discussion that there are elements of order in the amino acid sequence diversity of Ig V regions— V and C regions, framework and HV regions, groups, and subgroups. It was thought that if enough V_H and V_L amino acid sequences were determined, the mechanisms by which the sequence diversity arose might be deduced. Theories on the origin of V region sequence diversity fall into two major classes (Edelman and Gally, 1972):

1) Germ line theories generally held that all of the V_L and V_H regions a species could make are encoded in the germ line DNA of that species. Expression of an Ab by a somatic cell simply involved gene activation, transcription, and translation of one germ line L chain and one germ line H chain gene per cell and assembly into a complete Ig molecule.

2) Somatic theories held that the germ line of a species contained a small number of V_L and V_H genes. Additional sequence diversity was thought to be introduced into these genes in somatic cells by spontaneous mutation, by recombination, by gene conversion, by a special enzymatic mechanism which preferentially introduced variability into HV regions, or by a combination of the above.

Participants in the debate over the origin of Ab V region diversity agreed that the existence of C region allotypes made it likely that a single copy of each C region existed in the germ line DNA, and that some mechanism must exist at the DNA level to join the V region to the C region in somatic cells to create an intact gene coding for a complete Ig chain. Since all classes of H chains seemed to share a common set of V_H regions, and since C_H regions allotypes were inherited as a group, it was apparent that $C\mu$, $C\alpha$, the $C\gamma$ subclasses, $C\delta$ and $C\varepsilon$ must be closely linked. Combined data from man, mouse and rabbits indicated that kappa, lambda, and H chain loci were unlinked and must lie on different chromosomes. Also, $V\kappa$, $V\lambda$, and V_H regions were clearly distinguishable on the basis of amino acid sequence, and so it was clear that there must be a set of V regions linked to $C\kappa$, $C\lambda$, and the C_H genes which was for exclusive use by the linked C region locus (loci) in forming complete Ig chains. V region genes might then be joined in somatic cells to the linked C region gene by a deletion or translocation event at the DNA level to form a VC gene which codes for a complete Ig chain. This set of events was inferred and by no means proven, and it still did not say whether all V region sequence diversity was encoded in the germ line DNA, or whether much of the diversity was generated by somatic mechanisms from a relatively small number of germ line genes. It was apparent, however, that multiple germ line genes were required for H chain V regions and for V regions of each L chain type. Groups of $V\kappa$ regions in man and in the mouse, for example, are so dissimilar in framework sequence that there must be at least one germ line gene for each group. In the mouse, it would appear that there must be at least one germ line gene for each subgroup.

3. ANALYSIS OF Ig GENES AT THE DNA LEVEL

3.1. General considerations, and the mouse λ locus

The advent of the technology for cloning, sequencing, and heteroduplex analysis of germ line and somatic cell DNA within the past 5-7 years has provided considerable insight into the origins of Ab sequence diversity. Much V region sequence diversity is encoded in the germ line, but somatic mechanisms appear to contribute greatly to the sequence diversity observed in Ab V regions. The remainder of this article is a brief summary of what has been learned at the nucleotide level concerning mechanisms of Ab gene expression and somatic generation of Ab sequence diversity. The evidence for these mechanisms has been contributed largely by the work of S. Tonegawa, P. Leder, L. Hood and their coworkers.

One of the biggest suprises revealed by analysis of Ig genes at the DNA level was that V regions encoded in the germ line are not present in their entirety. For example, a DNA sequence coding for an entire L chain V region (approximately 107 amino acid residues) must be assembled in somatic cells by joining a V gene segment encoding the hydrophobic leader sequence and approximately 95 residues of V region to a J (or joining) segment which encodes the remaining 13 residues of the V region. This is illustrated in Figure 1 which depicts the simplest antibody gene locus known— that of the mouse λ_1 L chain. In the mouse, λ_1 chains account for less than 5% of Ab L chains, and V_λ sequence diversity is extremely limited as determined from myeloma and hybridoma V_λ amino acid sequences. The existence of λ_2 and λ_3 chains and their corresponding genes (Azuma et al., 1981; H.N. Eisen, personal communication) has complicated the simple view of the λ locus shown in Figure 1, but the present figure is sufficient for illustrative purposes. A single J_λ segment lies some 1200 nucleotides from the C_λ gene in the mouse germ line DNA (Bernard et al., 1978). When a somatic cell becomes committed to synthesizing a λ chain, the V segment is joined to the J segment with deletion of the intervening DNA to create a V_λ J_λ segment coding for a complete protein V_λ region. The entire segment consisting of sequences coding for hydrophobic leader, V_λ, J_λ and C_λ as well as non-coding intervening sequences is then transcribed into RNA. This primary RNA transcript is processed by an as yet uncharacterized enzymatic RNA splicing mechanism (Berget et al., 1977) to remove the non-coding intervening sequences, leaving a mature messenger RNA (mRNA) with contiguous hydrophobic leader, V_λ, J_λ and C_λ coding segments. This mRNA is translated in the cytoplasm, and as the protein is passed through the endoplsamid reticulum, the hydrophobic leader is proteolytically cleaved off, leaving a light chain as we know it to exist in an Ig molecule (Burstein and Schechter, 1977).

The joining of V to J appears to be brought about by specific nucleotide sequences which lie immediately to the right of V and to the left of J (Sakano et al., 1979; Max et al., 1979) as shown in a simplified form in Figure 2 (Gottlieb, 1980). The nucleotide sequence to the right of V is an imperfect palindrome, i.e., reading one strand left to right is nearly the same as reading the complementary sequence from right to left. This same palindromic heptamer is present to the left of J but inverted 180 degrees. and the two palindromes are termed an inverted repeat. If DNA in these regions is allowed to melt, and renaturation occurs involving inverted repeat sequencs within each DNA strand rather than between strands, the V and J coding sequences are brought very close to each other. Special DNA binding proteins and/or enzymes must exist to favor

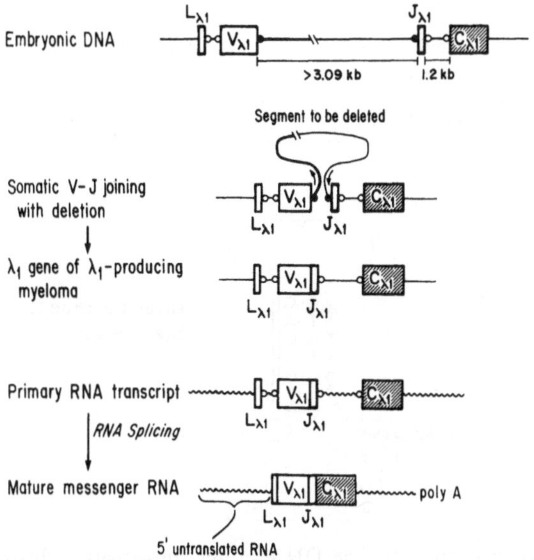

Figure 1: Rearrangement of embryonic DNA and splicing of RNA lead to production of a mature mouse λ_1 L chain messenger RNA. $L_{\lambda 1}$ = hydrophobic leader; —— = DNA; ~~~ = RNA: • = DNA rearrangement site; o = RNA splicing site (From Gottlieb, 1980).

this occurrence. Enzymes could then cleave and rejoin the nucleotide strands to yield contiguous VJ coding segments and to delete the inverted repeat stem structure.

The inverted repeat stem structure is actually more complex than that shown in Figure 2, and it involves homologous intrastrand base pairing in addition to the inverted palindromic heptamers. In the DNA segment separating V_K and J_K, 12 and 23 nucleotides away from the palindromes adjacent to V_K and J_K, respectively, are A + T rich nonamers which are so oriented as to contribute to the inverted repeat stem structure by intrastrand base pairing (Sakano et al., 1979). In fact, all joining events involving V, J and other DNA segments that contribute to V region sequence diversity (see below) appear to involve, at one and of the joining event, a palindromic heptamer and A + T rich nonamer separated by 12 nucleotides, and at the other end of the joining event, a palindromic heptammer and A + T rich nanomer separated by 23 nucleotides (Sakano et al, 1980, 1981). This 12/23 rule appears to hold for the joining of all DNA segments

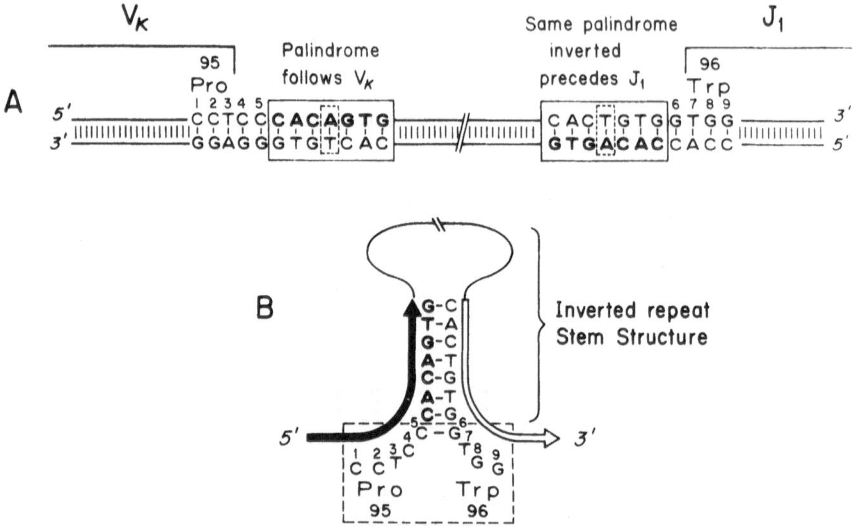

Figure 2. A possible mechanism for DNA rearrangement which permits V_K - J_K joining as well as somatic generation of sequence diversity in CDR3. (a) Double stranded DNA showing the palindrome following V_K and the same palindrome inverted preceding J1. The first complete codon in J1 is numbered 96 since in some instances it becomes residue 96 of a completed V region. Bases numbered 1-9 are discussed below. (b) Only the top DNA strand of part (a) is shown. Palindromic sequences following V_K and preceding J1 have annealed to form an inverted repeat stem structure and to bring bases 1-5 and 6-9 into proximity to allow site-specific recombination. The complementary DNA strand (not shown) probably forms a similar structure (From Gottlieb, 1980).

which contribute to L and H chain V regions, and it is likely that the inverted repeat DNA stem structures which result provide sites for DNA binding proteins and the enzymatic machinery required to effect DNA joining.

Comparison of different V_λ sequences (as well as V_K sequences) at the protein level has shown that the most highly variable residue is exactly at the site of V-J joining predicted by the inverted repeat stem structure mechanism, corresponding to the beginning of the third L chain HV region. Max et al. (1979) and Sakano et al. (1979) have suggested that if the precise site of breaking and rejoining is somewhat flexible, the rejoining event could result in an amino acid at the point of joining which is not present in either the germ line V or J segment. Thus new diversity in a CDR might be introduce by the somatic joining event.

3.2. The mouse κ light chain locus

The diversity of κ light chains in the mouse is far greater than that of λ chains, and the κ locus is much more complex. Amino acid sequence analysis of mouse $V_κ$ regions suggests that there may be approximately 50 $V_κ$ groups (Weigert and Riblet, 1978), each of which must be encoded by at least one germ line $V_κ$ segment (see above). The presence of subgroups within groups suggests that the germ line $V_κ$ locus may contain considerably more than 50 germ lines $V_κ$ segments.

Our present view of the mouse κ L chain locus (Figure 3) is based largely on the extensive library of amino acid sequences of Ig chains (Kabat et al., 1979), and on DNA studies by Max et atl. (1979), (Sakano et al., (1979) and others. Nucleotide studies revealed five J-like DNA segments, each coding for 13 amino acids, approximately 1500 nucleotides upstream from a single $C_κ$ coding segment in germ line DNA. One of

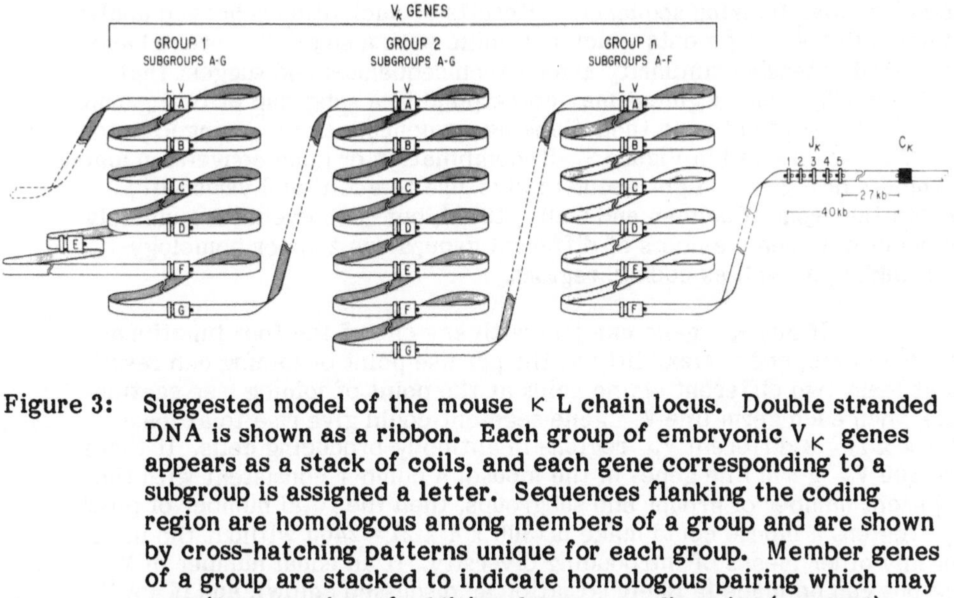

Figure 3: Suggested model of the mouse κ L chain locus. Double stranded DNA is shown as a ribbon. Each group of embryonic $V_κ$ genes appears as a stack of coils, and each gene corresponding to a subgroup is assigned a letter. Sequences flanking the coding region are homologous among members of a group and are shown by cross-hatching patterns unique for each group. Member genes of a group are stacked to indicate homologous pairing which may permit generation of additional sequence diversity (see text). Subgroup E of group 1 is folded out of the stack to indicate that pairing need not only involve adjacent genes. Note that reciprocal recombination between subgroups 1D and 1F would lead to a deletion. • = DNA rearrangement site; o = RNA splicing site; Note that no RNA splicing site follows J3, possibly explaining failure to observe the sequence of J3 in myeloma κ chains (Sakano et al., 1979; Max et al., 1979) (From Gottleib, 1980).

these J_κ -like segments appears to be non-functional since, unlike the other 4 J_κ segments, its sequence has never been observed in mouse κ proteins. Preceding each J_κ segment was a palindrome similar to that preceding J_λ, suggesting that an inverted repeat stem structure might be involved in V_κ -J_κ joining in somatic cells at the DNA level. Similar palindromes, inverted 180°, have been found following germ line V_κ genes, reinforcing this suggestion.

Evidence concerning the number of germ line genes which code for a particular group of mouse V_κ regions was obtained by hybridization of a V_κ probe specific for a myeloma L chain belonging to a particular V_κ group to a restriction enzyme digest of mouse germ line DNA which had been electrophoresed in agarose and transferred to nitrocellulose (a so-called Southern hybridization). Each V_κ group-specific probe hybridized to 6-10 restriction fragments, and probes for different V_κ groups hybridized to a <u>different set</u> of 6-10 fragments (Seidman et al., 1978a,b). Study of the nucleotide sequences flanking each V_κ gene indicated that V_κ segments hybridizing with particular V_κ probe shared similar flanking sequences, whereas V_κ gene segments of different V_κ groups had dissimilar flanking sequences. Results of nucleotide sequence analysis of germ line V_κ segments which hybridize with a single V_κ probe have indicated extensive similarity among such sequences and suggest that each such V_κ gene segment may correspond to a <u>subgroup</u> of that group. Similar V_κ segments and their flanking sequencs may be arranged in tandem, and this is reflected in Figure 3. Recombination or gene conversion among members of a group might occur to give rise to additional variability within the group (Edelman and Gally, 1969), but such events are unlikely to occur between members of different groups due to poor homology of flanking, as well as coding, regions.

If any V_κ gene can join with any one of the four functional J_κ segments, and if flexibility at the precise point of joining can result in at least two different amino acids at the point of joining (see section 3.1), then each germ line V_κ gene segment could give rise to at least (1 x 4 x 2) = 8 different V_κ regions in antibody-producing cells. If there are 300 V_κ germ line genes in the mouse, a number consistent with the apparent number of groups and subgroups, then the total number of possible V_κ regions a mouse could make is (300 x 4 x 2) = 2400 without the need for any other means of introducing diversity. If an equal number of V_H regions can be made (a likely situation as discussed below), and if any V_H can pair with any V_κ, then the number of antibodies a mouse could make would be (2400 x 2400) = 5.96×10^6 without any other means of introducing variablity. A repertoire of binding sites this large would appear to be of an order of magnitude approaching that necessary to specifically recognize the world of antigens.

3.3. <u>Generation of V_H region diversity in the mouse</u>

Amino acid sequence analysis of H chains and nucleotide sequence analysis of DNA segments determining V_H regions have reveaed a situation similar to, though more complex than, the V_κ locus. At least two DNA joining events appear necessary to generate a DNA segment coding for a complete V_H region. Analysis of mouse germ line DNA has revealed four J_H segments approximately 8000 nucleotides upstream from the C_μ gene (Newell et al., 1980; Sakano et al., 1980). However, to generate a gene coding for a complete V_H region, a V_H gene segment must join at the DNA level to a D_H segment, encoded somewhere between the V_H and J_H genes, and the D_H segment must join to one of the (at least) four J_H DNA segments. There is some evidence that a third joining event--joining of two different D_H segments— may further contribute to V_H region sequence diversity (Kurosawa et al., 1981).

The numbers of V_H and D_H segments encoded in germ line DNA are not known. However, it is easy to see that (a) if any V_H can join to any D_H (b) if D_H segments can join to other D_H segments; and (c) if D_H segments can join to any J_H segment; then the total number of possible V_H region sequences could be very large. Furthermore, if additional diversity can be generated by flexibility at the point of the joining event, as suggested for V_κ -J_κ and V_λ -J_λ joining, this number could be increased even further.

3.4. Is additional V region diversity introduced by somatic mutation?

Most of the sequence diversity observed in V_L regions can be accounted for by the germ line V and J DNA segments and the somatic mechanisms discussed above. Although there exists much less data on V_H, D_H, and J_H germ line DNA segments, it is likely that these may account in similar fashion for most of the V_H region sequence diversity observed. However, recent work of Gearhart et al,. (1981) and Crews et al. (1981) has suggested that somatic mutation in DNA coding for framework and HV regions may provide some of the amino acid sequence diversity observed in V_κ and V_H regions of IgG and IgA but not IgM antibodies. This suggestion is based in part on the findings that the V_H regions of phosphorylcholine-binding IgM myelomas which bear the T15 idiotype are identical in sequence to each other and to one of the several germ line genes which hybridize to a cloned cDNA probe specific for the T15 myeloma V_H region. The V_H regions of some T15 idiotype-positive IgG and IgA molecules contain variations in both framework and hypervariable regions which cannot be accounted for by V_H-D_H, D_H-J_H, or D_H-D_H joining, or by recombination or gene conversion involving the several germ line V_H segments identified using the T15 V_H probe. Analyses of V_κ sequences of IgM, IgG and IgA proteins bearing the T15 idiotype also revealed more sequence diversity in L chains of IgG and IgA proteins than in IgM (Gearhart et al., 1981). Increased V region sequence diversity in IgG as compared with IgM antibodies has been noted by Rodwell and Karush (1980).

The mechanism(s) by which this additional diversity is generated is (are) not known. It is possible that the signal which operates the switch of an antibody-producing cell from μ chain (and IgM) production to production of a different H chain (and Ig) class— possibly delivered by T cells-- activates a V gene-specific mutation mechanism that acts on both V_H and V_L region DNA to generate additional V region sequence diversity. Kim et al. (1981) have obtained evidence consistent with extensive somatic mutations confined to the coding and nearby flanking nucleotides of the rearranged V_H genes of the M167 and M603 myelomas— closely related to the germ line T15 V_H gene segment. Such mutations do not appear in the C_α gene of those α -chain producing myelomas, or in flanking regions more distant from the rearranged V_H genes. These workers suggest a somatic hypermutation mechanism which is highly localized in its site of execution, and which is highly restricted in its time of operation during B cell development. In addition, it has been suggested that mutations may accumulate at the basal mutation rate, and selection by antigen for higher affinity antibody, or escape from negative regulation by anti-idiotype may favor proliferation of cells which have accumulated such mutations (Crews et al., 1981). The result would be a shift from a repertoire of V region diversity resulting strictly from joining germ line-encoded V,D, and J gene segments, to a repertoire containing somatic mutants superimposed on the germ line theme. This extra somatic diversity would be imposed only on antibodies derived from antigen-stimulated B cells, allowing fine tuning of the response to a given antigen.

4. CONCLUSION

It should be apparent that the somatic mechanisms described above, operating on a relatively small library of germ line V, D, and J DNA segments, is capable of generating the enormous amino acid sequence diversity seen in antibody V regions. Nucleotide sequence analysis of DNA flanking the V,D, and J coding regions which have been identified have provided important clues concerning the mechanisms by which these germ line segments are brought together in somatic cells to form functional Ig genes. However, the DNA-binding proteins and the enzymatic machinery implicit in the proposed mechanisms have yet to be identified. It appears, however, that the debate over whether the origin of antibody V region sequence diversity is germ-line or somatic has nearly been resolved with the conclusion that extensive diversity results from combinatorial joining of germ line-encoded V,D, and J gene segments. A major question which remains is the extent to which, on top of the extensive diversity provided by combinatorial joining, additional level of somatic mutation, perhaps associated with the H chain switch, results in further fine-tuning of the antibody system. Evidence for this new level of V region diversity has been obtained, but whether its source will prove to be a hypermutation mechanism as proposed (Kim et al., 1981) or a manifestation of as yet unidentified germ line DNA segments remains to be discovered.

REFERENCES

Amzel, L.M., Poljak, R.J., Saul, F., Varga, J.M. and Richards, F.F., 1974. The three dimensional structure of a combining region - ligand complex of immunoglobulin NEW at 3.5 A resolution. Proc. Natl. Acad. Sci. USA 71:1427-1430.

Azuma, T., Steiner, L.A. and Eisen, H.N., 1981. Identification of a third type of λ -light chain in mouse immunoglobulins. Proc. Natl. Acad. Sci. USA 78:569-573.

Berget, S.M., Moore, C. and Sharp, P.A., 1977. Spliced segments at the 5' terminus of adenovirus 2 late mRN. Proc. Natl. Acad. Sci. USA 74:3171-3175.

Bernard, O., Hozumi, N. and Tonegawa, S., 1978. Sequences of mouse immunoglobulin light chains before and after somatic changes. Cell. 15:1133-1144.

Burstein, Y. and Schechter, I., 1977. Amino acid sequence of the NH_2-terminal extra piece segments of the precursors of mouse immunoglobulin λ_1-type and κ -type light chains. Proc. Natl. Acad. Sci. USA 74:716-720.

Crews, S., Griffin, J., Huang, H., Calame, K. and Hood, L. 1981. A single V_H gene segment encodes the immune response to phosphorylcholine: somatic mutation is correlated with the class of the antibody. Cell 25:59-66.

Edelman, G.M. and Gally, J.A., 1969, Antibody structure, diversity and specificity, Brookhaven Symp. Biol. 21:328-343.

Edelman, G.M. and Gally, J.A., 1972. The genetic control of antibody synthesis. Ann. Rev. Genet. 6:1-46.

Gearhart, P.J., Johnson, N.D., Douglas, R. and Hood, L., 1981. IgG antibodies to phosphorylcholine exhibit more diversity than their IgM counterparts. Nature 291:29-34.

Gottlieb, P.D., 1980. Immunoglobulin genes. Molec. Immunol. 17:1423-1435.

Hilschmann, N. and Craig, L.C., 1965. Amino acid sequence studies with Bence-Jones proteins, Proc. Natl. Acad. Sci. USA 53:1403-1409.

Kabat, E.A., Wu, T.T. and Bilofsky, H., 1979. Sequences of immunoglobulin chains, U.S. Department of Health, Education and Welfare, Public Health Service, National Institutes of Health, NIH Publication No. 80-2008, October 1979.

Kim, S., Davis, M., Sinn, E., Patten, P. and Hood, L. 1981. Antibody diversity: somatic hypermutation of rearranged V_H genes. Cell 27:573-581.

Kurosawa, Y., von Boehmer, H., Haas, W., Sakano, H., Truaneker, A. and Tonegawa, W., 1981. Identification of D segments of immunoglobulin heavy-chain genes and their rearrangement in T lymphocytes. Nature 290:565-570.

Max, E.E., Seidman, J.G. and Leder, P., 1979. Sequences of five potential recombination sites encoded closed to an immunoglobulin κ constant region gene. Proc. Natl. Acad. Sci. USA 76:3450-3454.

Newell, N., Richards, J.E., Tucker, P.W. and Blattner, F.R., 1980. J genes for heavy chain immunoglobulins of mouse. Science 209:1128-1131.

Rodwell, J.D. and Karush, G., 1980. Restriction in IgM expression. I. The V_H regions of equine anti-lactose antibodies. Molec. Immunol. 17:1553-1561.

Sakano, H., Huppi, K., Heinrich, G. and Tonegawa, S., 1979. Sequences at somatic recombination sites of immunoglobulin light chain genes: their implication for the mode of recombination, antibody diversity and evolution. Nature Lond. 280:288-294.

Saskano, H., Maki, R., Kurosawa, Y., Roeder, W. and Tonegawa, S., 1980. Two types of somatic recombination are necessary for the generation of complete immunoglobulin heavy-chain genes. Nature 286:676-683.

Sakano, H., Kurosawa, Y., Weigert, M. and Tonegawa, S., 1981. Identification and nucleotide sequence of a diversity DNA segment (D) of immunoglobulin heavy-chain genes. Nature 290:562-565.

Seidman, J.G., Leder, A., Edgell, M.H., Polsky, F., Tilghman, S.M., Tiemeier, D. and Leder, P., 1978a. Multiple related immunoglobulin variable-region genes identified by cloning and sequence analysis. Proc. Natl. Acad. Sci. USA 75:3881-3885.

Seidman, J.G., Leder, A., Nau, M., Norman, B. and Leder, P., 1978b. Antibody diversity. Science 202:11-17.

Weigert, M. and Riblet, R., 1978. The genetic control of antibody variable regions in the mouse. Springer Seminar Immunopath. 1:133-169.

Weigert, M., Gatmaitan, L., Loh, E., Schilling, J. and Hood, L. 1978. Rearrangement of genetic information may produce immunoglobulin diversity. Nature 276:785-790.

Wu, T.T. and Kabat, E.A., 1970. An analysis of the sequences of the variable regions of Bence-Jones proteins and myeloma light chains and their implications for antibody complementarity. J. Exp. Med. 132:211-250.

3

HEMATOPOIETIC TUMORS: NORMAL OR ABNORMAL MODELS OF LEUKOCYTE DIFFERENTIATION?

Noel L. Warner,*,** Lewis L. Lanier,*,** Edwin Walker,** Carlton Stewart,*** Gary Wood,+ and Charles Balch++

*Becton Dickinson Monoclonal Center, Mt. View, CA, 94043; **Department of Pathology, University of New Mexico, Albuquerque, NM 87131; *** Los Alamos National Laboratories, Los Alamos, NM 87545; +Department of Pathology, Stanford University Medical Center, Stanford, CA 94305; and ++Department of Surgery, University of Alabama, Birmingham, AL 35294

1. INTRODUCTION

One of the major problems in cellular immunobiology has been the issue of resolving the cellular bases of the many different specific functional properties mediated by the immune system. This is in essence a dual problem, as it involves both a detailed phenotypic characterization of all the differentiation lineages in the hematopoietic system, and the correlation of each cellular subset with specific functional properties. This problem is particularly compounded both by the wide heterogeneity in the differentiation stages of cells within hematopoietic tissues, and by the requirements for collaboration and interaction amongst different cell types in the mediation of specific immunological functions (Katz and Benacerraf, 1977). Analysis of the cell types involved in any given immune response is further complicated by the wide heterogeneity in cell populations when cell suspensions are obtained from in vivo tissue and blood sources.

It is into this complexity of cellular and functional heterogeneity that monoclonal antibodies have been applied in an attempt to more precisely resolve the specific cellular components that may be present, and in turn to use monoclonal antibodies as identification reagents for specific cell subsets with which specific functional properties may be associated. Previous studies in this area have attempted to use heterologous or allogeneic antisera as probes to define specific cell populations (Warner and

McKenzie, 1977). The results obtained with such antisera were frequently confusing. It was felt that this was principally due to the mixed populations of antibodies present in most antisera, and the extreme difficulty in absorbing them such to render the antisera effectively monospecific. Monoclonal antibodies should remove this major difficulty by providing a homogeneous and monospecific antibody population.

In attempting to deal with the problem of cellular heterogeneity that is encountered with studies using in vivo derived cell preparations, the use of homogeneous cloned tumor cell lines has become of increasing interest and value (Moller, 1979). The essential premise behind the use of tumor cell lines in the study of normal differentiation is that the oncogenic event and subsequent clonal expansion of the tumor does not involve any aberrant expression of the genes that are activated in normal leukocyte stages of differentiation. From many studies in this field reviewed elsewhere (Moller, 1979), the general view might be proposed that most hematopoietic tumor cell lines:

(i) are of clonal origin and represent the cellular clonal expansion of cells of a single differentiation stage;
(ii) are relatively restricted in their stage of differentiation, although some intra-tumor heterogeneity to varying degrees may frequently be detected. (Such intra-tumor heterogeneity may be associated with cell cycle associated expression of specific gene functions (Lanier and Warner, 1981) or may be due to tumor stem cells in the line which represent a more immature stage in the particular hematopoietic differentiation lineage. Alternatively, true differentiation related heterogeneity may exist within the cell line, perhaps in part due to the induction of the tumor cell lines thru agents associated or involved with the culture of such cells.); and
(iii) express cell surface components that are determined by specific genes that are expressed and representative of the corresponding normal cellular differentiation stage.

This latter point is particularly the one that is most frequently and seriously challenged, concerning the use of tumor cell lines as representative of normal differentiation stages and subpopulations. The alternative view is that since tumors may demonstrate aberrant control of cellular proliferation, the antigenic and functional characteristics of tumor cell lines may not reflect normal hematopoietic differentiation pathways. It is certainly evident that tumor cell lines, particularly those maintained either in in vivo transplantation or in tissue culture for extended periods, can develop subsequent mutational changes. Such variants, particularly those involving loss of gene function or expression, are clearly not normal, and not reflective of the normal differentiation state even though some functions relevant to the normal cell subpopulation from which they were derived may still be expressed. Perhaps the foremost example of this is the sequential loss of immunoglobulin gene expression by myeloma cells (Baumal and Scharff, 1973). Thus the detection of Bence-Jones protein (light chain only) producing myelomas does not necessarily indicate that a normal stage in B cell differentiation involves the secretion of light chains without heavy chains. However, this does not exclude the possibility that at least early following oncogenic transformation, the original parent myeloma line expressed a phenotype similar to normal plasma cells.

With specific exceptions of this type aside, we strongly reemphasize the view that when a particular tumor cell line is found to express a cell surface phenotype that has hitherto been undetected in normal cell populations, one should not assume that this is due to aberrant gene expression. Rather, we would propose that such situations should be exhaustively and extensively studied from the view point that the given tumor may represent either a particular hematopoietic cell subpopulation, or a specific stage in differentiation which is associated with a normal cell population that was previously undetected.

The emphasis of this report is thus to illustrate this view with six selected examples of tumor cell lines that initially were found to demonstrate bizarre unexpected cell surface phenotypes, and subsequently have turned out to be representative of normal cell populations.

2. CELL SURFACE PHENOTYPES

2.1. Fc Receptors

Perhaps one of the simplest historical examples to illustrate this general point is the expression of Fc receptors by T cells. In the early 1970's one of the markers proposed to be characteristic of B cells in distinction to T cells, was the expression of a cell surface receptor for antibody and antigen-antibody complexes (Warner, 1974). At that time Fc receptors were clearly demonstrated to be on macrophages and most B cells, but had not been detected on normal T cell populations. In characterizing a series of T cell tumors, two tumors in particular, S49 and WEHI-22, were both clearly shown to express Fc receptors (Warner et al., 1975). These receptors appeared to show a specificity very similar to that reported for B cells. It was proposed that such T cell tumors may represent rather uncommon types of lymphoid tumors, perhaps representing a small hitherto unrecognized population of normal lymphoid stem cells expressing both T and B cell functions. Subsequently however it was demonstrated that the expression of Fc receptors is a normal property shown by selected populations of normal activated T cells (Dickler, 1976). Thus the expression of Fc receptors on WEHI-22 and S49 need not imply anything abnormal about those tumor cell lines with respect to that expression. Since in all other properties these two tumors demonstrated exclusively

T cell specific cell surface components, the most likely explanation is that these two tumors are representatives of T cell subpopulations that normally express Fc receptors.

2.2. Thy-1

Possibly one of the most widely used cell surface markers for distinguishing T cells from B cells has been the Thy-1 marker (Reif and Allen, 1964). This marker is not exclusively specific for T cells, since it was widely recognized to be also expressed by cells of the brain and other non-lymphoid tissues (Golub, 1972). However this property does not detract at all from its value as a reagent for discriminating T cells from B cells. In setting aside this issue of the non-hematopoietic cell expression of Thy-1, the use of Thy-1 as a discriminating agent for T cells versus B cells has however somewhat slipped into the connotation that Thy-1 is thus a specific marker for T cells. With this as the prevailing background, it was therefore rather surprising to find that a well characterized myelomonocytic tumor WEHI-3, also expressed the Thy-1 antigen (Ledbetter and Herzenberg, 1979). This tumor is unequivocally in the myeloid series as has been firmly documented by a wide range of cell surface markers and specific macrophage related functions, including lysozyme production, colony stimulating factor production, phagocytosis and IL-1 production. The detection of Thy-1 on WEHI-3 had been confirmed using several different reagents as listed in Table I. Since most of the reagents used react with different epitopes on the Thy-1 molecule, it is most likely that the WEHI-3 associated Thy-1 antigen is similar or identical to the Thy-1 molecule produced by T cells. Preliminary studies in collaboration with Dr. J. Ledbetter have in fact demonstrated by 2D gel analysis of cell surface proteins from WEHI-3, that the molecule expressing the determinant detected by the 30-H12 monoclonal antibody is a Thy-1 like molecule, differing only slightly in its charge, perhaps related to different glycosylation. The molecule detected by the monoclonal antibody T30 is thought to be a Thy-1 related like molecule, and is also expressed by WEHI-3. It should be noted however that several sublines of WEHI-3 have been developed over the last decade, some of which lack the Thy-1 antigen (Lanier et al., 1982). The existence of sublines lacking this antigen suggest that its expression within the WEHI-3 related lineage may in fact reflect a stage specific related expression.

Is however, this expression of Thy-1 on WEHI-3 unique to this tumor? The results in Table II demonstrate that this is not abnormal amongst macrophage tumors. The issue therefore is whether this is a more universally occurring mutational variant to be found among macrophage tumors, or whether this reflects a normal differentiation lineage within the myeloid pathway. It has been conclusively shown that in the rat, immature cells within a variety of hematopoietic lineages including cells of the myeloid and B cell lineage, do express Thy-1 (Hunt, 1979).

TABLE I

Anti-Thy-1 reagents shown to react with WEHI-3 myelomonocytic leukemia[a]

REAGENT CODE	DESCRIPTION
30-H12	Rat anti-mouse hybridoma derived monoclonal antibody to Thy-1.2
53-3.1	Rat anti-mouse hybridoma derived monoclonal antibody to Thy-1 framework antigen.
BAΘ	Rabbit anti-mouse brain antiserum.
GαΘ	Goat anti-serum to purified Thy-1 molecules from ASL-1 leukemia. (Kindly provided by Dr. Ray Daynes.)
MTLA	Rabbit anti-mouse thymocyte specific serum absorbed with nude mouse spleen cells. (Kindly provided by Dr. George Gutman)
ΘC3H	AKR anti-C3H mouse thymocyte serum.

[a] All reagents were tested on WEHI-3, WEHI-7 (T lymphoma) and WEHI-279 (B lymphoma) in indirect immunofluorescence by flow cytometry. All reagents were strongly positive on WEHI-3 and the T lymphoma, but showed no detectable reactivity on B lymphomas.

TABLE II

Expression of differentiation antigens on murine monocyte/macrophage cell lines

CELL LINE	ANTIGEN DETECTED (HYBRIDOMA CLONE)						
	Thy-1.2 (30-H12)	Ly-1 (53-7.3)	Ly-2 (53-6.7)	Ly-3.2 (53-5.8)	E2 (30-EL)	DNL 1.9 (DNL 1.9)	Mac-1 (M1/70)
WEHI-3	+++	-	-	-	-	-	+
WEHI-265	±	-	-	-	-	-	+
WEHI-274	±	-	-	-	-	-	+
WEHI-78.2/3	+	-	-	-	-	-	+
HTX-1	+++	-	-	-	-	-	++
WEHI-78.1/4	-	-	-	-	-	-	+
PU.5(1R)	-	-	-	-	+	-	++
J774	-	-	-	-	-	-	++++
P38801	-	-	-	-	-	-	++

Cells were stained with an optimal amount of rat monoclonal antibody against the indicated antigen, followed by staining with FITC-goat anti-rat Ig. Analysis was performed using a FACS III system as described elsewhere (Lanier et al., 1981b; 1982).

Monoclonal antibodies to Thy-1.2, Ly-1, Ly-2, Ly-3 and E2 are reported in Ledbetter and Herzenberg (1979); Mac-1 by Springer et al., (1979) and DNL-1.9 by Dressner and Loken (1981).

This has not however been demonstrated in the mouse, and further studies in this regard are clearly warranted.

Studies of normal mouse bone marrow by flow cytometry have revealed the existence of a small subpopulation of Thy-1 bearing cells in the marrow. The size of these Thy-1$^+$ cells suggests they may be in the myeloid lineage. Since virtually all cells in this particular size range have also reacted with the monoclonal Mac-1, it is possible that there is a normal non-lymphoid hematopoietic cell population within the mouse bone marrow which expresses Thy-1. When pure macrophages were obtained by culturing normal mouse bone marrow in CSF (MGF), a subpopulation of cells ($\sim$10%) was found to express Thy-1. Further studies to identify this particular macrophage population and to demonstrate its functional association are in progress.

2.3. LEU-3

In screening a wide panel of human tumor cell lines with various lymphocyte specific monoclonal antibodies, an unexpected reaction was observed in that the tumor line U937 was found to react with the monoclonal antibodies Leu-3a, Leu-3b and OKT4. Since all three of these monoclonal antibodies are known to react with the same molecule expressed on a subset of T cells associated with helper function, and since the U937 tumor represents the clearest example of a human monocytic tumor cell line (Sundstrom and Nilsson, 1976), these results question whether the U937 line inappropriately expresses this molecule or alternatively whether this represents a normal monocytic subset. The results demonstrated in Fig. 1 show that the Fab fragment of Leu-3a monoclonal antibody specifically reacts with the cell line. Preliminary studies on the molecular characterization of the antigen expressing the Leu-3/T4 determinant indicate that it is a molecule of approximately 58,000 molecular weight similar to the molecular expressing these determinants produced by T cells.

Previous flow cytometry analyses of human peripheral blood monocytes have not shown that monocytes express Leu-3 or OKT4. However using a sensitive three-stage procedure for immunofluorescence, it has been possible to show that the light scatter gated human peripheral blood monocyte population reacts with anti Leu-3 but not anti Leu-2a (Fig. 2). Although the reaction with anti Leu-3 is weak, it does suggest that Leu-3 may be expressed in low density on the surface of human peripheral monocytes. More definitive information has been obtained by histological staining of frozen tissue sections and cytocentrifuge preparations using a three-stage system of monoclonal antibody/biotinylated goat anti-mouse IgG/Avidin horse radish peroxidase. The cell preparations studied included sinusoidal histiocytes of lymph nodes, tissue macrophages in lymph nodes and skin, epithelioid histiocytes in tuberculosis and sarcoidosis granulomas, Langerhans cells in the skin, histiocytosis X in lymph nodes, skin and soft tissues,

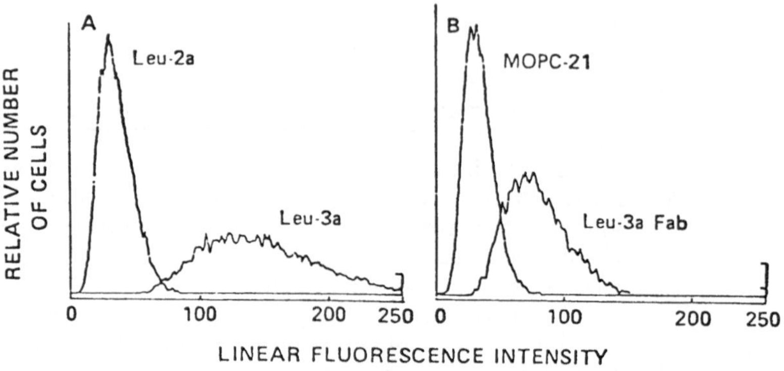

Figure 1: Expression of LEU-3 antigen on the surface of U937 monocytic tumor cells. All samples were treated with the indicated first stage reagents (anti-Leu-2a, anti-Leu-3a, MOPC-21 control myeloma protein and Fab fragment of anti-Leu-3a), followed by FITC goat anti-mouse immunoglobulin. All three intact monoclonal antibodies are of IgG isotype. Note that both intact and Fab fragment of anti-Leu-3a showed marked reactivity with the U937 cell line, whereas there was no staining with anti-Leu-2a or a control MOPC-21 myeloma protein.

and malignant histiocytosis in skin and spleen. All cell types stained for Ia antigen and were unstained with antibodies against Leu-1, Leu-2a, Leu-4, and Leu-4 determinants. In contrast all cells stained clearly with both Leu-3a, Leu-3b, and T4, except Langerhans cells and malignant histiocytois histiocytes which were stained in some but not all cases. In general, a diffuse cytoplasmic pattern of staining was obtained with all of these reagents. These results are reported in more detail elsewhere (Wood et al., 1982).

These results clearly demonstrate that although the expression of Leu-3/T4 within cells of the T cell lineage may be selective for a subset of cells including those with helper-inducer function, the expression of this molecule is not restricted solely to cells of the T cell lineage. It appears that in the normal hematopoietic differentiation pathway,

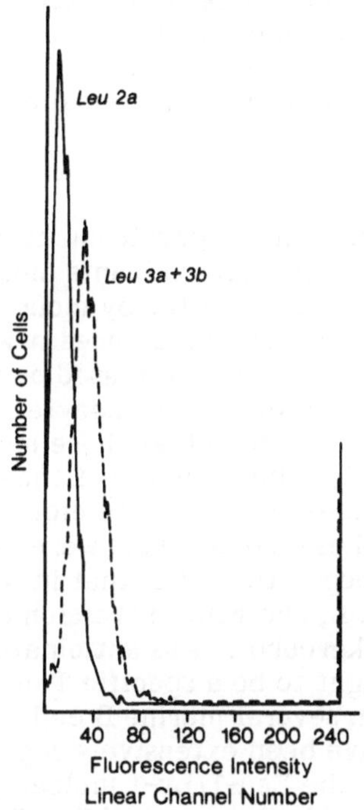

Figure 2: FACS IV analysis of the binding of anti-LEU-3 monoclonal antibody to human peripheral blood monocytes. A ficoll/hypaque preparation of human peripheral blood mononuclear cells was stained with either anti-Leu-2a monoclonal antibody or the multiclone anti-Leu-3a plus 3b antibodies. These samples were then stained with biotinylated goat anti-mouse immunoglobulin, followed by a third step reagent, fluorescein conjugated avidin. For analysis of the three step stain cell preparation, the light scatter profile was gated so that only the large scatter (monocyte) preparation was analyzed. Weak but significant staining of the monocytes was observed with the Leu-3a plus 3b reagent.

most cells of the macrophage/monocyte lineage produce a molecule expressing this epitope, but retain the molecule primarily as an intracytoplasmic molecule with only a low density expression on the cell surface. The expression of the molecule at a high density on the U937 cell surface may be inappropriate. Alternatively, it may subsequently turn out that there is a normal albeit rare stage in macrophage/myeloid differentiation where a relatively high density expression of this molecule occurs. The functional properties to be attributed to this molecule thus must relate to some functional property expressed by both macrophages and a subset of T cells, rather than specifically being associated only with the function of T cell help or induction. Again, however the point that we would wish to stress in this article is that an observation originally made with a tumor cell line is reflective of the existence of a normal cell expression.

2.4. LY-1/LEU-1

In the initial studies on the development of monoclonal antibodies to human T cells, one of the determinants most frequently recognized is a pan T cell determinant that is detected by monoclonal antibodies such as Leu-1, T1, T101, and 3.9. In all instances it was reported that these monoclonals detected a molecule expressed on all normal T cells, that was not present on normal B cells or monocytes. In most instances however it was also subsequently found that these monoclonal antibodies reacted with a proportion of lymphoid tumors of the CLL type (Wang et al., 1980). Since most of these tumors were clearly of the B cell lineage, it was initially viewed that these monoclonals were useful as pan T reagents when used with normal cell populations, but that in some lymphoid malignancies, specifically CLL, aberrant expression of the Leu-1 antigen occurred. With this as a background it was anticipated that the expression of Ly-1, previously thought to be a specific T cell marker in the mouse would also be found on several murine B cell tumors (Lanier et al., 1981a). These studies have been extensively reported elsewhere and unequivocally document that the Leu-1/Ly-1 molecules of mouse and man are respectively expressed both in normal T cells and on a subset of B cell tumors (Wang et al., 1980; Lanier et al., 1981b; Ledbetter et al., 1981).

Further studies have questioned whether expression of Ly-1 on murine B cell tumors is in fact reflective of a similar expression by at least a subset of normal mouse B cells. It is already apparent that murine B cells can be divided by flow cytometry analysis on the basis of mu and delta chain expression into three distinct cell populations. One of these subpopulations unequivocally expresses the Ly-1 antigen (Hardy et al., 1982). Thus, in the mouse, it is now firmly established that the initial observation of B cell tumors expressing Ly-1 is indeed reflective of the existence of a normal B cell subpopulation expressing this molecule. Further studies on the possible expression of a Leu-1/T1/T101/3. positive normal B cell subset are proceeding.

2.5. E2

In screening a series of xenogeneic monoclonal antibodies against murine lymphoid cells. one monoclonal was detected which reacted with erythrocytes and with a subset of normal mouse B cells (Ledbetter and Herzenberg, 1979). This monoclonal termed E2 reacted with about 20% of spleen cells. The majority if not all of these cells appeared to be of B cell derivation. In examining the wide range of tumor cell lines, it was also observed that approximately 50% of murine B cell tumors expressed E2 (Lanier et al., 1981b) (Table III). Thus these results demonstrate that E2 is a specific differentiation anigen useful for discriminating a subset of neoplastic and normal B cells, whereas normal T cells do not express this antigen. It might be noted that one macrophage tumor, PU5, also expressed E2 (Lanier et al., 1982) (Table II). Thus this result might be interpreted as inappropriate expression of this antigen by this particular macrophage tumor.

However, in studies of normal murine bone marrow cell populations by flow cytometry, it was observed that the E2 determinant was expressed on a subpopulation of cells which were virtually all positive with the macrophage specific monoclonal antibody Mac-1. In an attempt to subset type macrophage subpopulations within the bone marrow, a culture system was used which permitted the subsequent expansion only of cells of the macrophage lineage. These cells in turn could be further separated after approximately a week of culture on the basis of adherence, yielding both a non-adherent and an adherent population of cells. While the non-adherent population was highly enriched for macrophage progenitor cells, the adherent population was virtually 100% macrophages as defined by a variety of properties including the expression of the Mac-1 antigen. These cell populations are completely negative for B cell markers as defined by the expression of membrane immunoglobulin and DNL 1.9 (Dressner and Loken, 1981), and are negative for T cell markers as defined by T30, Ly-1 and Ly-2. It was clearly apparent that the non-adherent population was totally negative for the expression of E2 whereas the adherence positive subset clearly expressed E2 on virtually all cells (Fig. 3). Thus E2 is a determinant that is expressed on erythrocytes. on a subset of normal macrophages, a subset of normal B cells, and on a subset of macrophage and B cell tumors. The functional properties associated with this specific molecule and the definition of the molecular nature of this cell surface component are currently under investigation.

2.6. LEU-7

In this final example we would like to illustrate the types of results obtained with a monoclonal antibody that was developed against a tumor cell line thought to be of one particular cell lineage, yet when tested on normal cell populations reacts with another cell population whose differentiation lineage at the present time has yet to be clearly resolved. Leu-7 (HNK-1) is a monoclonal antibody developed by Drs. Abo and Balch against the human tumor cell line HSB-2 (Abo and Balch, 1981). This tumor has been previously characterized as a T cell tumor line and expressed Leu-1 but not Leu-2, 3, 4, 5, 6, DrW, or Ig. It has been reported to react weakly with OKT1, OKT4, and OKT10. This monoclonal antibody that was developed against this presumed T cell tumor line, reacts with approximately 15% of normal human peripheral blood lymphocytes (Abo and Balch, 1981). Included within the Leu-7 positive population of cells are all cells showing human NK activity.

TABLE III

Distribution of differentiation antigens on
T cell, B cell, and monocyte/macrophage tumors

	NUMBER POSITIVE/NUMBER EXAMINED				
	Thy-1.2	Ly-1	Ly-2	DNL1.9	sIg
B Cell Lymphomas	0/20	6/20	0/20	19/20	20/20
T Cell Lymphomas	18/18	11/16	11/16	0/7	0/18
Monocyte-Macrophage Tumors	6/13	0/13	0/13	0/7	0/13

A panel of cell lines in the B cell, T cell, and monocyte/macrophage tumors were stained with monoclonal antibodies against the indicated antigen, and were analyzed using a FACS III system.

A summary of the results are presented above. For a more detailed description see Lanier et al. (1981b, 1982).

Previous studies using light scatter sorting had however demonstrated that all human NK activity could be restricted to a population of no more than 3% of human peripheral blood mononuclear cells (Tai and Warner, 1981). Thus these two observations must imply that the Leu-7 antigen was expressed on at least two cell types in human peripheral blood, namely those cells with NK function, and an additional subset without NK function. Flow cytometry analysis of human peripheral blood lymphocytes stained with Leu-7 and/or Leu-2 have revealed that there is a subset of Leu-2 bearing cells that also expressed Leu-7 (Fig. 4). This has also been confirmed by two-color dual laser FACS analysis. Since Leu-2 and Leu-7 bind independently on the cell surface, additivity studies could be used to determine the proportion of cells expressing either one or both of these determinants. The results for two individuals shown in Table IV clearly demonstrate that the use of these two monoclonals define at least four populations of human lymphocytes. The Leu-2 negative, Leu-7 negative cells are predominantly if not exclusively Leu-3 positive, as are the helper inducer containing subset. Recent data demonstrates that the subset of the cells expressing Leu-7 but not Leu-2, contain virtually all of the cells with active NK function (Abo and Balch, unpublished observation). Thus NK cells might be more specifically defined as being Leu-2 negative Leu-7 positive. In addition Leu-7 therefore also subdivides

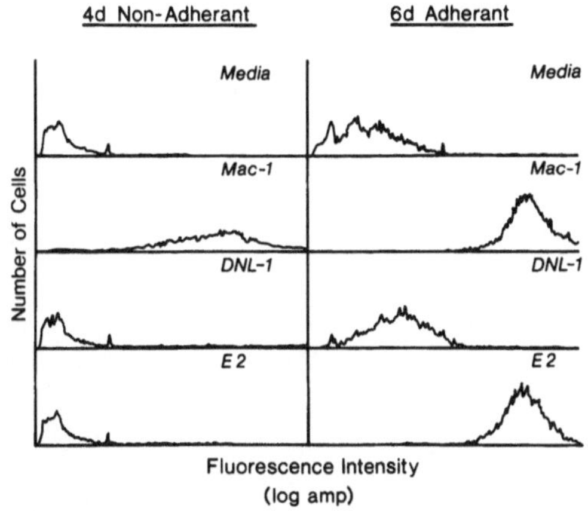

Figure 3: Flow cytometry analysis of cultured mouse bone marrow cell preparations. Normal mouse bone marrow cells were cultured in L-cell conditioned medium. The non-adherent cells were removed at four days and the adherent cells taken at six days after culture. The two cell preparations were then separately stained with either media or three monoclonal antibodies, Mac-1, DNL-1.9, or E2. All reagents were followed by staining with fluoresceinated mouse anti-rat immunoglobulin monoclonal antibody. The results of the 8 cell preparations are presented as histograms of number of cells versus fluorescence intensity using a 3 decade logarithm amplifier. Significant staining of the non-adherent preparation was observed only with the Mac-1 reagent, whereas the six day adherant preparation stained with both Mac-1 and E2. The control B-cell specific monoclonal, DNL-1.9, showed no reactivity with either preparation. These histograms were derived using the flow cytometry analytical system at Los Alamos National Laboratory.

the Leu-2 positive subset of T cells into both Leu-2 positive, Leu-7 negative, and Leu-2 positive, Leu-7 positive subpopulations. Studies are currently in progress in an attempt to delineate whether these two subpopulations have different functional properties. In preliminary experiments, two subsets of NK cells have been demonstrated on the basis of expression of Leu-4 and Leu-7. A subset which expresses the $Leu\text{-}4^-/Leu\text{-}7^+$ phenotype demonstrates efficient NK activity, whereas the $Leu\text{-}4^+/Leu\text{-}7^+$ subset may represent an immature population of NK cells (Abo et al., 1982).

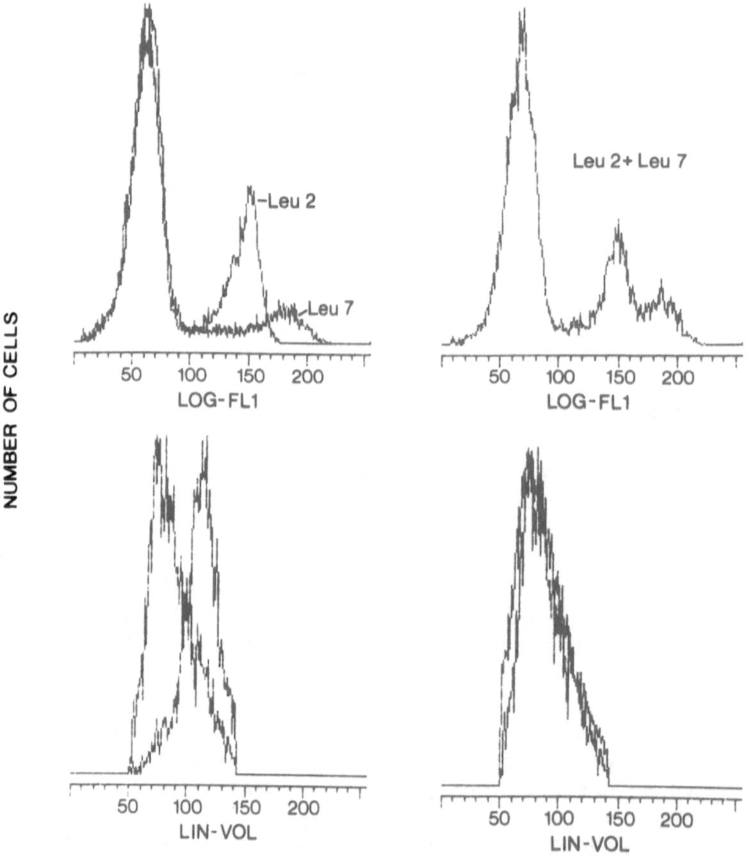

Figure 4: Flow cytometry analysis of human peripheral blood lymphocyte preparations stained with either LEU-2 or LEU-7 monoclonal antibodies alone or in combination, followed by fluoresceinated goat anti-mouse immunoglobulins. A ficoll/hypaque preparation of human peripheral blood mononuclear cells was stained with the monoclonal antibodies either alone (upper left) or in combination (upper right) followed by FITC goat anti-mouse immunoglobulins. Three preparations were then separately analyzed on the BD research analyzer. The figure in the upper left shows the fluorescence histogram as an overlay of the Leu-2 preparation and the Leu-7 preparation. Clearly resolved Leu-2 and Leu-7 populations were observed. In the upper right figure, the fluorescence histogram demonstrates the results of staining cells with both reagents. A mixed reactivity with a reduction in the Leu-2 stained peak was observed. This was due to a significant overlap of Leu-2 and Leu-7 bearing cells.

Thus a monoclonal antibody raised against a T cell tumor appears to define two subpopulations of normal peripheral cells: one population with a known function (natural killing), which may or may not be in the T cell lineage, and a second previously defined T cell subset. One possible explanation that is compatible with all of these observations is that the NK cells at immature stages may express both Leu-2 and Leu-7. During maturation, Leu-7 may increase in density, whereas several presumed T cell antigens such as Leu-2 and Leu-5 (E rosette positivity), may decrease in density to undetectable levels. The HSB-2 tumor might be reflective of a normal stage in natural killer lineage differentiation which is an immature stage, in that it lacks NK function, but expresses the T cell positive phenotype of being Leu-7 positive, pan T antigen positive. The lack of Leu-2 expression on the HSB-2 tumor may represent a loss of cell surface expression of Leu-2 by this tumor.

3. DISCUSSION

The above six examples thus firmly establish the view point that the finding of so called aberrant expression of cell surface antigens on given tumor cell lines, need not reflect an inappropriate gene expression by that tumor, but rather reflect an "inappropriate" state of knowledge of investigators at the time of the observation. In most of the above instances it was demonstrated that the particular cell surface phenotype initially observed on a tumor, was subsequently also found on normal lymphoid or myeloid cell populations. Thus studies of tumor cells may provide valuable clues as to the existence of normal cell populations

---

Figure 4: (continued)
The lower figures demonstrate the volume profile of the stain cell preparations. The bottom right histogram demonstrates the volume profile of Leu-2 stained cells versus Leu-3 stained cells (separate fluorescence histogram not shown). Both Leu-2 and Leu-3 cells demonstrate a virtually identical volume distribution.

The volume profile for the Leu-7 negative cells versus the Leu-7 positive cells in shown in lower left histogram. The Leu-7 negative lymphocytes (i.e., predominately Leu-2 or Leu-3 cells) demonstrate a volume distribution similar to the cells shown in the lower right histogram. In contrast, the right profile of significantly larger size cells is for the Leu-7 positive preparation. This clearly demonstrates that Leu-7 bearing cells are significantly larger and is compatible with their designation as large granular lymphocytes.

TABLE IV

Reactivity of anti-Leu-2a and Leu-7 on human peripheral blood lymphocytes

	PERCENT POSITIVE CELLS	
Observed Results	Experiment 1	Experiment 2
Anti-Leu-2a	32.4	47.5
Anti-Leu-7	22.3	26.2
Anti-Leu-2a + Leu-7	37.9	48.7
Calculated		
Leu-2^- Leu-7^-	62.1	51.3
Leu-2^+ Leu-7^-	15.6	22.5
Leu-2^+ Leu-7^+	16.8	25.0
Leu-2^- Leu-7^+	5.5	1.2

Observed results were obtained by indirect immunofluorescence staining of peripheral blood mononuclear cells. Analysis was performed using a FACS IV with light scatter gating on the lymphocyte fraction. Monoclonal reagents as first steps were used either singly or mixed. Neither of these reagents blocks the binding of the other, and hence permits calculation of the percent of cells of the four indicated combinations between these two reagents. FITC conjugated goat antimouse was used as the second antibody in these experiments.

which are present in relatively low frequency, but which may serve specific functions in the immune system. This is not to say that all tumor cell lines when studied at a particular time in their passage must reflect normal phenotypes. Mutational changes can occur in tumor cell lines which clearly are aberrancies. All in all however, these are probably usually due to either deletions of chromosomes or repression of gene function, rather than inappropriate gene derepression or expression. In many instances the loss of specific cell surface components may reflect some change in the biosynthetic pathway that is necessary for cell surface expression of the molecule. The expression of specific cell surface antigens on tumors however may indeed reflect a normal state of gene expression and therefore indicate the existence of a specific normal cell population.

REFERENCES

Abo, T., and Balch, C.M., 1981, A differentiation antigen of human NK and K cells identified by a monoclonal antibody (HNK-1), J. Immunol. 127:1024.

Abo, T., Cooper, M., and Balch, C., 1982, Characterization of human NK cells identified by the monoclonal HNK-1 (Leu-7) antibody, in: Natural Cell Medicated Immunity (R.B. Herberman, ed.) Academic Press, NY.

Baumal, R., and Scharff, M.D., 1973, Synthesis, assembly and secretion of mouse immunoglobulin, Trans. Rev. 14:163.

Dickler, H.B., 1976, Lymphocyte receptors for immunoglobulins, Adv. Immunol. 24:167.

Dressner, D., and Loken, M., 1981. DNL 1.9: A monoclonal antibody which specifically detects all murine B lineage cells, Eur. J. Immunol. 11:282.

Golub, E.S., 1972, Brain-associated stem cell antigen: An Antigen shared by brain and hematopoietic stem cells, J. Exp. Med. 136, 369.

Hardy, R.R., Hayakawa, K., Haaijman, J., and Herzenberg, L.A., 1982, B cell subpopulation identifiable by two-color fluorescence analysis using a dual-laser FACS, Ann. N.Y. Acad. Sci., In press.

Hunt, S.V., 1979, The presence of Thy-1 on the surface of rat lymphoid stem cells and colony-forming units, Eur. J. Immunol. 9:853.

Katz, D.H., and Benacerraf, B. (eds,), 1976, The Role of Products of the Histocompatibility Gene Complex in Immune Responses, Academic Press, New York.

Lanier, L.L., and Warner, N.L., 1981, Cell cycle related heterogeneity of Ia antigen expression on a murine B lymphoma cell line: Analysis by flow cytometry, J. Immunol. 126:626.

Lanier, L.L., Warner, N.L., Ledbetter, J.A., and Herzenberg, L.A., 1981a, Expression of Lyt-1 antigen on certain murine B cell lymphomas, J. Exp. Med. 153:998.

Lanier, L.L., Warner, N.L., Ledbetter, J.A., and Herzenberg, L.A., 1981b, Quantitative immunofluorescent analysis of surface phenotypes of murine B cell lymphomas and plasmacytomas with monoclonal antibodies, J. Immunol. 127:1691.

Lanier, L.L., Walker, E.B., Coffman, R., Loken, M., Springer, T., Richey, J., and Warner, N.L., 1982, Cell surface antigens and functional properties of murine monocyte/macrophage tumor cell lines, In preparation.

Ledbetter, J.A., and Herzenberg, L.A., 1979, Xenogeneic monoclonal antibodies to mouse lymphoid differentiation antigens, Immunol. Rev. 47:64.

Ledbetter, J.A., Evans, R.L., Lipinski, M., Cunningham-Rundles, C., Good, R.A., and Herzenberg, L.A., 1981. Evolutionary conservation of surface molecules that distinguish T lymphocyte helper inducer and cytotoxic/suppressor subpopulations in mouse and man, J. Exp. Med. 153:310.

Moller, G. (ed,)., 1979, Activation and regulation of immunoglobulin synthesis in malignant B cells, Immunological Reviews, Volume 48, Munksgaard, Copenhagen, Denmark.

Reif, A.E., and Allen, J.M.V., 1964, The AKR thymic antigen and its distribution in leukemias and nervous tissues, J. Exp Med. 120:413.

Springer, T.A., Galfre, G., Sacher, D.S., and Milstein, C., 1979, Mac-1: A macrophage differentiation antigen identified by monoclonal antibody, Eur. J. Immunol. 9:301.

Sundstrom, C., and Nilsson, K., 1976, Establishment and characterization of a human histiocytic lymphoma cell line (U-937), Int. J. Cancer 17:565.

Tai, A., and Warner, N.L., 1981, Characterization of the differentiation lineage of natural killer cells, Fed. Proc. 40:2711.

Wang, C.Y., Good, R.A., Ammirato, P., Dymburt, G., and Evans, R.L., 1980, Identification of a p69/70 complex expressed on human T cells sharing determinants with B-type chronic lymphatic leukemia cells, J. Exp. Med. 151:1539.

Warner, N.L., 1974, Membrane immunoglobulins and antigen receptors on B and T lymphocyes, Adv. Immunol. 19:67.

Warner, N.L., Harris, A.W., and Gutman, G., 1975, Membrane immunoglobulin and Fc receptors of murine T and B cell lymphomas, in: Membrane Receptors of Lymphocytes (M. Seligurann, J.L. Preud'homme, and F.M. Kourelsky, eds.), Elsevier North Holland, NY, p. 203.

Warner, N.L., and I.F.C. McKenzie, 1977, Surface antigens and receptors of normal and neoplastic lymphocytes: Nature and distribution, in: The Lymphocyte, Structure, and Function (J.J. Marchalonis, ed.), Marcol Dekker, N.Y., p. 433.

Wood, G., Warner, N.L., and Wamecke, R., 1982, Leu-3 antigen is expressed by cells of monocyte/macrophage and Langerhans lineage, In preparation.

4

DISSECTION OF HUMAN IMMUNOREGULATORY T LYMPHOCYTES: IMPLICATIONS FOR UNDERSTANDING CLINICAL DISEASE

Ellis L. Reinherz, Chikao Morimoto, Stefan Meuer and Stuart F. Schlossman

Division of Tumor Immunology, Sidney Farber Cancer Institute
Department of Medicine, Harvard Medical School, Boston, MA 02115

1. SUMMARY

Functional T lymphocyte subpopulations can be identified in humans by antibodies which detect stable glycoprotein antigens on their surface. Thus, inducer T lymphocytes bear an antigen termed T4 while suppressor T lymphocytes bear a different antigen termed T5. Immune homeostasis results from a delicate balance between inducer and suppressor subsets within the T cell circuit. Perturbation in subset dynamics may initiate a wide variety of immunopathologic disorders. An understanding of this circuit will be important for the elucidation of the pathogenesis of a number of diseases and should permit the orderly manipulation of the human immune response through modulation of selected T cell subsets.

2. INTRODUCTION

The precise dissection of cellular mechanisms and interactions involved in the generation of human T cell responses has been facilitated in recent years by advances in four areas: first, the development of in vitro methods for characterization and identification of human T lymphocyte subsets by cell surface markers; second, the development of new techniques for the isolation of highly purified subclasses of human T lymphocytes depending on cell surface markers; third, the development of in vitro techniques to discriminate functional properties and interactions of the isolated subsets of T lymphocytes and other cells (that is, B cells, null cells and macrophages): and fourth, the capacity to correlate normal and abnormal in vitro functional properties of T lymphocyte subpopulations with in vivo disorders of the immune response. These advances have made possible the elucidation of the major T lymphocyte subsets in man and their unique functional programs.

The genetic program of the human T lymphocyte is a complex one as it includes immunoregulation as well as the capacity to recognize specific antigens and to execute unique effector functions. Thus, T lymphocytes proliferate to soluble and cell surface antigens and polyclonal activators, including the mitogens phytohemagglutinin (PHA) and Concanavalin A (Con A) (Chess et al., 1974a; 1974b). They are responsible for cytotoxic killer activity in cell-mediated lympholysis (CML) (Sondel et al., 1975) and they produce a host of soluble factors (Geha et al., 1973; Rocklin et al., 1974) which effect a variety of cellular functions. Perhaps more importantly, T lymphocytes are involved in virtually all regulatory interactions, including helper and suppressor cell function (Reinherz and Schlossman, 1980a; 1980b).

In this review, we will focus upon the recent developments in understanding the differentiation history of T lymphocytes and their functional maturation. We will also provide evidence which supports the notion that, during differentiation, T cells diverge into functionally distinct subsets programmed for their respective inducer (helper) and cytotoxic/suppressor functions which can be defined by unique cell surface glycoprotein antigens.

3. RESULTS AND DISCUSSION

3.1. Differentiation of T lymphocytes

A thymic microenvironment is necessary for the differentiation of T cells in all species. It appears that precursor bone marrow cells (prothymocytes) migrate to the thymus gland, where they are processed, become functionally competent, and are then exported into the peripheral lymphoid compartment (Moore and Owen, 1967; Owen and Ritter, 1969; Owen and Raff, 1970; Stutman and Good, 1971). Moreover, the profound changes in cell surface antigens mark the stages of T cell ontogeny (Raff, 1971; Konda et al, 1973).

In man the earliest lymphoid cells within the thymus lack mature T cell antigens but bear antigens shared by bone marrow cells of several lineages (Reinherz et al., 1980a). This population accounts for approximately 10% of thymic lymphocytes and is reactive with two monoclonal antibodies, anti-T9 and anti-T10 (Stage I) (Fig. 1). Although these two antibodies are not specific for T cells (Reinherz et al., 1980a; Reinherz and Schlossman, in press) they are reactive with normal, activated, and malignant cells of non-T lineage, they are useful in providing an understanding of antigenic changes occurring during T cell ontogeny. With maturation, thymocytes lose T9, retain T10, and acquire a thymocyte distinct antigen defined by anti-T-6. Concurrently, these cells express antigens defined by anti-T5 and anti-T5/T8 (Stage III). The T4+, T5/8+, T6+, and T10+ thymocytes account for approximately 70% of the total thymic population. With further maturation, thymocytes lose the T6 antigen, maximally express the T1 and T3 antigens, and then segregate

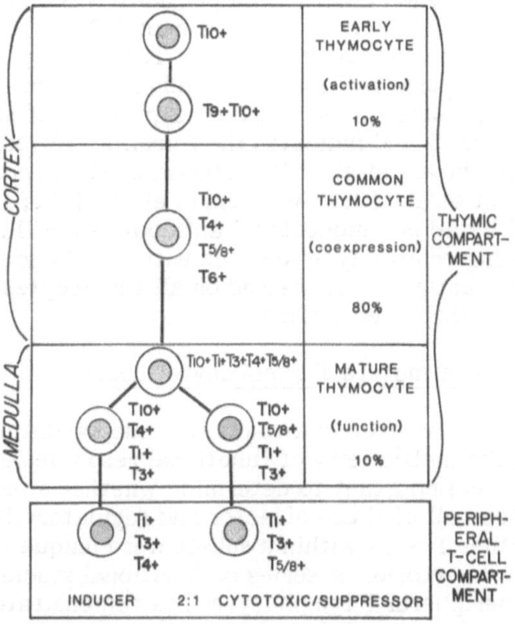

Figure 1. Stages of T cell differentiation in man.
Three discrete stages of thymic differentiation can be defined on the basis of reactivity with monoclonal antibodies. The most mature thymocyte population (Stage III) gives rise to the peripheral T cell inducer and cytotoxic/suppressor subsets. The cell surface antigens expressed during T cell ontogeny are shown.

into T4+ and T5+ subsets (Stage III). (T5+ cells also express the T8 antigen. For simplification, however, these cells are referred to as T5+ rather than T5+/8+.) Immunologic competence is acquired at this stage but is not fully developed until thymic lymphocytes are exported (Reinherz et al., 1979a). Outside the thymus the resting (T1+, T3+,T4+) and (T1+,T3+, T5+) subsets lack T10 and represent circulating inducer (helper) (Reinherz et al., 1979b; 1979c) and cytotoxic/suppressor populations (Reinherz and Schlossman, 1979; Reinherz et al., 1980b; 1980c), respectively.

Unlike the majority of the thymocytes which express little or only faint reactivity with anti-T1 and anti-T3 (Bhan et al., 1980), all of the circulating peripheral T cells are strongly reactive with T1+ and T3+. The T4 antigen is expressed on approximately 55-65% and the T5 (Reinherz et al., 1979c) is on 20-30% of peripheral T cells.

These two subsets correspond to TH_2- helper and TH_2+ cytotoxic/suppressor cells, respectively (Reinherz and Schlossman, 1979; Evans et al., 1978a). Moreover, unlike Stage II thymocytes, T4 and T5/T8 antigens are expressed on mutually exclusive subsets of mature T cells (Reinherz et al., 1980a; 1980b; 1980c). Table I lists the cell surface expression of antigens defined by monoclonal antibodies and indicates their preliminary biochemical characterization (van Agthoven et al., 1981; Martin et al., in press; Terhorst et al., 1980; Terhorst et al., 1981; McMichael et al., 1979; Ledbetter et al., 1981). Also listed is a monoclonal antibody, anti-T11 which defines an antigen comprising or closely associated with the sheep erythrocyte receptor. Since this antigen is expressed on all thymocytes and peripheral T cells, it will not be discussed further.

3.2. Functions of mature T lymphocyte subsets

Given the existence of two distinct subpopulations of peripheral T cells and the multiplicity of functional responses effected by T lymphocytes, it was important to determine whether an individual T lymphocyte possessed all of these effector and regulatory functions or, alternatively, whether T cells within a subset were unique with respect to their functional repertoire. A series of functional studies on isolated subpopulations of peripheral T lymphocytes has demonstrated that the latter hypothesis is correct and that the specific program of T cells is linked to the expression of a particular cell surface antigen (Fig. 2).

For example, only the T1+,T3+,T4+ population responded to soluble antigen (Reinherz et al., 1979b). In contrast, both subsets of cells show a maximal response to cell surface antigens (alloantigens). In additional studies, it was shown that the T1+,T3+,T4+ population responds maximally to PHA, while the T1+,T3+,T5+ subset shows a diminished response at all doses tested. Both populations of cells respond comparably to Con A.

As mentioned above, one of the major effector functions of human T lympocytes is the capacity to become sensitized to HLA-A, -B and -C locus antigens and to effect specific cell-mediated killing, It was found that only the T1+T3+T5+ subset (Reinherz et al., 1980b; 1980d) contained a cytotoxic effector population when separated after allogeneic activation in mixed lympocyte culture. The T1+,T3+,T4+ population, although capable of proliferating to alloantigen did not become cytotoxic when separated after sensitization (Reinherz et al., 1979b).

Perhaps the most important difference between these T cell subsets was evident from their differential regulatory effects on the immune response (Reinherz and Schlossman, 1980a; 1980b). The T1+,T3+,T4+ cells were shown to provide inducer (helper) function in the T-T, T-B, and T-macrophage interactions. For example, although the T1+,T3+,T4+ cells were not cytotoxic effectors when separated after allogeneic stimulation, they were required for optimal development of cytotoxicity within

TABLE I

Monoclonal antibodies to human T cell surface antigens

Monoclonal Antibodies	Approximate molecular weights of antigens 25-30		Cell surface expression (% reactivity with antibodies)			Trade designations[a]
	Non-reduced	Reduced	Thymocytes	T cells	Non-T cells	
Anti-T1	69K	69K	10[b]	100	0	OKT1, Leu1
Anti-T3	19K	19K	10[b]	100	0	OKT3, Leu4
Anti-T4	62K	62K	75	60	0	OKT4, Leu3a/3b, CC T4
Anti-T5	76K	30K + 32K	80	25	0	OKT5
Anti-T8	76K	30K + 32K	80	30	0	OKT8, Leu2a/2b, CC T8
Anti-T6[c]	49K	49K	70	0	0	NAI/34, OKT6
Anti-T9[d]	190K	94K	10	0	0	OKT9, 5E9
Anti-T10[d]	37K	45K	95	5	10	OKT10
Anti-T11	50K	50K	100	100	≤ 10	OKT11, Leu5, CC T11

[a]OK designations are from Ortho Pharmaceutical, Raritan, NJ. Leu designations are from Becton-Dickinson, Mountain View, CA. CC designations are from Coulter Electronics, Hialeah, FL. NAI/34 is from Accurate Chemical, NJ.

[b]10% of thymocytes express a high density of T1 and T3 antigens while the remaining thymocytes express little or faint reactions with anti-T1 and anti-T3.

[c]T6 is β2 microglobulin associated.

[d]T9 and T10 antigens are not T lineage specific and are found on normal and malignant populations of non-T cells. In addition, both antigens are expressed on a fraction of peripheral T cells following mitogen stimulation.

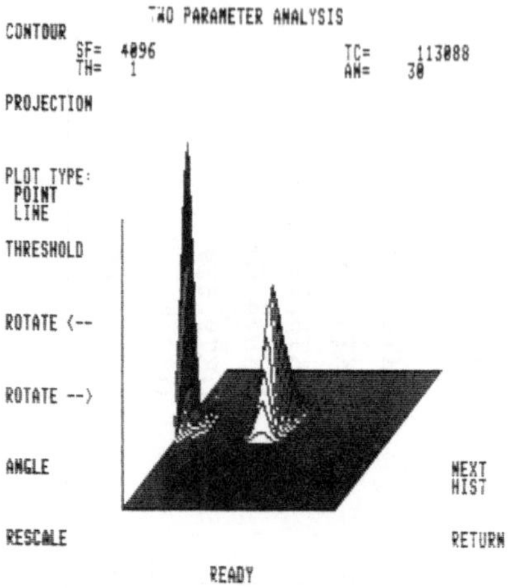

Figure 2. Immunofluorescence profile of normal human peripheral T cells with Anti-T4. As shown, anti-T4 reacts with 60% of the peripheral T cell population (right peak). The remaining 40% of T cells are unreactive with anti-T4 (left panel).

the T1+,T3+,T5+ effector population (Reinherz et al., 1979b). This is similar to findings in previous studies which showed that the TH_2+ population defined by heteroantisera contained the cytotoxic effector cell, while the Th2- population contained the helper cell for development of cytotoxicity (Evans et al., 1978).

Thus the T1+,T3+T4+ (T4+) T cells provided an inducer function in T-T interactions. In addition, T4+ T cells provided helper function in T-B interactions as well (Reinherz et al., 1979c; 1980c). Only the T4+ T cell subset provided the signals necessary to help autologous B cells to proliferate and differentiate into immunoglobulin-containing cells (Fig. 3). In contrast, the T1+,T3+,T5+ (T5+) T cells neither induced B cells to proliferate nor to differentiate. Moreover, the inducer role of the T4+ T cells for B cell immunoglobulin production was shown in both a pokeweed mitogen (Reinherz et al., 1979c) and an antigen-stimulated system (Reinherz et al., 1980e).

Prior studies demonstrated that antigen triggered T cell produced helper factors, including lympocyte mitogenic factor (Geha et al., 1973), which induced proliferation of all major lympocyte subclasses (T cell, B cell, null cell and macrophage) and, T cell replacing factor which initiated B cell immunoglobulin synthesis in the absence of T cells.

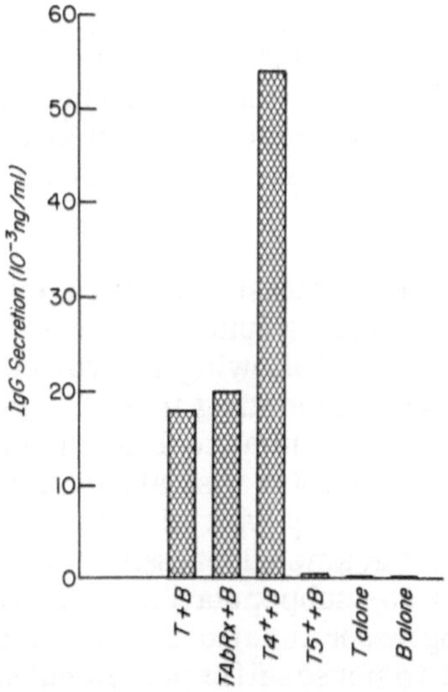

Figure 3. Influence of isolated T cell subsets on B cell IgG secretion. Isolated populations of T cells and T cell subsets, B cells, and autologous T-B combinations were cultured in the presence of PWM for 7 days. Subsequently, culture supernatants were harvested and IgG secretion quantitated by RIA. The T_{AbRx} population was treated with anti-T4 or anti-T5 and G/M FITC, but not fractionated into subsets. As shown, the T4+ T cell subset helps B cells secrete IgG and its inductive effect is significantly greater than the unfrac-T cell population. In contrast, the T5+ subset provides no inducer function for IgG secretion. Results are representative of four experiments performed.

In recent studies, it was found that only the T4+ subset made these non-specific helper factors (Reinherz et al., 1980e). The T cell subset restriction of these factors in man further stresses the importance of this subset of T cells in its inductive influence on the human immune response.

The above findings helped assign an inducer role to the T4+ population in T-T, T-B and T-macrophage interactions. Moreover, they provided additional evidence that a proliferative response to soluble antigen is restricted to the inducer population. The regulatory effects of the T4+ population do not appear limited to cells of lymphoid lineage. Since it is known that antigen stimulated T lympocytes produce helper factors which modulate erythroid stem cell production in vitro, it is probably that the T4+ population of lymphocytes is important in some aspects of erythroid differentiation (Lipton et al., 1980). Similarly, osteoclast-activating factor (Horton et al., 1972) and soluble factors inducing fibroblast proliferation and collagen synthesis have been shown to be derived from antigen-stimulated T lymphocytes (Cohen et al., 1979). These findings suggest a much broader biologic role for the T4+ inducer population in man.

The T5+ subset in contrast contains a mature population of cells with cytotoxic and suppressor function but not inducer functions (Reinherz et al., 1980b; 1980c). Following activation with Con A, T5 cells suppressed autologous T cells responding in MLC. In addition, this same T5+ population suppressed B cell immunoglobulin production (Fig. 4). It should be emphasized at this point that although both the T4+ and T5+ subpopulations proliferated equally well to mitogenic stimulation by Con A, only the T5+ population became suppressive. These results support the view that the T4+ and T5+ subpopulations are programmed for their respective helper and suppressor functions independent of their ability to discriminate and react to nonspecific polyclonal mitogens or antigens. Moreover, these results suggest that the programming of the specific cell function is linked to the expression of a particular cell surface phenotype and that such programming occurs before cell activation.

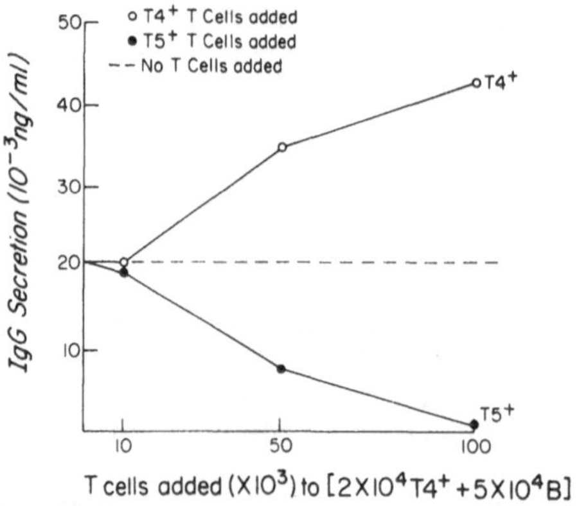

Figure 4. Suppressive effect of the T5+ T cell subset on autologous B cell IgG secretion. The capacity of T4+ and T5+ T cells to effect IgG secretion by a mixture of 2×10^4 T4+ T cells and 5×10^4 B cells was determined. T cell subsets were separated as described in Experimental Procedures. Subsequently, different numbers of T4+ (o) or T5+ (•) T cells were added to the constant

Further evidence substantiating this notion is the subsets' differential susceptibility to expression of Ia antigens following specific activation stimuli (Reinherz et al., 1979f). In several species, immunoregulatory activities are mediated by intercellular signals involving products of the I region of the major histocompatibility complex. Human Ia-like antigens were first defined by alloantisera (Winchester et al., 1975) and subsequently by heteroantisera (Humphreys et al., 1976) and monoclonal antibodies (Reinherz et al., 1979f). In man, Ia antigens are expressed on the surface of B cells, most monocytes and a subset of null cells, but are not detected on resting T cells (Reinherz et al., 1979f; Schlossman et al., 1976). Activation of human T cells results in de novo biochemical synthesis and cell surface expression of Ia antigens (Evans et al., 1978b). Thus, the appearance of Ia antigen on T cell subsets serves as a marker of specific T cell activation. Following alloactivation in MLC or stimulation by PHA and Con A, both T4+ and T5+ T cell subsets express Ia antigens. In contrast, when the T cell population is specifically activated by soluble antigens, only the T4+ subset expresses Ia. The appearance of Ia antigens on unique T cell subsets therefore depends upon the activation stimuli used and the ability of that individual subset to respond to a given stimulus.

The observation that only a fraction (∿ 40%) of T4+ T cells expressed Ia antigen upon activation suggested that the T4+ population might be heterogeneous. To test this possiblity, antigen activated T4+ T cells were separated into T4+Ia+ and T4+Ia-populations and characterized (Reinherz et al., 1981). It was found that the T4+Ia+ population contained the majority of proliferating T cells and that most of this proliferation was nonspecific. Elimination of the T4+Ia+ T cell subset with monoclonal anti-Ia and complement treatment diminished subsequent proliferation to both the triggering antigen and to unrelated antigens. In addition. the antigen induced T4+Ia+ subset alone produce nonspecific helper factors. Although the Ia- fraction represents 60% of the T4+ population, it showed minimal proliferation to soluble antigen and did not elaborate helper factors. Nonetheless, a mixture of both T4+Ia+ and T4+Ia- T cells was required for maximal Ig secretion by B cells since these two T4+ subsets worked in a synergistic fashion.

Figure 4. (Continued)
T4+ T-B mixture and supernatants harvested after 7 days in culture with PWM. As shown, T5+ cells were capable of suppressing IgG secretion while a comparable number of T4+ cells provided additional T cell help. Results are representative of four experiments performed.

Other studies have provided additional evidence to support the existence of heterogeneity within the T4+ subset. Specifically it was shown that approximately 35% of T4+ T cells and 10% of T5+ cells were reactive with an antibody found in the serum of many patients with active juvenile rheumatoid arthritis (Morimoto et al., 1981). These T4+JRA+

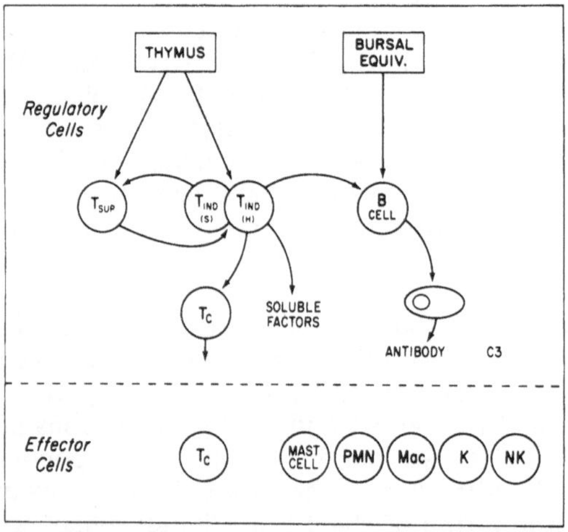

Figure 5. The human T cell circuit. Cellular and humoral responses are regulated by T4+ inducer (Tind) and T5+ suppressor (Tsup) T lymphocytes. One subpopulation of T4+ cells ($Tind_{(S)}$), reactive with JRA autoantibodies, incudes T5+ suppressor cell activation while a second subpopulation of T4+ cells ($Tind_{(H)}$) induces help for cytotoxic T cell (Tc) effector function, B cell differentiation and immunoglobulin production. Many of the effector cells illustrated above including NK, K, mast cells, polymorphonuclear leukocytes (PMN) and Macrophages (Mac) are under the influence of these regulatory cells and their products.

cells are required to induce the T5+ subset to mediate suppression of B cell Ig secretion. Thus T4+JRA+ T cells appear to be the inducer cells of suppression while T4+JRA-are the T inducers of help (Fig. 5). Yet to be determined is the relationship of T4+Ia+ lymphocytes to the T4+JRA+ subset.

Therefore the T4+ subset is analogous to the murine Lytl+2- subset which provides helper function through both specific and nonspecific signals and induces suppressor cell activation as well (Cantor and Boyse, 1977). In this regard, the T4+JRA+ subset has a counterpart in the Lytl+Qal+ murine subpopulations (Eardley et al., 1978) and the T4+Ia+ in the Lytl+2-Ia+ subset (Keller et al., 1980). The T5+ T subset in man appears to be analogous to the murine Lyt2,3 subset which mediates both cytotoxic and suppressor functions (Cantor and Boyse, 1977). Based on the evolutionary conservation of other cell surface molecules (that is, immunoglobulins, MHC encoded antigens, etc.), it is likely that the antigens defining the phenotypes of inducer and suppressor populations in man and mouse will be biochemically similar. There is already good evidence to indicate that the T5 antigen and the Lyt2,3 antigen are homologous (Terhorst et al., 1980). However, whether the T4 antigen and the Lytl antigen are similar is less certain. Nonetheless, given the observation that T4 and T5 are on reciprocal subsets in man with similar functions to the murine Lytl+2- and Lyt2+ populations, it is likely that an antigen equivalent to T4 exists on a murine Lytl+2- subset.

3.3. Clinical disorders of T lympocytes

In the present review, we have provided evidence to support the theory that it is possible to detect T lymphocyte subpopulations with unique biologic functions on the basis of their cell surface antigenic components. The application of this technology to human immunodiagnosis is just becoming appreciated. It is now possible to define the heterogeneity of T cell malignancies; diseases of T cell maturation and/or premature release; diseases associated with loss of T cells; diseases associated with imbalances of T cell subset restricted functions; and diseases associated with activation of T cell subsets.

Since immunologic functions are acquired only at the latest stage of intrathymic ontogeny, premature release of immunologically incompetent cells or aberrations of T cell maturation resulting in blocked differentation could lead to immunodeficiency. The development of probes that make it possible to define points along the differentiative pathway should help in the understanding of heterogeneity within congenital immunodeficiencies. In this regard, it has already become evident that patients with severe combined immunodeficiency may have thymocytes blocked in differentiation either at Stage I (T10+, or T9+T10+) or Stage III (T3+T4+T5/8+T10+) (Reinherz et al., 1981b). As expected, only those patients with the latter population express any T cell function (i.e. MLC proliferation).

Major immunologic abnormalities result from alterations in the mature T cell subsets. For example, some patients with acquired agammaglobulinemia lack the T4+ population and possess a T cell population that is incapable of triggering B cell synthesis of immunoglobulin. This specialized circumstance must be discriminated from that in the majority of patients with agammaglobulinemia, who have B cell abnormalities, but possess normal T cells (Reinherz and Schlossman, 1980a; 1980b; 1981b).

Circulating activated T4+ cells appear to result in different immunopathologic abnormalities including the formation of autoantibodies directed at red cells, white cells and platelets. Activated helper cells have been demonstrated in patients with active graft-versus-host disease (Reinherz et al., 1979d), and similar abnormalities have been seen in patients with sarcoid, scleroderma, and Sjogren's syndrome. Not only is there an increase in T4+ cells, but these activated T lymphocytes, unlike resting lymphocytes, express Ia-like (HLA-D related) antigens. The presence of activated T lymphocytes in human disease does not appear to be uncommon (Yu et al., 1980). Whether the activated T4+ T cells account for hyperglobulinemia, lymphocytosis dermal infiltration, granuloma formation, or fibrotic lesions in the diseases mentioned above is under investigation.

It is obvious that defects in immunoregulation could result from either a loss or persistent activation of the T5+ population. Loss of the T5+ T cells should result in unopposed inducer function, whereas activated T5+ cells should suppress the immune response. In patients with acute graft-versus-host disease, in which activated helper cells have been demonstrated, there is also a loss of suppressor cells (Reinherz et al., 1979d). A similar loss of T5+ cells has been seen in naturally occurring autoimmune diseases including systemic lupus erythematosus (Morimoto et al., 1980a), hemolytic anemia (Reinherz et al., 1979e), multiple sclerosis (Reinherz et al., 1980f), severe atopic eczema, hyper IgE syndrome and inflammatory bowel disease (Reinherz and Schlossman, 1980b). Moreover, the loss of the T5+ population may correlate temporally with the severity of clinical disease. The precise mechanism by which one population is lost or another activated is not clear. There is evidence from patients with lupus that autoantibodies are present in the serum and directed at the T5+ population (Morimoto et al., 1980b). Autoantibodies may selectively eliminate the suppressor population or moduclate its functional properties. Similarly, in studies of patients with juvenile rheumatoid arthritis, the loss of suppressor cell function correlates with increased B cell Ig secretion, the presence of autoantibodies to a T4+ subset which induces suppressor cell function, and increased disease activity (Morimoto et al., 1981).

In contrast, the presence of excessive numbers of activated suppressor cells results in severe immunodeficiency. For example, in a smaller number of patients with acquired agammaglobulinemia, activated T5+Ia+ cells were responsible for suppressing autologous B cell production of immunoglobulin (Reinherz et al., 1979e). Increased numbers of activated suppressor cells have also been seen after viral infections including those caused by Epstein-Barr virus and cytomegalovirus (Reinherz et al., 1980g).

In infectious mononucleosis, the self-limited increase in suppressor cells may account for transient immunologic hyporesponsiveness, but in patients with chronic graft-versus-host disease, persistent circulating suppressor cells cause prolonged immunologic incompetence (Reinherz et al., 1979d). Moreover, antigen-specific suppressor cells may result in human disorders. Patients with lepromatous leprosy have a T5+ population that can be specifically activated by lepromin (Mehra et al., 1980). In this case, activation of T5+ T cells is antigen specific; nevertheless, these activated suppressor cells may cause generalized immunosuppression. Presumably, the anergy seen in tuberculosis, systemic fungal infections, and protozoan infections may result from similar mechanisms. Finally it should be noted that an imbalance in the inducer; suppressor cell ratio is itself sufficient to result in diminished immunoglobulin production in vitro and agamma globulinemia in vivo. The finding of a relative increase in suppressor cells is a common finding in patients with acquired agamma-globulinemia and circulating B cells (Reinherz et al., 1981b).

3.4. Human T cell malignant diseases

The ability to define cell surface antigens that appear at specific stages of T cell differentation has, in addition, allowed for the orderly dissection of T cell malignant processes in human beings. In fact, these T cell diseases reflect the same degree of heterogeneity and maturation as is seen in normal T cell ontogeny (Reinherz et al., 1980a). For example, the tumor cells in most cases of acute T cell lymphoblastic leukemia arise from an early thymocyte or prothymocyte compartment (Stage I) whereas in only 20% of cases the cells are derived from a common thymocyte compartment (Stage II) and therefore bear the T6 antigen. To date, we have found only one T cell acute lymphoblastic leukemia tumor that arose from the most mature thymic compartment and expressed the T3 antigen (Stage III). Since normal thymocytes have not acquired mature T cell functions at either the level of the Stage I or II thymocytes, it is not surprising that the vast majority of acute lymphoblastic leukemia T cell in human beings have no demonstrable function.

The tumor populations from patients with T cells chronic lymphocytic leukemia, Sezary syndrome and mycosis fungoides are derived from the mature T cell compartment bearing the T1 or T3 antigens or both (Reinherz and Schlossman, in press; Boumsell et al., 1981). Therefore, it is not unexpected that some of these tumor cells display helper or suppressor functions. In this regard, all tumor populations from patients with Sezary syndrome bear the T1+T3+T4+ phenotype and may demonstrate the functional capacity to provide help for B cell Ig synthesis. T cell CLL tumor populations are of either the mature inducer or suppressor phenotypes. Since 6 of 8 T-CLL tumor populations studied to date have been of the helper phenotype, it would appear that the frequency of subset derivation corresponds to that expected from the normal ratio of inducer:suppressor cells (i.e., 2-3:1).

3.5. Immunotherapy by selective T cell subset manipulation

The ability to dissect normal lymphoid differentiation and to define the biology of lymphocytes in health and disease will allow for an understanding of disorders of the human immune response. In addition to their immunodiagnostic utility, the reagents described in this review may be potent therapeutic tools themselves. It is also likely that certain drugs will have slective effects on individual T cell subsets such that they could be utilized to alter the ratio of T inducer and T suppressor cells.

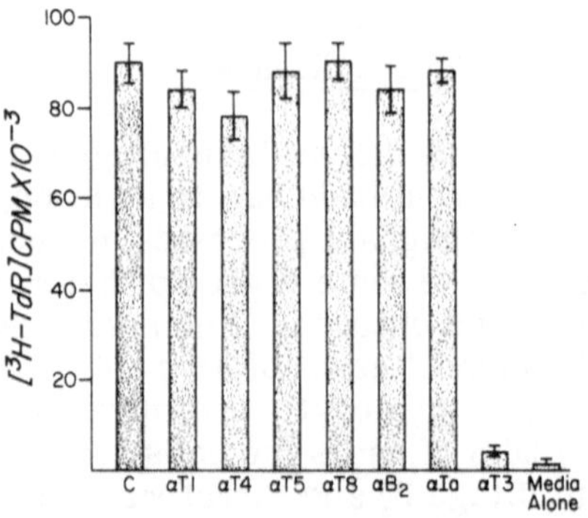

Figure 6. Unique capacity of Anti-T3 to inhibit MLC proliferation. The effect of anti-T3 and other monoclonal antibodies to inhibit MLC proliferation was determined by incubating responding T cell and mitomycin treated stimulator cell mixtures with antibodies during the in vitro culture period. Results are expressed as ^{3}H-TdR incorporation.

Finally, it is important to note that these monoclonal antibody-defined cell surface antigens may themselves represent important cell surface receptors. For example, anti-T3 alone was found in earlier studies to block proliferative responses of T lymphocytes to both soluble and cell surface antigens (Fig. 6) (Reinherz et al., 1980h). As few as 10^4 molecules of anti-T3 per cell appeared to inhibit these responses when added early in the culture period.

In addition to abrogating antigen-stimulated T cell proliferation responses, anti-T3 blocked the ability of T cells to provide help to autologous B lymphocytes in a T-dependent, PWM-driven system (Fig. 7). Thus, in the presence of anti-T3, Ig secretion by the T-B mixture was reduced to the level of Ig secretion found in B cells alone. Similar to the inhibition of T cell proliferation, anti-T3 inhibition of PWM-driven Ig secretion occurred during an early phase of cell-cell interaction. The capacity of anti-T3 to inhibit both T cell proliferation itself as well as T-B cooperation suggests that anti-T3 defines an important T cell interaction molecule. Both the failure of all other T cell specific antibodies to block these functions and the fully develped expression of the T3 antigen late in thymic differentiation at the time of acquisition of immunologic competence suggest that T3 is an important molecule (Reinherz et al., 1980h).

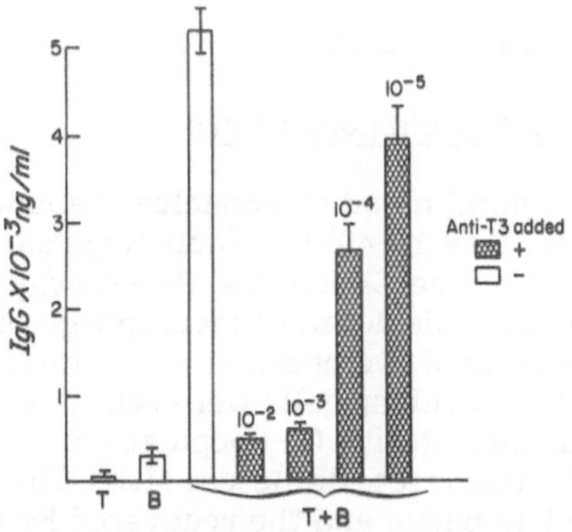

Figure 7. Inhibition by Anti-T3 cell help in PWM-driven B cell IgG secretion. The capacity of anti-T3 to effect IgG secretion by a mixture of T and B cells was determined. Anti-T3 was added to cultures in dilutions ranging from 10^{-2} to 10^{-5}. A shown, IgG synthesis is significantly inhibited by anti-T3 even at a dilution of 10^{-4}.

Furthermore, other T lymphocyte surface molecules may be critical for a variety of T cell effector functions. Recently, several monoclonal antibodies, anti-T5, anti-T8 and anti-$T8_A$, were found to define antigens on cytotoxic effector T cells (unpublished data; Reinherz et al., 1980a; 1980b). More importantly, as shown in Table II, anti-$T8_A$ markedly inhibited cell mediated lympholysis, anti-T8 partially effected CML, and anti-T5 had no effect. Immunoprecipitation studies and competitive binding experiments indicated that anti-T8 and anti-$T8_A$, like anti-T5, defined a related 33,000 dalton molecular weight antigen. Taken together, these results suggest that the single epitope marked by anti-$T8_A$ is critical for the cytolytic mechanism.

Additional studies in the murine system indicate that regulatory molecules (i.e., IJ), like effector molecules, are expressed on the cell surface membrane of some T lymphocytes. It is likely that antibodies to such determinants may serve to inhibit suppressor T cell function. In this regard, abrogation of suppressor cells has been shown to markedly alter the host immune response to autologous tumor cells.

Utilization of soluble regulatory products from T lymphocytes as well as manipulation of T cells and their cell surface receptor may be of paramount importance in controlling immune function. Given the recent evolution of T-T hybridomas in murine systems and the possibility of developing similar hybridomas in humans, it is conceivable that homogeneous helper and suppressor products could be produced. These and other strategies should prove useful to the host if manipulations are based on sound principles of immunobiology.

4. EXPERIMENTAL PROCEDURES

4.1. Isolation of lymphoid populations

Human peripheral blood mononuclear cells were isolated from healthy volunteer donors (age 15-40) by Ficoll-Hypaque density gradient centrifugation (Pharmacia Fine Chemicals, Piscataway, NJ). Unfractionated mononuclear cells were depleted of macrophages by adherence to plastic as previously described (Reinherz et al., 1979a). Subsequently, macrophage depleted mononuclear cells were separated into E rosette positive (E+) and E rosette negative (E-) populations with 5% sheep erythrocytes (Microbiological Associates, Bethesda, MD). The rosetted mixture was layered over Ficoll-Hypaque and the recovered E+ pellet treated with 0.155M NH_4Cl to lyse erythrocytes. The T cell population obtained was greater than 95% E+ and greater than 94% reactive with monoclonal antibody anti-T3 which defines an antigen present on all mature peripheral T lymphocytes. The E- population was further fractionated into B and null cell populations. Normal human macrophages were obtained from mononuclear populations by adherence techniques (Reinherz et al., 1979a). Suspensions of human thymocytes were made from fragments of thymus obtained at cardiac surgery from infants (age 2 months - 4 years).

4.2. Production and characterization of monoclonal antibodies

Monoclonal antibodies were derived by standard techniques after immunization of BALB/cJ mice (Jackson Laboratories, Bar Harbor, ME) with either human thymocytes, T cells or T cell chronic lymphatic leukemia human cells. Hybridoma cultures containing antibodies reactive with E+ cells or thymocytes but not B cells were selected, cloned and recloned by limiting dilution methods in the presence of feeder cells (Reinherz and Schlossman, 1980a). Malignant ascites were then developed and utilized for analysis. Immunoprecipitation studies were performed as previously described (van Agthoven et al., 1980; Terhorst et al., 1980; 1981.).

TABLE II

Effect of pre-incubation of CTL with monoclonal antibodies

		% specific lysis		
Monoclonal antibody	E/T ratio:	40/1	20/1	10/1
Control		28 ± 2	29 ± 1	21 ± 1
Anti-T5		30 ± 2	29 ± 2	22 ± 2
Anti-T8		16 ± 2	14 ± 2	15 ± 1
Anti-$T8_A$		2 ± 1	2 ± 1	1 ± 1

4.3. Analysis and separation of lymphocyte populations with a fluorescence activated cell sorter

Cytofluorographic analysis of cell populations was performed by means of indirect immunofluorescence with fluorescein conjugated $F(ab')_2$ goat anti-mouse $F(ab')_2$ on a fluorescence activated cell sorter (FACS-I) or Epics V as previously described (Reinherz et al., 1979a). Background fluorescence reactivity was determined with a control ascites obtained from mice immunized with nonsecreting hybridoma clones. For analysis, all monoclonal antibodies were utilized in antibody excess at dilutions of 1:250 to 1:1000.

4.4. Antigen stimulation of T lymphocytes

T lymphocytes were cultured at 1 x 10^6 per ml in final culture medium RPMI 1640 with 10% human AB serum, 200mM L-glutamine, 25mM HEPES buffer (Microbiological Associates), 0.5% sodium bicarbonate and 1% penicillin streptomycin with an equal volume of either tetanus toxoid (Massachusetts Department of Public Health Biological Laboratories, Boston, MA) at a 1:200 dilution, mumps C fixation antigen (Microbiological Associates) at a 1:20 dilution, 1 x 10^5 allogeneic B lymphoblastoid (Laz 156) cells, autologous E-cells, or medium alone. Specifically, 1 x 10^5 cells in 0.1 ml of culture media were placed in round bottomed microtiter plates, (Falcon, Oxnard, CA) and supplemented with 0.1 ml of soluble antigen, alloantigen or autologous E- cells as described above. Microplates were incubated in a 95% air, 5% CO_2 humid atmosphere for 5 days and cultures were then pulsed with 0.15 µCi of ^{3}H-TdR (specific activity, 1.9 CI/mM; Schwartz-Mann, Orangeburg, NY) per well.

After a 16 hour incubation with a radiolabelled thymidine, the cells were harvested using a MASH II apparatus (Microbiological Associates) and the ^{3}H-TdR incorporation measured in a liquid scintillation counter (Packard Instrument Co., Downers Grove, Il). Each experimental group was assayed in triplicate and expressed as mean specific count above background per minute (CPM) ± the standard deviation of the mean.

4.5. Detection of in vitro secretion of IgG

Unfractionated and separated populations of lymphocytes were cultured in round bottomed microtiter culture plates (Falcon) at 37°C in a humid atmosphere with 5% CO_2 for 7 days in RPMI 1640 supplemented with 20% heat inactivated fetal calf serum (Microbiological Associates), 0.5% sodium bicarbonate, and 1% penicillin streptomycin. To determine the effect of subpopulations of T cells on secretion of IgG by autologous plasma cells, 5 x 10^4 unfractionated T cells, unfractionated T cells treated with antibody and G/M FITC, or purified T4+ and T8+T cell subsets were added to 5 x 10^4 B cells or 2 x 10^4 T4+ T cells and 5 x 10^4 B cells in a volume of 0.1 ml of pokeweed mitogen (Grand Island, Grand Island, NY) at a 1:50 dilution. On day 7, cultures were terminated, supernatants harvested, and IgG secretion into supernatant determined by solid phase immunoassay utilizing a monoclonal antibody directed at the Fc portion of the human gamma heavy chain (anti-gamma Fc) (Gift of V. Raso, Sidney Farber Cancer Institute) as previously described (Reinherz et al., 1980h).

REFERENCES

Bhan, A.K., Reinherz, E.L., Poppema, S., McCluskey, R. and Scholossman, S.F. 1980, Location of T cell and MHC antigens in the human thymus. J. Exp. Med. 152:771-782.

Cantor, H. and Boyse, E.A. 1977, Regulation of cellular and humoral immune responses by T cell subclasses. Cold Spring Harbor Symp. Quart. Biol. 41:23-32.

Chess, L., MacDermott, R.P. and Schlossman, S.F. 1974a, Immunologic functions of isolated human lymphocyte subpopulations. I. Quantitative isolation of human T and B cells and response to mitogens. J. Immunol. 113:1113-1121.

Chess, L., MacDermott, R.P. and Schlossman, S.F. 1974b, Immunologic functions of isolation human lymphocyte subpopulations. II. Antigen triggering of T and B cells in vitro. J. Immunol. 113:1122-1127.

Cohen, S., Pick, E. and Oppenheim, J.J. 1979, Biology of the Lymphokines. (New York: Academic Press).

Eardley, D.D., Hugenberger, J., McVay-Boudreau, L., Shen, F.W., Gershon, R.K. and Cantor, H. 1978, Immunoregulatory circuits among T cell sets. II. T helper cells induce other T cell sets to exert feedback inhibition. J. Exp Med. 147:1106-1115.

Evans, R.L., Lazarus, H., Penta, A.C. and Schlossman, S.F. 1978a, Two functionally distinct subpopulations of human T cells that collaborate in the generation of cytotoxic cells responsible for cell-mediated lympholysis. J. Immunol 129:1423-1428.

Evans, R.L., Faldetta, T.J., Humphreys, R.E., Pratt, D.M., Yunis, E.J. and Schlossman, S.F. (1978b). Peripheral human T cells sensitized in mixed leukocyte culture synthesize and express Ia-like antigens. J. Exp. Med. 148:1440-1445.

Geha, R.S., Schneeberger, E., Rosen, F.S. and Merler, E. 1973, Interaction of human thymus derived and non-thymus derived lymphocytes in vitro. Induction of proliferation and antibody synthesis in B lymphocytes by a soluble factor released from antigen-stimulated T lymphocytes. J. Exp. Med. 138:1230-1247.

Howard, F.D., Ledbetter, J.A., Wong, J., Bieber, C.P., Stinson, E.B. and Herzenberg, L.A. 1981, A human T lymphocyte differentiation marker defined by monoclonal antibodies that block rosette formation. J. Immunol. 126:2117-2122.

Horton, J.E., Raiza, L.G., Simmons, H.A., Oppenheim, J.J. and Mergenhagen, S.E. 1972, Bone resorbine activity in supernatant fluid from cultured human peripheral blood leukocytes. Science 177:793-795.

Humphreys, R.E., McCune, J.M., Chess, L., Herrman, H.D., Malenka, D.J., Mann, D.L., Parham, P., Scholossman, S.F. and Strominger, J.L. 1976, Isolation and immunologic characterization of a human B lymphocyte specific cell surface antigen. J. Exp. Med. 144:98-122.

Kamoun, M.P., Martin, P.J., Hansen, J.A., Brown, M.R., Siadan, A.W. and Nowinski, R.C. 1980, Identification of human T lymphocyte surface proteins associated with the E Rosette receptor. J. Exp. Med. 153:207-213.

Keller, D.M., Swierhosz, J.E., Marrock, P. and Kappler, J.W. 1980, Two types of functionally distinct syngergizing helper T cells. J. Immunol. 124:1350-1359.

Konda, S., Stockert, E. and Smith, R.T. 1973, Immunologic properties of mouse thymus cells: membrane antigen patterns. Cell. Immunol. 7:275-289.

Ledbetter, J., Evans, R.L., Lipinski, M., Cunnningham-Rundles, C., Good, R.R. and Herzenberg, L.A. 1981, Evolutionary conservation of surface molecules that distinguish T lymphocytes inducer and cytotoxic, suppressor subpopulations in mouse and man. J. Exp. Med. 153:310-320.

Lipton, J.M., Reinherz, E.L., Kidisch, M., Jackson, P.L., Schlossman, S.F. and Nathan, D.G. 1980, Mature bone marrow erythroid burst-forming units (BFU-E) do not require T cells for induction of erythropoietic dependent differentiation. J. Exp. Med. 152:350-360.

Martin, P.J., Hanson, J.A., Nowinski, R.C. and Brown, M.A. 1980, A new human T cell differentiation antigen: unexpected expression on chronic lymphocytic leukemia cells. Immunogen. 11:429-439.

McMicheal, A.J. Pilch, Galfre, G., Mason, D.Y., Fabre, J.W. and Milstein, C. 1979, A human thymocyte antigen defined by a hybrid myeloma monoclonal antibody. Eur. J. Immunol. 9:205-210.

Mehra, V., Mason, L.H., Rothman, W., Reinherz, E.L., Schlossman, S.F. and Bloom, R.R. 1980, Delineation of a human T cell subset responsible for lepromin-induced suppression in leprosy patients. J. Immunol 125:1183-1188.

Moore, M.A.S. and Owen, J.J.T. 1967, Experimental studies on the development of the thymus. J. Exp. Med. 126:715-725.

Morimoto, C., Reinherz, E.L., Steinberg, A.D., Schur, P.H., Mills, J.A. and Schlossman, S.F. 1980a, Alterations in immunoregulatory T cell subsets in active SLE. J. Clin. Invest. 66:1171-1174.

Morimoto, C., Reinherz, E.L., Abe, T., Homma, M. and Schlossman, S.F. 1980b. Characterization of anti-T cell antibodies in systemic lupus erythematosus (SLE): Evidence for selective reactivity with normal suppressor cells defined by monoclonal antibiodies. Clin. Immunol. and Immunopathol 16:464-484.

Morimoto, C., Reinherz, E.L., Borel, Y., Mantzourais, E., Steinberg, A.D. and Schlossman, S.F. 1981, An autoantibody to an immunoregulatory inducer population in patients with juvenile rheumatoid arthritis (JRA). J. Clin. Invest. 67:753-761.

Owen, J.T.T. and Ritter, M.A. 1969, Tissue interactions in the development of thymus lymphocytes. J. Exp. Med. 129:431-437.

Owen, J.T.T. and Raff, M.C. 1970, Studies on the differentiation of thymus derived lymphocytes. J. Exp. Med. 132:1216-1232.

Raff, M.C. 1971, Surface antigenic markers for distinguishing T and B lymphocytes in mice. Transplant. Rev. 6:52-80.

Reinherz, E.L. and Schlossman, S.F. 1979, Con A-inducible suppression of MLC: evidence for mediation by the TH2+ T cell subset in man. J. Immunol. 122:1335-1341.

Reinherz, E.L. and Schlossman, S.F. 1980a, The differentiation and function of human T lymphocytes. Cell 19:821-827.

Reinherz, E.L. and Schlossman, S.F. 1980b, Regulation of the immune response: inducer and suppressor T lymphocyte subsets in man. N. Engl. J. Med. 303:370-373.

Reinherz, E.L. and Schlossman, S.F.. The derivation of human T cell malignancies. Cancer (In press).

Reinherz, E.L., Kung, P.C., Goldstein, G. and Schlossman, S.F. 1979a, A monoclonal antibody with selective reactivity with functionally mature thymocytes and all peripheral human T cells. J. Immunol. 123:1312-1317.

Reinherz, E.L., Kung, P.C., Goldstein, G. and Schlossman, S.F. 1979, Separation of functional subsets of human T cells by a monoclonal antibody. Proc. Natl. Acad. Sci. USA 76:4061-4065.

Reinherz, E.L., Kung, P.C., Goldstein, G. and Schlossman, S.F. 1979c, Further characterization of the human inducer T cell subset defined by monoclonal antibody. J. Immunol. 123:2894-2896.

Reinherz, E.L., Parkman, R., Rappeport, J.M., Rosen, F.S. and Schlossman, S.F. 1979d, Abberrations of suppressor T cells in human graft-versus-host disease. N. Engl. J. Med. 300:1061-1068.

Reinherz, E.L., Rubinstein, A.J., Geha, R.S., Rosen, F.S. and Schlossman, S.F. 1979e, Abnormalities of immunoregulatory T cells in disorders of immune function. N. Engl. J. Med. 301:1018-1022.

Reinherz, E.L., Kung, P.C., Pesando, J.M., Ritz, J., Goldstein, G. and Schlossman, S.F. 1979f, Ia determinants on human T cell subsets defined by monoclonal antibody: activation stimuli required for expression. J. Exp. Med. 150:1472-1482.

Reinherz, E.L., Kung, P.C., Goldstein, G., Levey, R.H. and Schlossman, S.F. 1980a, Discrete stages of intrathymic differentiation: analysis of normal thymocytes and leukemia lymphoblasts of T lineage. Proc. Natl. Acad. Sci. USA 77:1588-1592.

Reinherz, E.L., Kung, P.C., Goldstein, G. and Schlossman, S.F. 1980b, A monoclonal antibody reaction with the human cytotoxic/suppressor T cell subset previously defined by a heteroantiserum termed TH_2. J. Immunol 124:1301-1307.

Reinherz, E.L., Morimoto, C., Penta, A.C. and Schlossman, S.F. 1980c, Regulation of B cell immunoglobulin secretion by functional subsets of T lymphocytes in man. Eur. J. Immunol. 10:570-572.

Reinherz, E.L., Hussey, R.E. and Schlossman, S.F. 1980d, Absence of expression of Ia antigen on human cytotoxic T cells. Immunogen. 11:421-426.

Reinherz, E.L., Kung, P.C., Breard, J.M., Goldstein, G and Schlossman, S.F. 1980e, T cell requirements for generation of helper factor in man: analysis of the subsets involved. J. Immunol. 124:1883-1887.

Reinherz, E.L., Weiner, H.L., Hauser, S.L., Cohen, J.A., Distaso, J.A. and Schlossman, S.F. 1980, Loss of suppressor T cells in active multiple sclerosis: Analysis with monoclonal antibiodies. N. Engl. J. Med. 303:125-129.

Reinherz, E.L., O'Brien, C., Rosenthal, P. and Schlossman, S.F. 1980g, The cellular basis for viral-induced immunodeficiency: Analysis by monoclonal antibody. J. Immunol. 125:1269-1274.

Reinherz, E.L., Hussey, R.E. and Schlossman, S.F. 1980h. A monoclonal antibody blocking human T cell function. Eur. J. Immunol. 10:758-762.

Reinherz, E.L., Morimoto, C., Penta, A.C. and Schlossman, S.F. 1981a, Subpopulations of the T4+ inducer T cell subset in man: evidence for an amplifier population preferentially expressing Ia antigen upon activation. J. Immunol. 126:67-70.

Reinherz, E.L., Copper, M.D., Schlossman, S.F. and Rosen, F.S. 1981b, Abnormalities of T cell maturation and regulation in human beings with immunodeficiency disorders. J. Clin. Invest. 68:699-705.

Rocklin, R.E., MacDermott, R.P., Chess, L., Schlossman, S.F. and David, J.R. 1974. Studies on mediator production by highly purified human T and B lymphocytes. J. Exp. Med. 140:1303-1316.

Schlossman, S.F., Chess, L., Humphreys, R.E. and Strominger, J.L. 1976, Distribution of Ia-like molecules on the surface of normal and leukemic cells. Proc. Natl. Acad. Sci. USA 73:1288-1292.

Sondel, P.M., Chess, L., MacDermott, R.P. and Schlossman, S.F. 1975, Immunologic functions of isolated human lymphocytes subpopulations. III. Specific allogeneic lympholysis mediated by human T cells alone. J. Immunol. 114:982-987.

Stutman, O. and Good, R.A. 1971, Immunocompetence of embryonic hematopoietic cells after traffic to thymus. Transplant. Proc. 3:923-925.

Terhorst, C., van Agthoven, A., Reinherz, E.L. and Schlossman, S.F. 1980, Biochemical analysis of human T lymphocyte differentiation antigens specific for the inducer subset (T4) and the cytotoxic/suppressor subset (T5). Science 209:520-521.

Terhorst, C., van Agthoven, A., LeClair, K., Snow, P., Reinherz, E.L. and Schlossman, S.F. 1981, Biochemical studies of the human thymic differentiation antigens T6, T9 and T10. Cell 23:771-780.

van Agthoven, A., Terhorst, C., Reinherz, E.L. and Schlossman, S.F. 1981, Characterization of T cell surface glycoprotein T1 and T3 present on all human peripheral T lymphocytes and functional mature thymocytes. Eur. J. Immunol 11:18-21.

Winchester, R.J., Fu, S.M., Wernet, P., Kunkel, H.G., Dupont, B. and Jerslid, C. 1975, Recognition by pregnancy serums of non-HLA alloantigens selectively expressed on B lymphocytes. J. Exp. Med. 141:924-929.

Yu, D.T.Y., Winchester, R.J., Fu, S.M., Gibofsky, A., Ko, H.S. and Kunkel, H.G. 1980, Peripheral blood Ia-positive T cells: increased in certain diseases and after immunization. J. Exp. Med. 151:91-100.

5

IMMUNE (γ) INTERFERON AND LYMPHOTOXIN PRODUCTION BY T CELL LINES AND HYBRIDS

Nancy H. Ruddle and Barbara S. Conta

Department of Epidemiology and Public Health, Yale University Medical School, P.O. Box 3333, New Haven, Ct. 06520

1. SUMMARY

T cell functions have been immortalized in antigen-specific continual cell lines and hybrids. C57BL/6 mice were sensitized to produce delayed hypersensitivity by injection of NP-chicken gamma globulin (NP-CGG)[1] subcutaneously in the tail. T cells have been maintained continuously for several months in the presence of CGG, x-irradiated syngeneic spleen cells and T cell growth factor (TCGF). These TCGF maintained cells produce lymphotoxin (LT), as assayed by killing of innocent bystander A9 or L929 cells, and immune interferon (IFN- γ), as assayed by inhibition of vesicular stomatitis virus cytopathic effects in L929 cells. T cells (3×10^5/ml) grown with CGG produce 256 IU IFN/ml after 96 hours of culture. Addition of CON A boosts this to 2048 IU/ml. LT of high titers was simultaneously produced by these cells. Growth and lymphokine production can be dissociated in these lines. CON A, normally mitogenic for primary T cells, though inhibitory for many T cell tumors, inhibits growth of T cell lines while stimulating LT and (IFN- γ) production.

Hybrids have been made between BW5147, an AKR $HGPRT^-$ T lymphoma, and T cell lines described above, by means of polyethylene glycol induced

---

[1] Abbreviations: NP-CGG, nitrophenyl chicken gamma globulin; HGPRT, hypoxanthine guanine phosphoribosyl transferase; HAT, hypoxanthine-aminopterin-thymidine; LT, lymphotoxin; MIF, migration inhibitory factor; TCGF, T cell growth factor; IL-2, interleukin-2; IFN, interferon; CON A, concanavalin A; PHA, phytohemagglutin; CGG, chicken gamma globulin; Hy, hemocyanin; OVA, ovalbumin; GPI, glucose phosphate isomerase; TG, thioguanine; and Ou, oubain.

fusion and HAT selection. One such hybrid, Hyb 82C10, produced LT constitutively, but did not produce IFN. Other T cell hybrids, for example (Hyb 64 C11), made with BW5147 and T cells prepared directly from ovalbuminova (OVA) sensitized mice have also made high titers of LT constitutively. In these and all other hybrids isolated so far, lymphokine production ceased abruptly within 8 weeks, possibly due to chromosome segregation.

This work shows that T cell hybrids and IL-2 maintained T cell lines can be homogeneous sources for lymphokines for further molecular characterization and functional analysis.

2. INTRODUCTION

The goal of our work is an indepth analysis of the thymus derived lymphocyte (T cell) and its products. T cell functions, products and molecular biology can now be analyzed through the use of continual T cell lines and hybrids. As demonstrated by Goldsby and his colleagues (1977), primary T cells can be immortalized within the milieu of a proliferating AKR Thy 1.2 $HGPRT^-$ lymphoma, BW5147, through the technique of somatic cell hydridization and selection in HAT medium. Constitutive traits such as isozymes, Thy 1 and H-2 antigens of both parents, and antigen specific activity of the primary parent can be expressed by such hybrids (Ruddle, 1978a; Whitaker and Ruddle, 1980; Ruddle et al., 1980b), and expression of many such activities is regulated by culture phase (Ruddle et al., 1980a). We (Goodman et al., 1980) and others (Kapp et al., 1980; Taniguchi et al., 1979) have demonstrated constitutive production of antigen specific suppressor factors by such hybrids. The scheme for preparing antigen specific, functional T cell hybrids is depicted in Figure 1. A hybrid cell maintaining expression of all functions of the primary T cell parent is not always achieved, as chromosomes may segregate from T cell hybrids with concommitant loss of traits. This situation also occurs in B cell hybridomas (Köhler et al., 1977) and all types of somatic cell hybrids (reviewed in Giles and Ruddle, 1973).

Lymphokines are T cell products which are induced by mitogen or specific antigen whose final activity is neither antigen specific nor H-2 restricted, though in some cases the target is species restricted. Their production is positively correlated with delayed hypersensitivity. Lymphotoxin (LT) and migration inhibitory factor (MIF) are produced predominantly by Ly 1 T cells, though production by Ly 2 cells can also occur (Eardley et al., 1980; Kuhner et al., 1980; Newman et al., 1978; and Adelman et al., 1980). Other lymphokines include T cell growth factor (TCGF) or interleukin 2 (IL-2) and immune interferon (IFN-γ). IL-2 production has been induced in T cell hybrids by Concanavalin A (CON A) (Harwell et al., 1980) and by specific antigen (Kappler et al., 1981). Low level production of IFN-γ has been reported in murine T cell lines maintained in TCGF (Marcucci et al., 1981). Thus far, there have been no reports of LT production by T cell lines or hybrids, nor IFN-γ production in antigen specific lines or in any T cell hybrids.

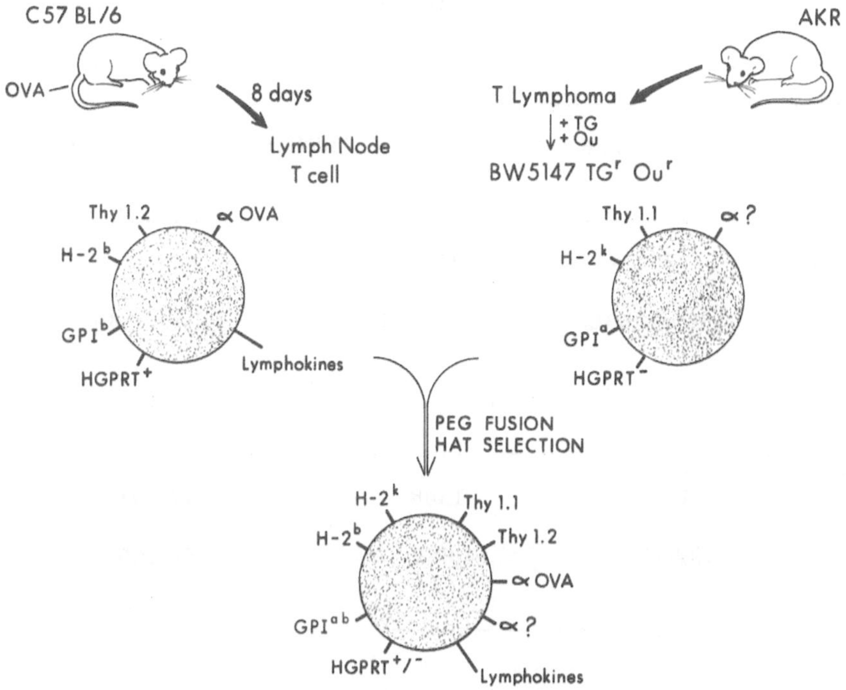

Figure 1. Schema for preparation of T cell hybrids.

The specific goal of the work reported here was the production of T cell lines and hybrids making LT and/or IFN-γ so that we might compare the regulation, physiology and biochemistry of these two lymphokines. Preparations of each frequently contain activities associated with the other (Ware and Granger 1979; Blalock et al., 1980; Ito and Buffett, 1981), so we wished to be able to clearly distinguish them and to obtain substantial quantities of homogeneous IFN-γ and LT, which are both of potential importance as therapeutic agents in cancer.

3. RESULTS

3.1. Growth factor lines

3.1.1. Properties. The growth factor maintained lines described here are all Thy 1.2 positive, surface immunoglobulin negative by immunofluorescence. Their growth is stimulated by the presence of specific antigen (CGG) (Table I). At high concentrations of TCGF (>10%) the antigen dependency for growth is not as stringent, though still apparent. The cells of most of the lines studied here have a doubling time of approximately 35 hours in the presence of antigen, 10% TCGF, and x-irradiated syngeneic spleen cells. Cells may be grown up to concentrations of 4 x 10^5/ml in microtiter wells, but their growth in bulk cultures

TABLE I

Antigen Specific Stimulation of DNA Synthesis in Line 19[a]

cpm 6 DAYS AFTER ADDITION OF			
TCGF	TCGF	TCGF	TCGF
—	spleen cells	spleen cells	spleen cells
—	—	Hy	CGG
536[b]	1450	1348	17,901
2157[c]	19,575	14,924	59,195

[a]T cells were isolated from C57BL/6 mice 8 days after sensitization with 100 μg NP-CGG subcutaneously in the tail, and maintained with TCGF, CGG, and syngeneic x-irradiated spleen cells.

[b]2.5% TCGF

[c]25% TCGF

is less reliable. The cell lines described here have been maintained in culture from 4 to 8 months. The cells can be frozen and recovered from liquid nitrogen.

3.1.2. Production of LT and IFN. We have been able to detect LT and IFN in supernatants of certain T cell lines grown in the presence of specific antigen. Supernatants sampled from cultures of Line 32 seeded at 1×10^5 cells/ml contain both activities within 24 hours of plating and activity increases to maximal levels at approximately 96 hours (Table II). The addition of CON A elicits both an earlier peak and a higher titer of IFN (2048 IU/ml in the experiment depicted in Table II). The lymphotoxin titer is also higher in supernatants of cells grown in the presence of mitogen (Conta et al, 1983).

TABLE II

LT and IFN Production by T cell line 32 in the presence of antigen, CON A and PHA[a]

Culture Condition	LT % Cytotoxicity	IFN IU/ml
Control + CGG	0	<4
Control + CGG + CON A	0	<4
CGG	81%	256
CGG + CON A	90%	2048
CGG + PHA	94%	1024

[a]Supernatants were taken 96 hours after addition of TCGF, x-irradiated syngeneic spleens and CGG or CGG and CON A (5 μg/ml) or PHA (20 μg/ml). Control supernatants were generated in the absence of T cell line 32.

Not all TCGF lines make and release both LT and IFN. Cells of one line, Line 19, are cytotoxic to innocent bystander cells, but only release LT into supernatants if the T cells are grown in the presence of CON A. IFN is also released into the supernatants only if the cells are treated with antigen. The supernatants of some sublines of Line 32 described above contain only LT and not IFN, though the reverse, i.e., production of IFN in the absence of LT, has not been detected as yet.

The IFN produced by the T cell lines studied here fulfills criteria for immune (γ) IFN. It is produced by T cells, is elicited by specific antigen or mitogen, and is acid labile. Dialysis of supernatants against 0.1 M glycine-HCl at pH2 for 4 hours and then against PBS eliminates a portion (75%) of the activity. Control dialysis at pH7 only marginally affects it (Conta et al, 1983).

Lymphokine production and cellular proliferation are not necessarily coupled in these T cell lines. In many lines the antigen dependency for proliferation is more pronounced than for lymphokine production. Furthermore, though CON A and PHA are both mitogenic for, and induce lymphokines from, normal T cells, they inhibit the growth of the T cell lines described here (Figure 2) while at the same time they stimulate LT and IFN production. Similar CON A induction of a more differentiated phenotype, while inhibiting growth, has been observed in the case of a T cell hybrid whose productionof IL-2 is stimulated by CON A (Harwell et al., 1980). PHA and CON A have previously been reported to inhibit the growth of T cell tumors, and this has even been suggested as one means of determining the cell of origin of a lymphoid tumor (Ralph, 1973; Ralph and Nakoinz, 1973). In this respect, the T cell lines described here are more similar to malignant than to normal cells.

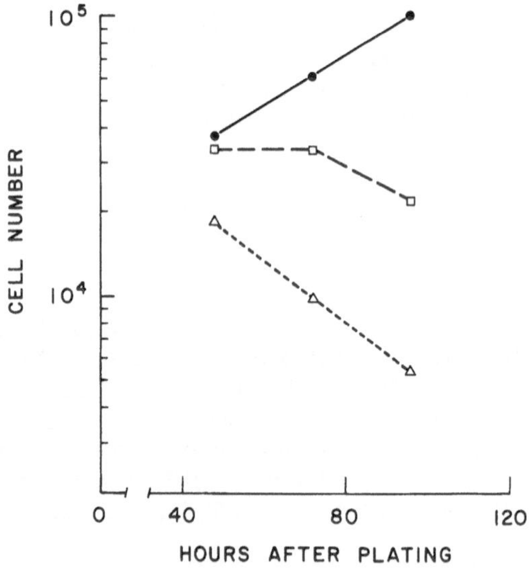

Figure 2. Growth curve of T cell Line 32. Cells were grown in 0.2 ml in the presence of 10% TCGF, 8 x 10^5/ml x-irradiated spleen cells/ml, and the following additives; ●——●CGG, 250 μg/ml; □---□ CGG 250 μg/ml, PHA 20 μg/ml; Δ ---- Δ CGG 250 μg, CON A 5 μg/ml. Cell number refers to the number of viable T blast cells/well at the indicated times.

3.2. T cell hybrids

3.2.1. Analysis. As is the case with all T cell hybrids we have previously described, the T cell hydrids described here have been subjected to analysis of a minimum of 3 traits coded by genes on different chromosomes, in order to confirm their identification as true hybrids, and not revertants, and to establish that T cell functions of both parents could be expressed. These traits are described in Table III.

3.2.2. LT Production. Previous experiments in which T cells were removed from mice 8 days after sensitization with antigen in vivo, and then immediately hybridized with BW5147, did not result in the isolation of hybrids making LT consitutively or after induction with feeder cells and antigen. Hybrid lines have been obtained which produce LT constitutively when the primary parent cells were obtained from mice sensitized for 8 days in vivo with 50 μg OVA and complete Freund's adjuvant and then restimulated in vitro for 17 hours before hybridization. However expression of LT continues for only a brief period and then ceases. Hyb 64C11, whose LT production is depicted in Figure 3, is an example of such a hybrid. This early cessation of LT production may be due to chromosome segregation. Hyb 64C11 has been subjected to extensive nucleic acid analysis with cloned immunoglobulin gene probes. This hybrid, which stopped producing LT within 1 month of hybridization, when analyzed 3 months later still retained at least one copy of the C57BL/6 primary parent chromosome 12 and made a unique Cμ RNA not made by BW5147 (Zuniga, D'Eustachio and Ruddle, 1982). This suggests that the presence of chromosome 12, which contains structural genes for immunoglobulin heavy chain production, is not sufficient to allow LT production.

Hybrids were prepared between BW5147 and cells of one of the T cell growth factor lines (19). One of the (Hyb 82C10) released LT constitutively into the supernatant and maintained this production for 8 weeks. It did not produce IFN, nor could this be induced by the addition of antigen, spleen cells, CON A, or any combination of these agents.

4. DISCUSSION

We have presented evidence that IL-2 maintained T cell lines can produce LT and IFN when stimulated by specific antigen or mitogen. Furthermore, a portion the IFN produced by these T cell lines after antigen or mitogen induction is acid labile, consistent with its identification as immune or IFN-γ. It is apparent that individual T cell lines exhibit different properties. Line 19 cells themselves can kill A9 cells in the presence of antigen, but release LT (and IFN-γ) into the medium only upon induction with mitogen. Line 32 produces and releases both LT and IFN-γ when stimulated by antigen, and these titers are elevated considerably when mitogen is present. Certain sublines of 32 release only LT, and not IFN into the medium after antigen stimulation. It is probable that

TABLE III

Analysis of T cell hybrids

Parameter	Chromosome	Method
Hypoxanthine guanine phosphoribosyl transferase	X	Growth in HAT medium
Isozymes		
Glucose phosphate isomerase	7	Starch gel electrophoresis
Isocitrate dehydrogenase	1	Starch gel electrophoresis
Antigens		
H-2	17	Quantitative absorption of cytotoxicity Direct cytotoxicity
Thy-1	9	Quantitative absorption of cytotoxicity Immunofluorescence
Lymphokines		
Lymphotoxin		Killing A9 or L929 cells
Interferon		Inhibition of cytopathic effect of vesicular stomatitis virus
Nucleic Acids		
DNA	6,12,16	Southern blot analysis with cloned immunoglobulin gene probes
RNA		Northern blot analysis with cloned immunoglobulin gene probes

these uncloned lines undergo selection during maintainance in culture, and thus exhibit different properties. We are in the process of cloning the lines in order to determine if a single cell is able to release both lymphokines simultaneously. Our results with T cell hybrids, such a Hyb 82C10, as well as sublines of 32 which make only LT and not IFN, suggest that, within the limits of our detection, the production of LT is not dependent upon the simultaneous synthesis of IFN, and that these two activities are physically separable.

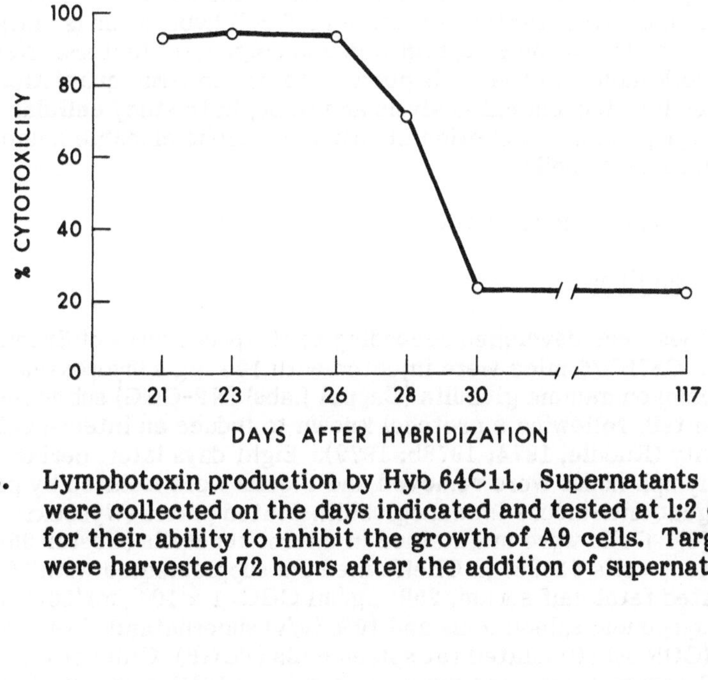

Figure 3. Lymphotoxin production by Hyb 64C11. Supernatants were collected on the days indicated and tested at 1:2 dilution for their ability to inhibit the growth of A9 cells. Target cells were harvested 72 hours after the addition of supernatants.

The IFN- γ levels reported here (2048 IU/3 x 10^5 cells/ml) are significantly higher than those in the only other report of IFN- γ production by T cell lines. The highest level obtained in that publication at equivalent CON A levels (5 μg/ml) was 800 IU/1 x 10^6 cells/ml (Marcucci et al., 1981). Levels of human IFN-γ comparable to those reported here have been detected in supernatants of a mitogen treated human T cell tumor as well (Nathan et al., 1981).

The consistent finding of an abrupt cessation of LT production in T cell hybrids indicates that this trait is considerably less stable than others we have studied (Ruddle, 1982). Though T cells are not usually considered to be targets of LT activity, it may be that production of LT is slightly detrimental to T cell growth. Cells which have lost LT production, perhaps through chromosome segregation, might possess even a slight growth advantage. This putative chromosome segregation may permit the determination of chromosome assignments of individual lymphokines.

The work presented here indicates that the technology is available for the preparation of T cell lines and hybrids which produce lymphokines. Each system possesses individual problems. Difficulties in cloning, adaptation to growth in bulk culture, the requirement for the presence of spleen cells, and TCGF preparations which may contain other lymphokines are all disadvantages in the use of T cell lines. The presence of the BW5147 genome is a complicating factor in the use of T cell hybrids, as is marker instability due to chromosome segregation and epigenetic factors. Nevertheless, our work indicates that it is possible to obtain large quantities of lymphokines for biochemical analysis and to begin to study cellular regulation of lymphokine production in antigen-specific clonable homogeneous populations of cells.

5. EXPERIMENTAL PROCEDURES

5.1. T cell lines

Lines were developed according to the procedures of Sredni et al., (1980). C57BL/6 mice were injected with 100 µg 4-hydroxy-3-nitrophenyl-chicken gamma globulin (Cappel Labs) (NP-CGG) subcutaneously in the tail, following a protocol known to induce an intense delayed hypersensitivity (Ruddle, 1974; 1978b; 1979). Eight days later, periaortic and inguinal lymph nodes were removed and T cells were isolated by panning on rabbit or goat anti-mouse immunoglobulin coated plates (Wysocki and Sato, 1978). Cells were maintained in microtiter wells (Costar 96-well flat bottom wells #3596) in Click's medium supplemented with 10% heat inactivated fetal calf serum, 250 µg/ml CGG, 1×10^5 /ml/2000 R x-irradiated syngeneic spleen cells and 10% (v/v) supernatants from Concanavalin A (CON A) stimulated rat spleen cells (TCGF) (Gillis et al., 1978). TCGF preparations were induced with 5 µg CON A/ml and 1×10^7 rat spleen cells/ml, in RPMI plus 5% HI FCS, and 5×10^{-5} M 2-mercaptoethanol. Supernatants were pooled at 24 and 48 hours, and screened for LT and IFN activity. All preparations used here contained less than 10% LT activity and 4 IU IFN.

5.2. T cell hybrids

Hybrids were made as described previously (Ruddle, 1978a) by means of polyethylene glycol (Baker 1540 MW, 40% weight/volume) induced fusion of BW5147, an AKR HGPRT$^-$ lymphoma, with primary T cells or T cell lines. Hybrids were selected in HAT medium and maintained in DME plus 10% calf serum. All hybrids have been tested for constitutive and facultative markers characteristic of each parent. These have included isozymes, Thy 1, H-2, (Ruddle 1978aa; Ruddle et al., 1980a) and the presence of polymorphic regions of chromosome 12 (Zuniga, D'Eustachio and Ruddle, 1982).

5.3. Lymphotoxin assay

Cells or supernatants were analyzed for LT activity essentially as described previously (Ruddle, 1979; Eardley et al., 1980). T cells, or their supernatants, after incubation with specific antigen, were added to innocent bystander A9 or L929 cells. Seventy two hours later the surviving fibroblasts were enumerated with a Coulter counter at a setting which excluded the (smaller) lymphocytes and included the (larger) target cells. Cytotoxicity was calculated as:

$$\%/\text{Inhibition} = 100 \times \frac{\text{No. A9 cells in cultures with supernatants from test cells}}{\text{No. A9 cells in cultures with Click's medium}}$$

The effect of antigen, TCGF, and spleen cells on A9 growth was analyzed in experiments where applicable.

5.4. Interferon Assay

IFN activity was analyzed as described by Dahl and Degre' (1972) and measured as protection of L929 cells from the cytopathic effect (CPE) of vesicular stomatitis virus (VSV). Interferon standard (βIFN-L929 origin) or test supernatants were added to L929 cells. Eighteen hours later, 8×10^3 plaque forming units (PFU) of VSV (Indiana strain, obtained through the generosity of Dr. Peter Lengyel) was added to each well. Twenty four hours later viral cytopathic effects (ranging from 3+ to -) were scored blind. Quantitation of interferon international units (IFN-IU/ml) was determined by comparison with an NIH reference standard (NIH #G-002-904-511).

Acknowledgements

This work was supported by NCI grant CA 16885 and grant RR05543 awarded by the Biomedical Research Grant Program, DRR, NIH. NHR is the recipient of an American Cancer Socity Faculty Research Award FRA-196. We thank Belinda Beezley for excellent technical assistance and Vera Wardlaw for help in manuscript preparation.

NOTE ADDED IN PROOF:

Results obtained after cloning of line 32 indicate that not only do these cells produce LT and IFN- γ, but they release IFN-α , β as well (Conta et al., 1983).

REFERENCES

Adelman, N., Ksiazek, J., Yoshida, T., and Cohen S., 1980, Lymphoid sources of murine migration inhibition factor, J. Immunol. 124:825-830.

Blalock, J.E., Georgiades, J.A., Langford, M.P., and Johnson, H.M., 1980, Purified human immune interferon has more potent anticellular activity than fibroblast or leukocyte interferon, Cell Immunol. 49(2):390-394.

Conta, B.C., Powell, M.B. and Ruddle, N.H., 1983, Lymphotoxin and immune interferon production by T cell lines, J. Immunol. 130: in press.

Dahl, H., and Degré, M., 1972, A microassay for mouse and human interferon, Acta. Path. Microbial. Scand. Section B 80:863-870.

Eardley, D., Shen, F.W., Gershon, R.K., and Ruddle, N.H., 1980, Lymphotoxin production by subsets of T cells, J. Immunol. 124:1199-1202.

Giles, R.E., and Ruddle, F.H., 1973, Production and characterization of proliferating somatic cell hybrids, in: Tissue Culture, Academic Press, New York, 475-500.

Gillis, S., Ferm, M.M., Ou, W., and Smith, K.A., 1978, T cell growth factor: Parameters of production and a quantitative microassay for activity, J. Immunol. 120:2027-2032.

Goldsby, R.A., Osborne, B.A., Simpson, E., and Herzenberg, L.A., 1977, Hybrid cell lines with T-cell characteristics, Nature 267:707-708.

Goodman, J.W., Lewis, G.K., Primi, D., Hornbeck, P., and Ruddle, N.H., 1980, Antigen-specific molecules from murine T lymphocytes and T cell hybridomas, Molecular Immunol. 17:933-945.

Harwell, L., Skidmore, B., Marrack, T., and Kappler, J., 1980, Concanavalin A-inducible, interleukin-2-producing T cell hybridomas, J. Exp. Med. 152:893-904.

Ito, M., and Buffett, R.F., 1981, Cytocidal effect of purified human fibroblast interferon on tumor cells in vitro, J. Nat. Canc. Inst. 66:819-825.

Kapp, J.A., Araneo, B.A., and Clevinger, B.L., 1980, Suppression of antibody and T cell proliferative responses to L-glutamic acid^{60}-L-alanine^{30}-L-tyrosine^{10} (GAT) by a specific monoclonal T cell factor, J. Exp. Med. 152:235-240.

Kappler, J.W., Skidmore, B.J., White, J., and Marrack, P., 1981, Antigen-inducible, H-2 restricted, IL-2 producing hybridomas: Lack of independent antigen and H-2 recognition, J. Exp. Med. 153:1198-1214.

Köhler, G., Pearson, T., and Milstein, C., 1977, Fusion of T and B cells, Somatic Cell Genet. 3:303-312.

Kühner, A.L., Cantor, H., and David, J.R., 1980, Ly phenotype of lymphocytes producing murine migration inhibitory factor (MIF), J. Immunol. 125:1117-1119.

Marcucci, F., Waller, M., Kirchner, H., and Krammer, P., 1981, Production of immune interferon by murine T cell clones from long-term cultures, Nature 291:79-81.

Nathan, I., Groopman, J.E., Quan, S.G., Bersch, N., and Golde, D.W., 1981, Immune (γ) interferon produced by a human T lymphoblast cell line, Nature 292:842-844.

Newman, W., Gordon, S., Hammerling, U., Senik, A., and Bloom, B.R., 1978, Production of migration inhibition factor (MIF) and an inducer of plasminogen activator (IPA) by subsets of T cells in an MLC, J. Immunol. 120:927-931.

Ralph, P., 1973, Retention of lymphocytic characteristics of myelomas and Θ + lymphomas: Sensitivity to cortisol and phytohemagglutinin, J. Immunol. 110:1470-1475.

Ralph B., and Nakoinz, I., 1973, Inhibitory effects of lectins and lymphocyte mitogens on murine lymphomas and myelomas, J. Nat. Canc. Inst. 51:883-890.

Ruddle, N.H., 1974, Cytotoxicity reactions mediated by antigen-activated rat and mouse lymphocytes, in: Mechanisms of Cell-Mediated Immunity (McCluskey and Cohen), Wiley, Chichester, pp. 401-407.

Ruddle, N.H., 1978a, T cell hybrids with specificity for individual antigens, Curr. Top. Microbiol. Immunol. 81:203-211.

Ruddle, N.H., 1978b, Delayed hypersensitivity to soluble antigens in mice. I. Analysis in vivo, Int. Archs. Allergy Appl. Immun. 57:560-566.

Ruddle, N.H., 1979, Delayed hypersensitivity to soluble antigens in mice. II. Analysis in vitro, Int. Archs. Allergy Appl. Immun. 58:44-52.

Ruddle, N.H., 1982, T cell hybrids; their functions, products, and regulation, Lymphokines 5:49-76.

Ruddle, N.H., Beezley, B.B., and Eardley, D.D., 1980a, Regulation of self-recognition by T cell hybrids, Cellular Immunol. 55:42-55.

Ruddle, N.H., Beezley, B.B., Lewis, G.K., and Goodman, J.W., 1980b, Antigen specific T cell hybrids. II. T cell hybrids which bind azobenzenearsonate, Molec. Immunol. 17:925-931.

Sredni, B., Tse, H., and Schwartz, R., 1980, Direct cloning and extended culture of antigen-specific MHC-restricted proliferating T lymphocytes, Nature 283:581-583.

Taniguchi, M., Saito, T, and Tada, T., 1979, Antigen-specific suppressive factor produced by a transplantable I-J bearing T-cell hybridoma, Nature 278:5704, 555-558.

Ware, C.F., and Granger, G.A., 1979, A physicochemical and immunologic comparison of the cell growth inhibitory activity of human lymphotoxins and interferons in vitro, J. Immunol. 122(5):1763-1770.

Whitaker, R.B., and Ruddle, N.H., 1980, Antigen specific T cell hybrids. I. Ovalbumin binding T cell hybrid, Cellular Immunol. 55:56-65.

Wysocki, L.J., and Sato, V.L., 1978, "Panning" for lymphocytes: A method for cell selection, Proc. Natl. Acad. Sci. USA. 75:2844-2848.

Zuniga, M.C., D'Eustachio, P., and Ruddle, N.H., 1982, Immunoglobulin heavy chain gene rearrangement and transcription in murine T cell hybrids and T lymphomas, Proc. Natl. Acad. Sci. USA. 79:3015-3019.

PART III.

MONOCLONAL ANTIBODIES AS PROBES OF ANTIGEN STRUCTURE

6

PARTIAL STRUCTURAL CHARACTERIZATION OF HUMAN HLA-DR SUBSETS USING MONOCLONAL ANTIBODIES

Carolyn Katovich Hurley, Gabriel Nunez, Robert Winchester*, Olja Finn**, Ronald Levy**, Peter Stastny and J. Donald Capra

The University of Texas Health Science Center at Dallas, Department of Microbiology and Internal Medicine, 5323 Harry Hines Boulevard, Dallas, Texas 75235 *Hospital for Joint Diseases, New York, New York **Howard Hughes Medical Institute Laboratories and the Department of Medicine, Stanford University Medical Center, Stanford, California

1. SUMMARY

Monoclonal antibodies have been used to isolate and characterize human HLA-DR[1] antigens. Using sequential immunodepletion experiments, four subsets of DR antigens have been identified. Partial amino-terminal amino acid sequencing has identified all four subsets as DR molecules with sequences homologous to the murine I-E alloantigens. In addition, sequence data provide evidence for primary structural differences in the DR beta chains of two of the subsets suggesting the presence of multiple DR beta chain loci.

2. INTRODUCTION

The human histocompatibility locus, HLA-D, like its counterpart, the murine I region, controls a variety of immune responses including T cell-B cell collaboration, mixed lymphocyte reactions, kidney and bone marrow graft rejection and disease susceptibility (for review, see Winchester and Kunkel, 1979; Uhr et al., 1979). In addition, both HLA-D and I regions encode a polymorphic set of antigens, termed DR and Ia, found primarily on the surface of B cells and macrophages. These antigens consist of two polypeptide chains of molecular weight 34,000 daltons (alpha) and 28,000 daltons

[1]DR, HLA-DR; MHC, major histocompatability complex; Ig, immunoglobulin; TBS, TRIS buffered saline; SDS, sodium dodecyl sulfate.

(beta) with the majority of the polymorphism residing in the beta chain (Uhr et al., 1979; Kaufman et al., 1980). In the mouse, there are two structurally distinct sets of I region antigens, I-A and I-E, which contain different alpha and beta chains. A large body of evidence suggests that three of these chains (A_{alpha}, A_{beta} and E_{beta}) are encoded in the I-A subregion of the murine major histocompatibility complex (MHC); the fourth chain (E_{alpha}) is encoded in the I-E subregion of the complex (Uhr et al., 1979). In man, DR antigens have been characterized in several laboratories (Springer et al., 1977; Allison et al., 1978) and the evidence argues strongly that they are homologues of the murine I-E alloantigens.

Recent attention has been focused on the polymorphism of the DR antigens within an HLA homozygous B cell line. Studies using two-dimensional gel electrophoresis have shown that both alpha and beta chains expressed in these lines show a high degree of microheterogeneity (Shackelford and Strominger, 1980). Not all of this variability can be accounted for by differences in glycosylation suggesting the presence of either additional post-translational modifications or the presence of nonallelic DR antigens (i.e., multiple DR loci).

Several structural studies using both conventionally derived antisera and monoclonal antibodies have been directed at dissecting this heterogeneity in the DR-like molecules (Shackelford et al., 1981a; Accolla et al., 1981; Quaranta et al., 1981; Nadler et al., 1981). In this paper, several monoclonal anti-DR antibodies were used to isolate four subsets of DR-like antigens from the human DR homozygous cell line NEW. These subsets are homologous to the murine I-E antigens and appear to be structurally distinct from one another.

3. RESULTS AND DISCUSSION

3.1. Two monoclonal antibodies (L203 and L227) recognize subsets of DR antigens

Sequential immunoprecipitations utilizing the monoclonal antibodies L203 and L227 were performed using the DR homozygous cell lines NEW, ER4, MAT, and PRIESS. Glycoprotein pools were depleted of reactivity to one of the monoclonals using several sequential immunoprecipitations; the ability of the second monoclonal to react with the depleted pool was then tested. Figure 1 shows the SDS gel profiles of the antigens precipitated from the cell line NEW by this procedure. L203 precipitated a bimolecular complex in the approximate molecular weight range of 27,000 to 38,000 daltons (Figure 1A). Following several precipitations with L203, the reactivity to that monoclonal antibody was completely depleted (Figure 1B). L227 was then added to the depleted glycoprotein pool and a bimolecular complex was again precipitated (Figure 1C). The converse reactions were identical. L227 precipitated a bimolecular complex (Figure 1D); following L227 depletion (Figure 1E), L203 was still able to react with a similar set of antigens (Figure 1F). It was observed that varying amounts of L203 and L227 reactive molecules were precipitated from different cell lines and that the amount precipitated from a single cell line varied from experiment to experiment (data not shown).

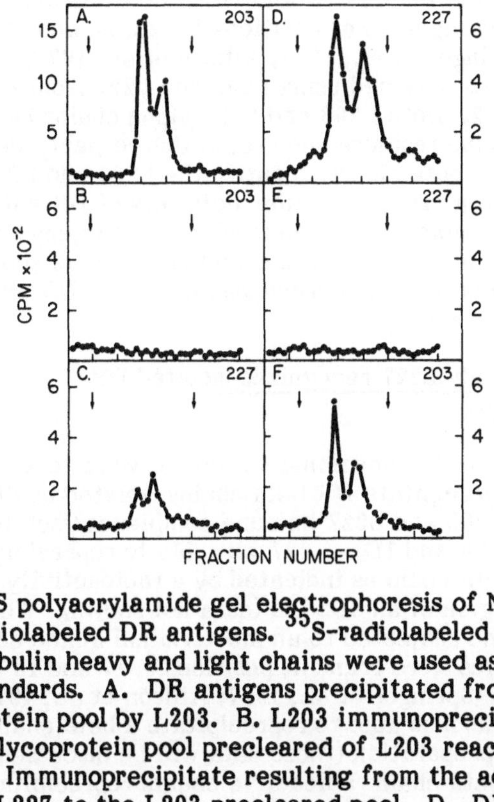

Figure 1. SDS polyacrylamide gel electrophoresis of NEW internally radiolabeled DR antigens. ^{35}S-radiolabeled immunoglobulin heavy and light chains were used as internal standards. A. DR antigens precipitated from a glycoprotein pool by L203. B. L203 immunoprecipitate of a glycoprotein pool precleared of L203 reactivity. C. Immunoprecipitate resulting from the addition of L227 to the L203 precleared pool. D. DR antigens precipitated by L227. E. L227 immunoprecipitate of a glycoprotein pool precleared of L227 reactivity. F. Immunoprecipitate resulting from the addition of L203 to the L227 precleared pool.

These experiments indicate that L203 and L227 recognize different subsets of DR antigens in a variety of B cell lines. Because each line is presumed to be homozygous at its DR locus, the subsets recognized by the monoclonal antibodies cannot represent allelic variations. The subsets are, therefore, either products of different DR loci or different modifications of the product of a single locus. These results confirm the observations of Lampson and Levy (1980) made using heterozygous cell lines.

3.2. Both sets of DR antigens are homologous to murine I-E antigens

To prove that the complexes precipitated by L203 and L227 are DR antigens, partial amino-terminal sequences were determined on the ^{3}H-phenylalanine labeled preparations isolated as described in Figures 1A and 1D. Phenylalanine residues have previously been reported at positions 12, 22, 24 and 26 in the DR alpha chain (Springer et al., 1977; Allison et al., 1978). The alpha chain sequences obtained using L203 and L227 isolated material are shown in Figure 2. Both L203 and L227 alpha chains have the expected phenylalanine residues and are, in this respect, identical to each other. These data clearly identify the L203 and L227 reactive complexes as DR antigens. In addition, both sets of DR antigens express sequences homologous to the murine I-E antigens (Uhr et al., 1979). Other amino acid residues examined (leucine and tyrosine) gave sequences concordant with previously published DR and I-E sequences (data not shown).

3.3. The antibody L227 recognizes isolated DR beta chains

L203 and L277 monoclonal antibodies were tested for their ability to precipitate DR beta chains isolated by SDS gel electrophoresis of L203 and L227 immune complexes (isolated as depicted in Figures 1A and 1D). L227 was able to reprecipitate the isolated L227 beta chain as indicated by a radioactivity peak in the beta chain region of an SDS gel (data not shown). When this peak was isolated and subjected to amino-terminal amino acid sequencing, phenylalanine residues were found at positions 7, 17 and 18 expected for a DR beta chain (Springer et al., 1977; Allison et al., 1978) (data not shown). L227 was also able to reprecipitate a beta chain from an L203 beta chain preparation (these data will be discussed below). The percentage of beta chain radioactive counts reprecipitated in both cases was only a small percentage of the total counts (approximately 10-20%). The antibody L203 showed no reactivity toward either isolated L203 or L227 alpha or beta chains suggesting that the L203 antibody may recognize a determinant formed by the combination of the alpha and beta chains or a determinant lost during isolation.

3.4. Both monoclonal antibodies react with a third population of DR antigens

Analysis of the radioactive counts precipitated in the depletion experiments described above showed that the number of counts in the immune precipitates of the clearing steps of the experiments (as described in Figures 1A and 1D) was always much greater than the number of counts precipitated in the second stage of the experiment (as described in Figures 1C and 1F). For example, in one experiment, preclearing the PRIESS glycoprotein pool with L203 precipitated a total of 4543 cpm; however, L203 added as a second step could only precipitate 1101 cpm in an equivalent pool precleared with L227. This result was suggestive of the presence of a third population of DR antigens which is reactive with both L203 and L227.

To examine this question, three aliquots of the same glycoprotein pool were cleared of reactivity to either L203, L227 or a combination of equal amounts of L203 and L227. The total amount of monoclonal antibody and the total amount of facilitating antibody used was the same for each pool. If a third population exists, the sum of the radioactive counts in the immune precipitates from the pools cleared with L203 (representing the L203 unique molecules plus the cross-reactive molecules) and L227 (representing the L227 unique molecules plus the cross-reactive molecules) should be greater than the number of counts preciptated using the combination of L203 and L227. The results of these experiments determined using two different cell lines, NEW and PRIESS, are shown in Table I. In the majority of experiments, the combination of the sums of each individual monoclonal antibody was significantly greater than the sum of the mixture of antibodies. For example, in one experiment (Table I, Line 1) the combination of the L203 and L227 sums precipitated 2419 cpm (A + B) while the sum of antibodies precipitated only 1826 cpm (column C), thus, providing evidence for the existence of a third population of cross-reactive molecules.

A third piece of evidence supporting the existence of a cross-reactive population arises from reprecipitation experiments of denatured beta chains. As described earlier, the antibody L227 can reprecipitate isolated beta chains from either L203 or L227 preparations. This suggests that L227 can recognize a fraction of the L203 precipitated beta chains.

These data suggest the presence of a third population of DR antigens, a cross-reactive population, which is reactive with both monoclonal antibodies L203 and L227. The beta chains from the cross-reactive population and the L227 reactive population would then share a determinant recognized by the L227 antibody.

TABLE I

A Comparison of the Cpm Precipitated by L203, L227 and a Combination of L203 Plus L227

(Cpm in Immune Precipitates)

Cell Line	A (L203)	B (L227)	A+B	C (L203+L227)	(A+B)-C
NEW	1407	1012	2419	1826	593
NEW	1056	987	2043	1547	496
PRIESS	4543	1562	6105	4223	1882
PRIESS	4521	1839	6360	4273	2087
PRIESS	3782	1952	5734	4186	1548

This determinant would not be present on the L203 unique beta chains. The proportion of DR antigens represented by the cross-reactive population varies from experiment to experiment suggesting that the expression of these populations varies.

3.5. Amino acid sequence analysis of the L203 and L227 unique populations

The L203 unique population was isolated from a glycoprotein pool that had been precleared of reactivity to the L227 antibody (see Figure 8). The L227 population was isolated from a pool that had a low proportion of cross-reactive molecules. ^{3}H-phenylalanine labeled polypeptide chains were isolated as described and analyzed by amino acid sequencing. The results show that both L203 and L227 unique populations are DR (I-E like) antigens and, at this level of analysis, are very similar. The alpha chain sequences of both L203 and L227 reactive molecules are identical to those

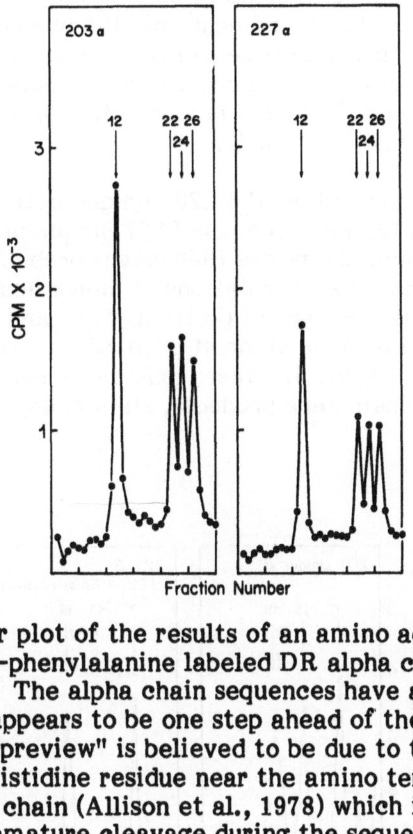

Figure 2: Linear plot of the results of an amino acid sequence of ^{3}H-phenylalanine labeled DR alpha chains from NEW. The alpha chain sequences have a minor component that appears to be one step ahead of the major sequence. This "preview" is believed to be due to the presence of a histidine residue near the amino terminus of the alpha chain (Allison et al., 1978) which is susceptible to premature cleavage during the sequencing procedure.

sequences shown in Figure 2. The L203 and L227 unique beta chains, however, differ from one another (Figure 3). The L203 beta chains, besides containing the expected phenylalanine residues at positions 7, 17 and 18, contain a phenylalanine at position 13. Phenylalanine is absent in the L203 beta chain isolated from a DRw4 cell line and may represent an allelic variation. An identical sequence has been observed for the L203 beta chain isolated from a population that contains the cross-reactive molecules. The L227 beta chains contain the expected phenylalanine residues at positions 7, 17 and 18; however, the yield of counts at position 18 is much greater than expected. This suggests that there may be molecules which lack phenylalanines at 7 and 17 and bear only the phenylalanine at 18. The minor peak in the L227 sequence observed at position 13 is due to contamination with the cross-reactive population. Unlike the L203 molecules, the L227 unique beta chain does not contain a phenylalanine at position 13.

These data argue that the L203 unique beta chains are structurally distinct from the L227 unique beta chains. In addition, the L227 beta chains are themselves probably heterogenous containing at least two populations of molecules, one set containing phenylalanine residues at positions 7, 17 and 18 and a second set containing a phenylalanine at position 18. This implies that in the DRw7 homozygous line there exists at least three different although related beta chain gene products, all homologous to the murine I-E beta chain.

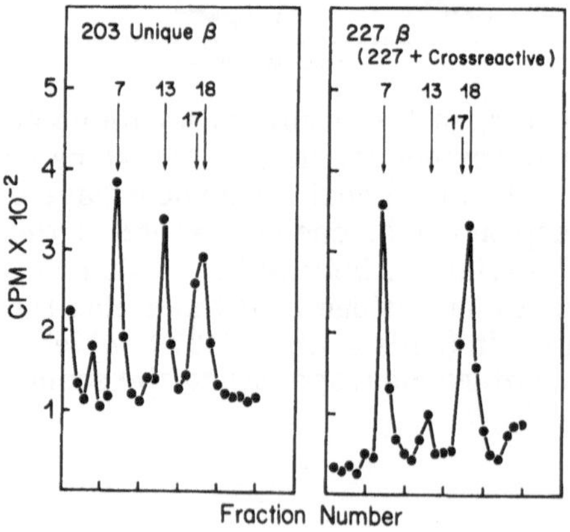

Figure 3. Linear plot of the results of an amino acid sequence of ^{3}H-phenylalanine labeled DR beta chains from NEW. The L203 beta chain was isolated from a pool precleared of L227 reactivity. The L227 beta chain was isolated from a pool containing a minor proportion of the cross-reactive beta chain.

3.6. The monoclonal antibodies IIIE3 and IVG1 recognize the same subset of DR antigens

When sequential immunodepletion experiments were carried out using the monoclonal antibodies IIIE3 and IVG1 on the homozygous cell line NEW, it was determined that both antibodies recognized the same subset of antigens (Figure 4). IIIE3 precipitated a bimolecular complex (Figure 4A). When all reactivity to IIIE3 was removed (Figure 4B) and IVG1 was added, no additional material was precipitated (Figure 4C). The converse experiment showed the same result. These experiments suggest that IIIE3 and IVG1 recognize the same population of DR antigens. It is not known whether the two monoclonal antibodies recognize the same epitope on those antigens.

3.7. IIIE3 recognizes a subset of L203 and L227 antigens

IIIE3 was compared to L203 and L227 using sequential immunodepletion experiments (data not shown). When an L203 depleted glycoprotein pool was precipitated with IIIE3, no antigens were precipitated. In the reverse experiment, IIIE3 depletion followed by L203, L203 precipitated additional DR antigens indicating that the IIIE3 reactive molecules were a subset of the L203 reactive antigens.

In a similar comparison to L227, removal of L227 reactive material did not remove all reactivity to IIIE3. The converse showed a similar result—removal of IIIE3 reactivity did not entirely remove L227 reactive material. Based on these results and on the relationship between the L203 and L227 subsets, the model shown in Figure 5 is proposed. In this model, the IIIE3 subset includes the L203-L227 crossreactive population of DR antigens as well as an additional subset of L203 molecules.

3.8. Amino acid sequence analysis of the IIIE3/IVG1 antigens

The amino-terminal sequence of the ^{3}H-phenylalanine labeled alpha and beta chains isolated from the cell line NEW using an antibody affinity column containing both IIIE3 and IVG1 antibodies was determined. The alpha chain sequence was identical to the sequences shown in Figure 2. The IIIE3/IVG1 beta chain sequence is shown in Figure 6 compared to the L203 unique beta chain sequence. Phenylalanine residues are found at positions 7, 13, 17 and 18— positions identical to the L203 sequence. This is consistent with the suggestion that the IIIE3/IVG1 molecules are a subset of the L203 population. This homogeneous sequence also suggests that the overlap or cross-reactive population alone has a sequence identical to the L203 unique sequence.

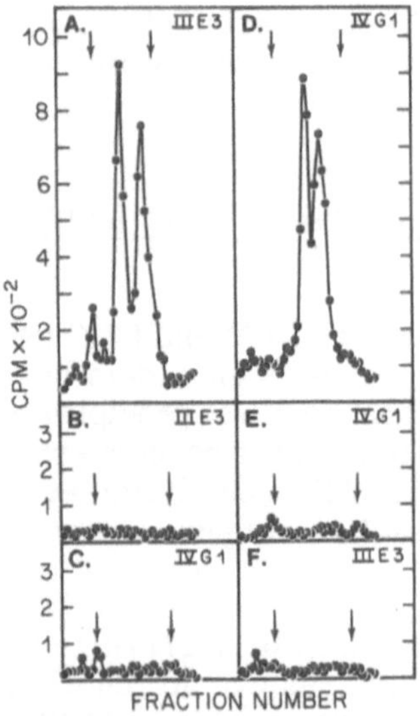

Figure 4. SDS polyacrylamde gel electrophoresis of NEW internally radiolabeled DR antigens. ^{35}S-radiolabeled immunoglobulin heavy and light chains were used as internal standards. A. DR antigens precipitated from a glycoprotein pool by IIIE3. B. IIIE3 immunoprecipitate of a glycoprotein pool precleared of IIIE3 reactivity. C. Immunoprecipitate resulting from the addition of IVG1 to the IIIE3 precleared pool. D. DR antigens precipitated by IVG1. E. IVG1 immunoprecipitate of a glycoprotein pool precleared of IVG1 reactivity. F. Immunoprecipitate resulting from the addition of IIIE3 to the IVG1 precleared pool.

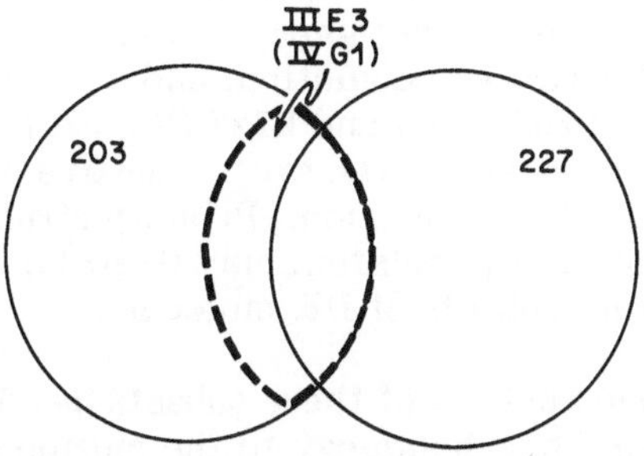

Figure 5: **Schematic representation of the subsets of the NEW DR antigens.**

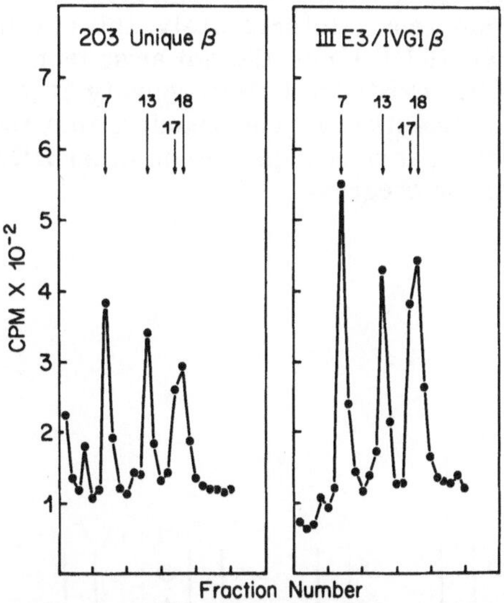

Figure 6: **Linear plot of the results of an amino acid sequence of ^{3}H-phenylalanine labeled DR beta chains from NEW. The L-203 beta chain was isolated from a pool precleared of L227 reactivity. The IIIE3/IVG1 beta chain was directly isolated from a glycoprotein pool.**

4. CONCLUSION

Using a series of monoclonal antibodies, four subsets of DR antigens have been isolated and partially characterized. The L203 antibody recognizes three sets of antigens— a unique L203 reactive population, a IIIE3/L203 reactive population, and a cross-reactive population reactive with L203, IIIE3 and L227 (Figure 7). The L227 antibody, in addition to recognizing the cross-reactive population, also recognizes a unique L227 population. Primary structural evidence suggests that this unique L227 population may itself be heterogenous consisting of two different subsets of DR molecules.

Amino acid sequence analysis of these subsets has defined them as DR antigens that show homology to the murine I-E antigens. The evidence suggests that a single alpha chain locus is associated with several structurally different beta chains. Evidence has been presented for at least three of these beta chains. Therefore, in contrast to the situation in the mouse, the human genome appears to encode several different I-E like antigens.

This premise is based on the supposition that the DR subsets defined by the monoclonal antibodies do not represent allelic differences, that is, that the DRw7 NEW line is homozygous at the loci encoding the isolated antigens. Unfortunately, this cell line, although typed as homozygous at its DR locus, did not arise from a consanguinous mating and the possibility exists for heterozygosity at loci other than that typed by the alloantisera. This requires that the structural analyses described in this paper be repeated on other DRw7 cell lines. These studies are in progress.

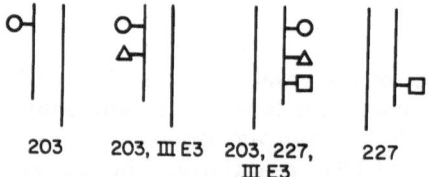

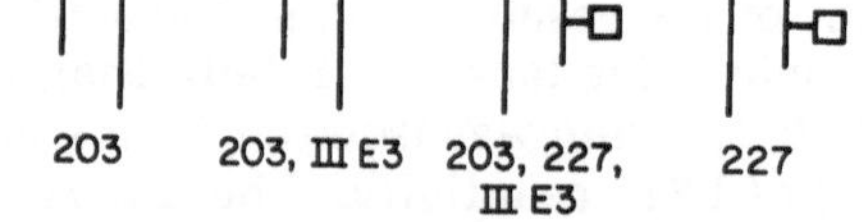

Figure 7. Structural models. Schematic representation of the determinants recognized by L203, L227 and IIIE3 (IVG1).

These data do not totally agree with the data or interpretation presented for L203 and L227 by Shackelford et al. (1981b) who suggested that L203 and L227 recognize various processing stages of a single DR molecule. We believe that different observations may arise depending on the choice of the cell line. Indeed, Shackelford briefly describes a cell line which exhibits two DR gene products differentially recognized by L203 and L227. It is likely that the expression of DR subsets varies among cell populations and probably varies during the cell cycle.

It has been the advent of monoclonal antibodies that has allowed the elucidation of DR antigen subsets. Previously, the only available antisera were either alloantisera which were available in small quantities and poorly precipitated DR antigens, and heteroantisera which did not have the fine specificity needed to characterize this complex locus. The idea of multiple DR or I-E loci has been suggested before (Shackelford et al., 1981a; Markert and Cresswell, 1980; Accolla et al., 1981; Lafuse, personal communication) but the data presented here using monoclonal antibodies provide the first primary structural evidence that these multiple DR subsets represent multiple I-E like gene products with distinct amino acid sequences.

5. EXPERIMENTAL METHODS

5.1. Cells

Epstein-Barr virus transformed human B cell lines, NEW (DRw7), ER4 (DRw4) and MAT (DRw3) were described by Hansen et al. (1979). The B cell line PRIESS (DRw4) was obtained from J. Kaufman, Harvard University. The cell lines are homozygous at their HLA-DR loci. The cell lines were cultured in RPMI 1640 (Microbiological Associates) supplemented with fetal calf serum, newborn calf serum, glutamine, sodium pyruvate, antibiotics and antimycotic.

5.2. Monoclonal antibodies and antisera

Monoclonal antibodies L203 and L227 were described by Lampson and Levy (1980). Monoclonal antibodies IIIE3 and IVG1 were generated by immunizing eight-ten week old BALB/c mice intraperitoneally with 10^7 NEW cells once a week for three weeks. After resting for 33 days, the mice were boosted with 10^7 cells intravenously and three days later the spleens were fused with the SP2/0-Ag14 cell line (Shulman et al., 1978) using the method of Kennett et al. (1978). Supernatant from the resultant clones was tested on the immunizing cell line using a direct binding assay (Williams, 1973) and immunoprecipitation.

Spent culture media containing the antibodies were concentrated five to ten times and directly used for immunoprecipitations. The immunoglobulin from ascites generated from these cell lines was isolated using a 50% ammonium sulfate cut. Goat anti-mouse immunoglobulin was a gift from E. Vitetta, Univesity of Texas Health Science Center at Dallas.

5.3. Isolation of radiolabeled DR antigens

Cells (1×10^8) were cultured at 1×10^7 cells/ml for six hours in radiolabeling media containing 10% fetal calf serum and two mCuries of each ^{3}H-amino acid labeled (New England Nuclear or Amersham). The labeling media was Spiner's salt solution (GIBCO) supplemented with nonessential amino acids, sodium pyruvate, glutamine, $CaCl_2$, vitamins, antibiotic-antimycotic, and all of the essential amino acids excluding the amino acid to be radiolabeled.

At the end of the labeling period, the cells were centrifuged from the media and washed once in Tris-buffered saline (TBS) (0.01 M Tris, 0.15 NaCl, pH 7.5). The cells were lysed in 10 ml 0.5% Nonidet P40 (NP40) (Shell) in TBS containing 1 mM phenylmethylsulfonylfluoride during a 30 minute incubation on ice; nuclei and debris were removed by centrifugation.

5.4. Lentil lectin affinity chromatography

Lentil lectin was isolated (Hayman and Crumpton, 1972) and coupled to Sepharose 4B (Pharmacia) (March et al., 1974). Five ml of the cell lysate was applied to a 6 ml lentil lectin affinity column and the nonadherent fraction eluted with TBS containing 0.25% NP40, 1 mM $CaCl_2$. The adherent material (glycoprotein pool) was eluted by 0.3 M alpha-methylmannoside in the column buffer. The adherent pool was concentrated by vacuum dialysis to 1-2 mls. Five to 20 percent of the acid precipitable counts from the lysate were recovered in this pool.

5.5. Immunoprecipitation

To remove radiolabeled immunoglobulin (Ig), the glycoprotein pool was precleared using protein A-bearing Staphylococcus aureus cells (Pansorbin, Calbiochem) (Cullen and Schwartz, 1976) that had been extensively prewashed with TBS containing 0.25% NP40 and 0.1% SDS. The Ig-depleted pool was then incubated with monoclonal antibody for 15 minutes at 37 °C and one hour to overnight at 4 °C. Goat anti-mouse immunoglobulin was then added and the same incubation protocol followed. The complexes were removed by centrifugation and washed with TBS containing 0.25% NP40 and with TBS.

5.6. Antibody affinity columns

Immunoglobulin preparations from ascites were dialyzed against 0.1 M sodium bicarbonate and coupled to Sepharose 4B (March et al., 1974). Briefly, 2 mgs of immunoglobulin were coupled to 1 ml of Sepharose which had been activated with cyanogen bromide at a ratio of 0.2 gm CNBr per ml of resin.

Glycoprotein pools isolated from 1-2 x 10^8 cells were passed over a 6 ml column and the column washed with three-four column volumes of TBS containing 0.25% NP40. The DR antigens were specifically eluted using 0.05 M diethylamine containing 0.25% NP40 (Parham, 1979). The eluate was dialyzed against TBS and concentrated. Lysozyme was added as a carrier protein to all of the eluates and some of the fall-through pools. Repassage of these separated fractions through the antibody affinity columns included a one hour incubation step prior to the column wash. A diagram of one such isolation is shown in Figure 8. This scheme illustrates the isolation of the L203 unique population and the cross-reactive population. The L227 unique population was isolated in a similar manner. The proteins were isolated from the pools by precipitation with acetone.

5.7. SDS polyacrylamide gel electrophoresis

Immune complexes or acetone precipitated proteins were boiled for five minutes in Laemmli sample buffer containing 5% mercaptoethanol (Laemmli, 1970). The samples were electrophoresed on 12.5% SDS polyacrylamide gels (Cullen

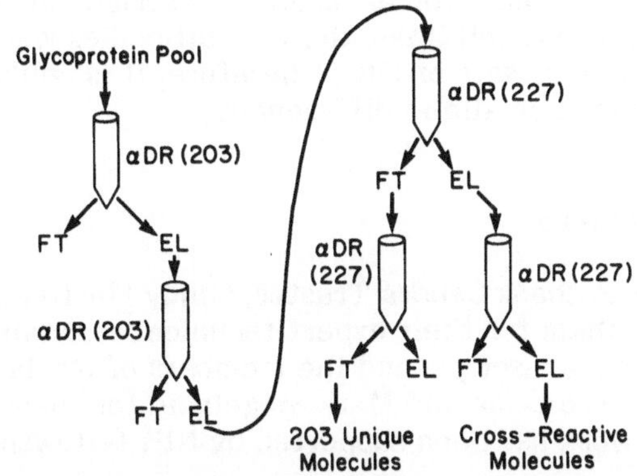

Figure 8. Antibody affinity column protocol used for the isolation of unique L203 and cross-reactive molecules.

et al., 1976). In some cases, an ^{35}S-labeled Ig was added to the sample as an internal standard. The gels were fractionated into a 0.05% SDS solution and aliquots counted to determine the position of the alpha and beta polypeptides. For preparative isolations, the fractions were incubated overnight at room temperature to elute the polypeptides from the gel fragments; the appropriate aliquots were then pooled and passed through a Millipore 0.45 micron filter to remove gel particles before lyophilization. The lyophilized alpha and beta chains were solubilized in 1 ml H_20 containing 1 mg lysozyme as a carrier and dialyzed overnight against H_20 at room temperature.

5.8. Amino acid sequence determination

Isolated alpha and beta chains labeled with a single ^3H amino acid were sequencer on a Beckman 890C sequenced modified with a cold trap (McCumber et al., 1980) using either 0.1 M Quadrol or DMAA programs. Polybrene (Tarr et al., 1977; Klapper et al, 1977) and occasionally an ^{35}S-labeled internal standard were added to the sequencer cup. The butyl chloride fractions were dried down and scintillation fluid was added to each fraction. The radioactivity in each fraction was determined using a liquid scintillation counter.

NOTE ADDED IN PROOF

The cell line NEW from our lab has been retyped and has been found to be a DR heterozygote. Allelic differences as an explanation for our results are, however, difficult to support since L203 and L227 recognize monomorphic determinants present on all cell lines tested. In immunodepletion experiments carried out on several different DR homozygous cell lines, the two antibodies were found to recognize different subsets of DR. Therefore, it is unlikely that these antibodies recognize allelic differences.

ACKNOWLEDGEMENTS

We would like to thank Sandra Traster, Cindy Martinez, Ana Spain and Sandy Graham for their expert technical assistance, Kathy Able for typing this manuscript, and the members of our laboratory especially Drs. Richard Cook and Mark Siegelman for their advice and support. This work has been supported by NIH fellowship 5 F32 AI05840.

REFERENCES

Accolla, R.S., Gross, N., Carrel, S., and Corte, G., 1981, Distinct forms of both alpha and beta subunits are present in the human Ia molecular pool, Proc. Natl. Acad. Sci. USA 78: 4549-4551.

Allison, J.P., Walker, L.E., Russell, W.A., Pellegrino, M.A., Ferrone, Reisfeld, R.A., Frelinger, J.A., and Silver, J., 1978, Murine Ia and human DR antigens: Homology of amino-terminal sequences, Proc. Natl. Acad. Sci. USA 75: 3953-3956.

Cullen, S.E. and Schwartz, B.D., 1976, An improved method for isolation of H-2 and Ia alloantigens with immunoprecipitation induced by protein A-bearing Staphylococci, J. Immunol. 117: 136-142.

Cullen, S.E., Freed, J.H., and Nathenson, S.G., 1976, Structural and serological properties of murine Ia alloantigens, Transpl. Rev. 30: 236-270.

Hansen, J.A., Fu, S.M., Antonelli, P., Kamoun, M., Hurley, J.N., Winchester, R.J., Dupont, B., and Kunkel, H.G., 1979, B-lymphoid cell lines derived from HLA-D homozygous donors, Immunogenetics 8: 51-64.

Hayman, M.J. and Crumpton, M. J., 1972, Isolation of glycoproteins from pig lymphocyte plasma membrane using Lens culinaris phytohemagglutinin, Biochem. and Biophys. Res. Commun. 47: 923-930.

Kaufman, J.F., Andersen, R.L., and Strominger, J.L., 1980, HLA-DR antigens have polymorphic light chains and invariant heavy chains as assessed by lysine-containing tryptic peptide analysis, J. Exp. Med. 152:37s-53s.

Kennett, R.H. Denis, K.A. Tung, A., and Klinman, N.R., 1978, Hybrid plasmacytoma production: Fusions with adult spleen cells, monoclonal spleen fragments, neonatal spleen cells and human spleen cells, Curr. Topics. Microbiol. Immunol. 81: 77-91.

Klapper, D.G., Wilde, C.E., and Capra, J.D., 1978, Automated amino acid sequence of small peptides utilizing polybrene, Analyt. Biochem. 85: 126-131.

Lampson, L.A. and Levy, R., 1980, Two populations of Ia-like molecules on a human B cell line, J. Immunol. 125: 293-299.

Laemmli, U.K., 1970, Cleavage of structural proteins during the assembly of the head of bacteriophage T4, Nature 227: 680-685.

March, S.C., Parikh, I., and Cuatrecasas, P., 1974, A simplified method for cyanogen bromide activation of agarose for affinity chromatography, Anal. Biochem. 60: 149-152.

Markert, M.L. and Cresswell, P., 1980, Polymorphism of human B-cell alloantigens: Evidence for three loci within the HLA system, Proc. Natl. Acad. Sci. USA 77: 6101-6104.

McCumber, L.J., Qadeer, M., and Capra, J.D., 1980, Comparison of solid phase and spinning cup methodologies III: Experiences with polybrene, a .1 M Quadrol program, and a modestly modified automated spinning cup sequencer, in: Methods in Peptide and Protein Sequence Analysis (C. Birr, ed.), Elsevier/North Holland Biomedical Press, Amsterdam, pp. 165-172.

Nadler, L.M., Stashenko, P., Hardy, R., Tomaselli, K.J., Yunis, E.J., Schlossman, S.F., and Pesando, J.M., 1981, Monoclonal antibody identifies a new Ia-like (p29,34) polymorphic system linked to the HLA-D/DR region, Nature 290: 591-593.

Parham, P., 1979, Purification of immunologically active HLA-A and -B antigens by a series of monoclonal antibody columns, J. Biol. Chem. 254: 8709-8712.

Quaranta, V., Pellegrino, M.A., and Ferrone, S., 1981, Serologic and immunochemical characterization of the specificity of four monoclonal antibodies to distinct antigenic determinants expressed on subpopulations of human Ia-like antigens, J. Immunol. 126: 548-552.

Shackelford, D.A. and Strominger, J.L., 1980, Demonstration of structural polymorphism among HLA-DR light chains by two-dimensional gel electrophoresis, J. Exp. Med. 151: 144-165.

Shackelford, D.A., Mann, D.L., van Rood, J.J., Ferrara, G.B., and Strominger, J.L., 1981a, Human B-cell alloantigens DC1, MT1 and LB12 are identical to each other but distinct from the HLA-DR antigen, Proc. Natl. Acad. Sci. USA 78: 4566-4570.

Shackelford, D.A., Lampson, L.A., and Strominger, J.L., 1981b, Analysis of HLA-DR antigens by using monoclonal antibodies: Recognition of conformational differences in biosynthetic intermediates, J. Immunol. 127: 1403-1410.

Shulman, M., Wilde, C.D., and Kohler, G., 1978, A better cell line for making hybridomas secreting specific antibodies, Nature 276: 269-270.

Springer, T.A., Kaufman, J.F., Terhorst, C., and Strominger, J.L., 1977, Purification and structural characterization of human HLA-linked B-cell antigens, Nature 268: 213-218.

Tarr, G.E., Beecher, J.F., Bell, M., and McKean, D., 1978, Polyquart-ernary amines prevent peptide loss from sequenators, Anal. Biochem. 84:622-627.

Uhr, J.W., Capra, J.D., Vitetta, E.S., and Cook, R.G., 1979, Organization of the immune response genes, Science 206: 292-297.

Williams, A.F., 1973, Assays for cellular antigens in the presence of detergents, Eur. J. Immunol. 3: 628-632.

Winchester, R.J. and Kunkel, H.G., 1979, The human Ia system, in Advances in Immunology, Volume 28 (F.J. Dixon and H.G. Kunkel, eds.), Academic Press, New York, pp. 221-292.

7

MONOCLONAL ANTIBODIES FOR ANALYSIS OF HLA ANTIGENS: FURTHER STUDIES WITH THE W6/32 ANTIBODY

Peter Parham, Harry T. Orr* and John S. Golden

Department of Structural Biology, Stanford University, Stanford, CA 94305. *The Biological Laboratories, Harvard University, 16 Divinity Avenue, Cambridge, MA 02138; Immunobiology Research Center, University of Minnesota, Minneapolis, MN 55455

1. INTRODUCTION

Many mouse monoclonal antibodies against HLA antigens have been described and their applications to tissue-typing, biological and biochemical research are being explored. A number of our studies revealed certain problems associated with an ignorance of the basic immunochemical properties of the monoclonal reagents we were using. For example:

(1) Certain HLA-A,B,C antibodies effectively blocked HLA-A,B,C[1] restricted killing of influenza virus infected cells (McMichael et al., 1980). Our initial experiments suggested that antibodies against some determinants but not others would block. This proved misleading because efficient blocking only occured if all the HLA target molecules were covered with antibodies. Blocking was therefore correlated with avidity rather than specificity of an antibody, thus leading to a very different interpretation of the results.

(2) Sections of thymus (Fig. 1) showed two distinct patterns when stained with different HLA antibodies (Rouse et al. 1981). Epithelial cells in medulla and cortex stained with some antibodies whereas others only stained cells in the medulla. It was unclear whether this was due to differences in the properties of the antibodies or if it represented differential expression of HLA antigenic determinants.

---

[1] To be brief HLA-A,B,C antigens will subsequent be referred to as HLA antigens.

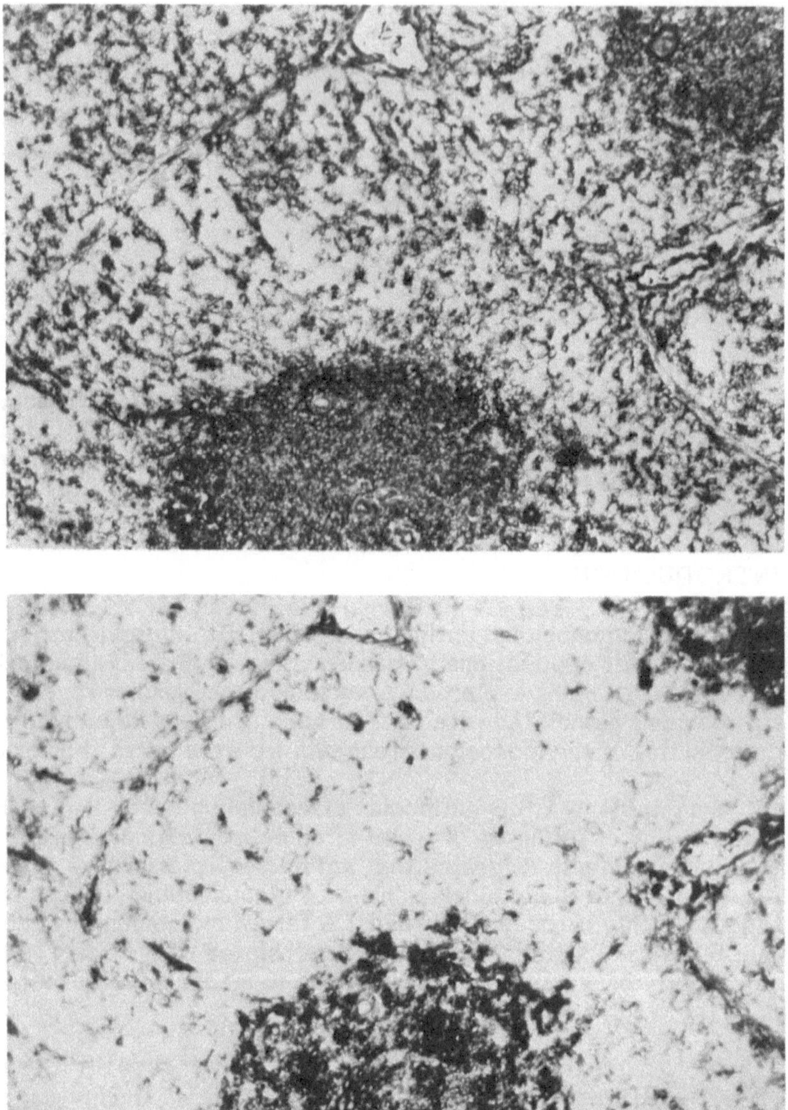

Figure 1. Top panel: Frozen section of 4 year old human thymus stained by peroxidase as described by Rouse et al., 1981 with W6/32 (monoclonal anti-HLA monomorphic determinant). The medulla is intensely stained with a confluent pattern; the cortex demonstrates intense dendritic staining. Bottom panel: Parallel section of same human thymus as in top panel, stained with PA2.1 (monoclonal anti-HLA-A2 polymorphic determinant, Parham and Bodmer 1978). Medulla is stained with a confluent pattern but the cortex demonstrates very little staining.

(3) The interaction with solubilized antigen of antibodies identified by serological screening assays involving cells as targets were not predictable due to the possibility of bivalent and/or univalent binding to cells (Mason and Williams 1980; Parham 1981; Parham and Brodsky, 1981).

In order to understand these details, we have begun to characterize more fully the binding properties of some monoclonal HLA antibodies and their proteolytic fragments. Much of the work described here concerns the characterization and applications of the W6/32 antibody. This was the first monoclonal HLA-A,B,C antibody described and has seen widespread use (Barnstable et al., 1978). Some properties of this antibody have been described (Parham et al. 1979; Parham 1979) and its relationship to other monomorphic HLA antibodies analyzed (Brodsky and Parham, 1982).

2. RESULTS

2.1. Interaction of W6/32 antibody with HLA antigen

2.1.1. Binding of W6/32 to cells at equilibrium. Trucco et al. (1980) determined the binding constants of their monoclonal HLA antibodies for cells. We have used their method with some modifications to investigate the interaction of W6/32 with various cells. One mg aliquots of purified W6/32 were iodinated with 1 mCi of isotope to a specific activity of $\simeq 1x10^6$ cpm/ μg. Between 20 and 30% of the material in these preparations was serologically active and radioactivity was assumed to be similarly distributed between active and inactive protein. Dilutions of antibody, starting at 2 μg total antibody, were incubated with a fixed number of cells having $\simeq$ 50 ng HLA in a volume of 1 ml for 4 hours at 4°C. 1.5 ml microfuge tubes were used and rotated gently and continuously to provide even distribution of cells. The cells were pelleted by a 0.5 min centrifugation and supernatants removed by aspiration. Bound antibody was determined by assaying the pellets for radioactivity, back-ground values were provided by control tubes containing Daudi cells which express no HLA. Background values were typically 1% of the input and could be accounted for in terms of the residual liquid. The results of a typical experiment are shown in Fig. 2. As

[2] K, binding constant; n, antigenic site number; Met, methionine; Phe, phenylalanine; Tyr, Tyrosine; DTT, dithiothreitol; PMSF, phenylmethylsulphonylfluoride; Staph. A., formalin-fixed, heat-killed Staphylococcus aureus Cowan 1, prepared by the method of Kessler (1975); SDS-PAGE, Sodium Dodecyl Sulphate Polyacrylamide Gel Electrophoresis; PTH, phenylthiohydantoin; HPLC, High Pressure Liquid Chromatography; TCA, trichloroacetic acid; PBS, phosphate buffered saline.

previously shown (Trucco et al., 1980), graphs of bound/free against bound radioactivity gave reasonable straight lines. The gradient and intercept were used to calculate a binding constant, (K) and a site number (n) respectively. A summary of results is in Table I. Although the data within any given experiment was in good agreement the results of independent experiments showed considerable variation which we assume to be due to the state of the cells on different days. This problem was alluded to by Trucco et al. (1980). Thus as can be seen in Table I any given determination of K or n is only good to within a factor of two. It is also clear that significant differences in K and n were found for the different B cell lines examined. K was always

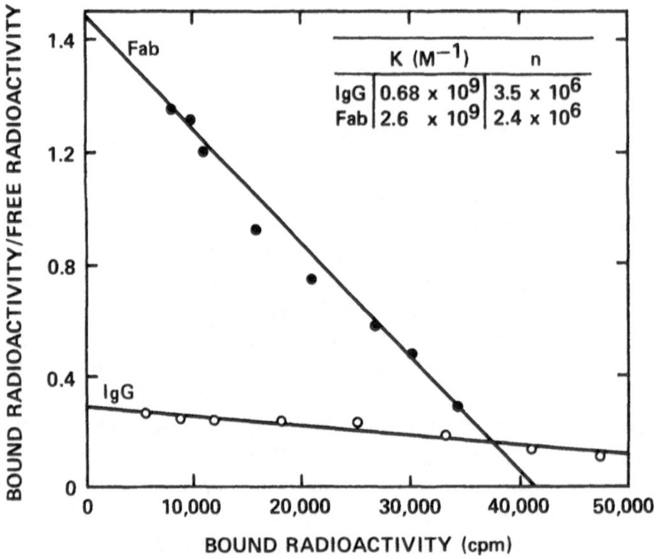

Figure 2. LB (A28, B40) or Daudi (no HLA-A,B,C) cells (1.1×10^5) were incubated in a volume of 1 ml with different concentrations of ^{125}I-W6/32-Fab or ^{125}I-W6/32-IgG at 4°C with continuous but gentle rotation for 4 hours at which point equilibrium was attained. After centrifugation in the Eppendorf centrifuge for 1 minute the supernates were removed by aspiration and the cells assayed for bound radioactivity. Specific binding was obtained by subtracting Daudi from LB values. Unbound radioactivity was calculated by subtraction of the bound radioactivity from the total input of active, i.e., bindable, radioactivity which was previously determined by titrating increasing numbers of cells against a fixed and limiting amount of antibody until a plateau of binding was reached. The quotient of bound and free radioactivity was plotted against bound radioactivity and straight lines fitted by the method of least squares. The binding constant (K) and the number of antigenic

in the range 10^8 -10^9 M^{-1} and n was in the range 2 - 8 x 10^6. Peripheral blood lymphocytes and thymocytes gave similar values of K but values of n that were 10 and 50 fold less respectively. This is in good agreement with our previous results using different methods (Parham et al., 1979; Brodsky et al., 1979).

2.1.2. Comparison of the binding of W6/32 IgG and Fab to LB cells at equilibrium. A preparation of ^{125}I-W6/32 IgG was divided in two and one half degraded with papain to give Fab and Fc fragments. They were separated from residual IgG by Sephadex G-200 chromatography as shown in Fig. 3. The K and n values for the binding of IgG and Fab to LB cells were then determined in the same experiments and typical values are shown in Fig. 2. ^{125}I-W6/32 Fab bound to cells with a binding constant (K) that was 4-10 times greater than ^{125}I-W6/32 IgG. Comparable values for the number of antigenic sites (n) were obtained for the two preparations.

Figure 2 (Continued)

sites per cell (n) were then calculated from the equations.

$$K = -\frac{\text{grad. SV} \times 10^9}{6.67}$$ for IgG assuming a molecular weight of 150,000

$$K = -\frac{\text{grad. SV} \times 10^9}{20.01}$$ for FAB assuming a molecular weight of 50,000

$$n = \frac{\text{intercept V} \times 10^{21}}{1.66\ \text{NK}}$$

where S = specific activity in cpm/ μg, V = volume in ml, N = number of cells. For the experiments shown here S was 1.3 x 10^6 cpm/ μg for ^{125}I-W6/32-=IgG and an identical value was assumed for Fab. The distribution of radioactivity between partially separated Fab and Fc peaks on Sephadex G-200 chromatography and the percentage of bindable radioactivity for the IgG and Fab preparation indicated this was a reasonable assumption.

TABLE I

Binding constants (K) and antigenic site number (n) calculated from analysis of the equilibrium binding of ^{125}I-W6/32 to various cells

Cell	$K \times 10^{-9}$ (M^{-1})	$n \times 10^{-6}$	Cell number
LB	0.75	2.7	5×10^5
LB	0.93	3.3	5×10^5
LB	0.56	2.9	5×10^5
LB	0.50	3.2	5×10^5
JY	0.18	3.7	5×10^5
JY	0.25	3.8	5×10^5
IDF	0.89	5.0	5×10^5
IDF	0.63	5.5	5×10^5
MFF	0.37	3.7	4×10^5
HOM-2	0.19	8.1	2×10^5
665	0.38	2.2	5×10^5
MST	0.54	2.7	5×10^5
PBL I	0.93	0.34	1.8×10^6
PBL II	0.86	0.25	1.6×10^6
Thymocyte I	0.47	0.082	6.9×10^7
Thymocyte II	0.62	0.069	1.4×10^7
mean:B cell line	0.51	3.9	
mean: PBL	0.90	0.30	
mean:Thymocyte	0.55	0.076	

Cell lines used were LB (A28,28:B40,40), JY (A2,2:B7,7), IDF (A26,28*:38,18) MFF(A2,28,B14,5) HOM-2(A3,3:B27,27) 665(A1,-:B8,27), MST(A3,3:B7,7). Peripheral blood lymphocytes (PBL) and thymocytes were of unknown HLA types.

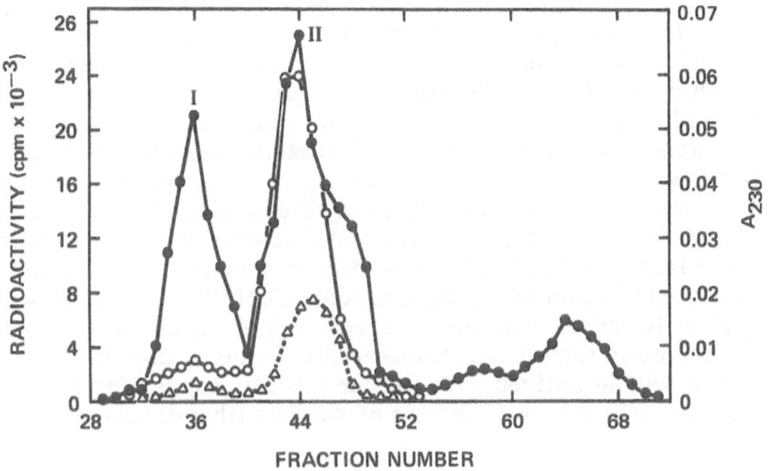

Figure 3. Sephadex G-200 chromatography of papain degraded ^{125}I-W6/32. ^{125}I-W6/32, 220 μg, was digested with 10 μg papain in 1.3 ml of 2 mM cysteine, 1 mM EDTA, 0.1 M acetate pH 5.5 for 3 hours at 37 °C with a second addition of papain at 90 minutes. The digest was then applied to a column of Sephadex G-200 (1.6 x 150 cm) equilibrated with PBS. The column was eluted with the same buffer, 2 ml fractions were collected and assayed for: (1) total radioactivity in 5 μl (●——●); (2) bindable radioactivity in 5 μl (o——o); (3) Absorbance at 230 nm (Δ—— Δ). SDS-PAGE and autoradiography showed that undegraded IgG eluted in peak I, Fab and Fc in peak II. Fractions 43-47 were pooled for use as a ^{125}I-W6/32-Fab preparation.

2.1.3. Kinetics of binding of W6/32 IgG and Fab to LB cells. The rates of association and dissociation of ^{125}I-W6/32 IgG and Fab were compared (Mason and Williams, 1980). As seen in Fig. 4, Fab associates about 10 times faster and dissociates about twice as fast as IgG. The dissociation rates for both species were extremely slow and precluded an accurate measurement of the dissociation constant. These results support the observed difference in the binding constant and show it is mainly due to an increased rate of association of Fab.

2.1.4. Binding of W6/32 to soluble HLA antigens. W6/32 IgG was radioiodinated to high specific activity ($\simeq 10^7$ cpm/ μg) and the inhibition of its binding to LB cells with various purified HLA antigens compared. As shown in Fig. 5, three different HLA preparations equivalently inhibited the binding in comparison to β_2-m which had no effect. In the same experiment a polymorphic antibody, MB40.4, was only inhibited by B7 and B40 preparations and not by A2 or β_2-m. In agreement with previous results using an indirect binding assay the binding of W6/32 to cell surface HLA was found to be effectively inhibited by similar quantities of solubilized HLA. In Fig. 5, 50% inhibition of binding occurred when approximately equal amounts of cellular and soluble HLA were present. This was not so for MB40.4 which required 16-50 times as much soluble antigen for effective inhibition. These results indicate that W6/32 binds to cells and to soluble antigen with similar affinity, suggesting that under these conditions, bivalent interactions of W6/32 with cell surfaces does not occur.

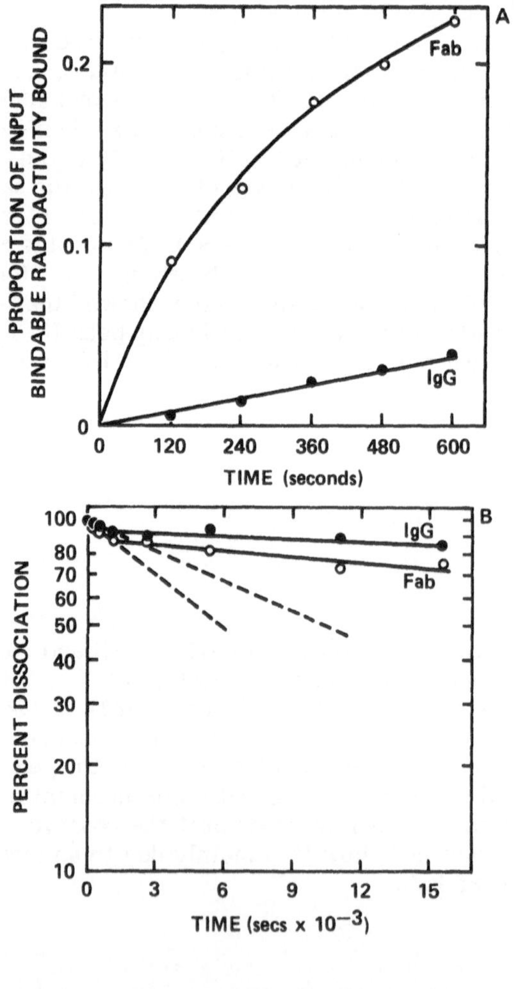

Figure 4A. Association of ^{125}I-W6/32-IgG and ^{125}I-W6/32-Fab to LB cells.

At two-minute intervals 5 x 10^5 LB (A28,28:B40,40) or Daudi (no HLA-A,B,C) cells were added to 1 ml of ^{125}I-W6/32-IgG or Fab and rotated at 4°C. After 10 minutes all samples were centrifuged for 0.5 mins and the supernatants removed by aspiration. Specific bound radioactivity (LB - Daudi) is plotted, as a proportion of the total bindable input (256,000 cpm for IgG, 104,000 cpm for Fab), against time. The specific activity(s) of both preparations was 1.3 x 10^6 cpm/μg. Forward rate constants were calculated from the equation, k_{+1} = gradient of graph/initial antigen concentration. The initial HLA-A,B,C concentration was calculated assuming 10^6 molecules per cell.

k_{+1} IgG = 2.7 x 10^4 M^{-1} S^{-1}

k_{+1} Fab = 3.0 x 10^5 M^{-1} S^{-1}

←

Figure 4B. Dissociation of ^{125}I-W6/32-IgG and Fab from LB cells.

4 x 10^6, LB or Daudi cells were incubated with ^{125}I-W6/32-IgG or Fab and then washed three times by centrifugation. Aliquots of 5 x 10^5 cells were resuspended in 1 ml containing 20 μg/ml non-radioactive W6/32, rotated at 4°C. At various times the cells were centrifuged, the supernatants removed and the cells assayed for residual bound radioactivity. The $\log_{10}$ of the percentage of dissociation is plotted against time. The radioactivity bound at t = 0 was 27,000 cpm for IgG, 34,000 cpm for Fab. The dissociation appeared biphasic and dissociation constants were calculated from both parts of the curves using the equation k_{-1} = 0.693/half time of dissociation. The values obtained were

	I	II
k_{-1}Fab	0.73 x $10^{-4}S^{-1}$	4.8 x $10^{-6}S^{-1}$
k_{-1} IgG	1.2 x $10^{-4}S^{-1}$	9.6 x $10^{-6}S^{-1}$

A comparison of the binding constant (K) calculated from k_{+1}/k_{-1} compared to the experimentallydetermined values suggests the initial rates of dissociation, given by I may be the more accurate estimates. These curves are shown extrapolated (---) in the figure.

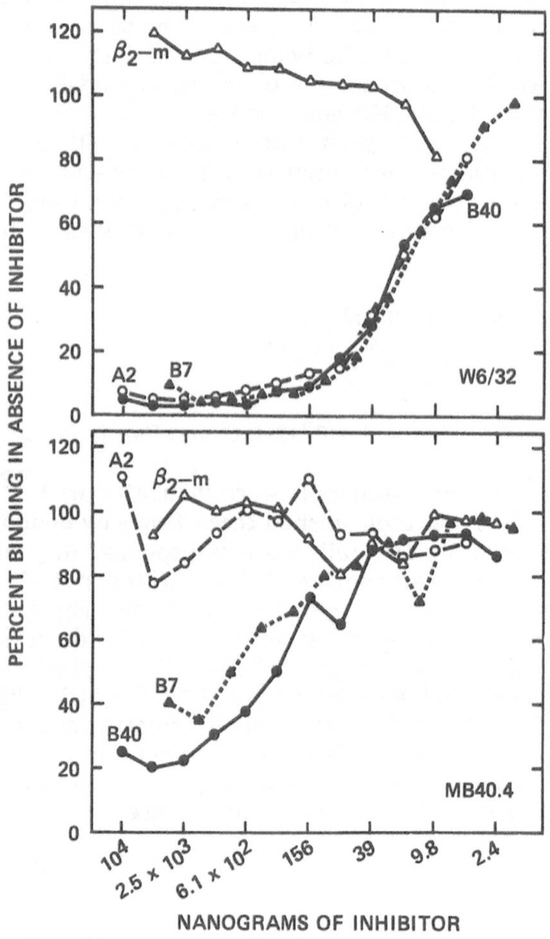

Figure 5. 10 μl of buffer of dilutions of inhibitors; β_2-m (Δ - Δ) HLA-A2 (o—o), HLA-B7 (▲ — ▲), HLA-B40 (●——●), were incubation with 5 μl (10^6 cpm) of ^{125}I-W6/32 or ^{125}I-MB40.4 for 30 mins at 4°C. 5 x 10^5 LB cells (25 μl) were then added and intubation continued with shaking for 1 hours at 4°C. The cells were washed and counted for bound radioactivity. Results are expressed as the percentage of binding in the absence of inhibitors.

The method of Frankel and Gerhard (1979) was used to measure the binding constant of ^{125}I-W6/32 to purified papain solublized HLA-A2 antigen adsorbed onto the wells of plastic microtitre plates. Three determinations gave values of 0.33, 1.1 and 1.5 x 10^9 M^{-1}, comparable to those found in the cellular binding assays. The amount of HLA bound per well was determined and the molecular density of HLA on the plastic was sufficiently low that no significant bivalent binding of W6/32 could have occurred. These determinations therefore represent the affinity of a single W6/32 binding site for an HLA molecule.

2.1.5. Conclusions. The utility of W6/32 as a reagent for immunoprecipitation, affinity chromatography and assay of soluble HLA antigens is indicative of relatively high affinity. The experiments described here in which we have investigated the interaction of W6/32 with monovalent (soluble) and polymeric (cell-surface) antigen show that the affinity of a single W6/32 combining site for HLA is about 10^9 M^{-1}. The values obtained for other monomorphic HLA antibodies were of the same order of magnitude (Trucco et al., 1980).

Under most conditions of analysis no evidence was obtained for bivalent interaction of W6/32 with cell surfaces. This contrasts to the behavior of a polymorphic HLA-A2 monoclonal antibody (BB7.2) which required bivalent binding for stable interaction (Parham and Brodsky, 1981). Interestingly, a number of anti-rat lymphocyte $F(ab')_2$ fragments were shown to distribute between monovalent and bivalently bound forms under conditions of antibody saturation (Mason and Williams 1980). When equilibrium binding experiments were done under conditions of low antibody concentration using ^{125}I-W6/32 preparations of high specific activity (s $\simeq$ 10^7 cpm/ µg) biphasic Scatchard plots were seen in which the low affinity part of the curve gave K values in agreement with the other experiments. The high affinity region could be the result of bivalent binding only seen at very low antibody concentrations ($\sim 10^{-12}$ M). The general lack of bivalent binding could be due to the presence of non-specific light chain in the W6/32 preparation such that all IgG molecules have only a single active combining site. Alternatively the iodination procedure may inactivate the W6/32 combining site. The bivalently bound state might also be energetically unfavorable due to strain induced in either antigen or antibody by bivalent binding.

A significant finding of this study was that Fab fragments of W6/32 bind to cells with higher binding constant than IgG. This is primarily a function of an increased association constant. It is unlikely that differences in diffusion constant are responsible for this affect as the binding reactions are orders of magnitude slower than diffusion. The cell surface is a complex microenvironment of proteins, lipids and carbohydrates and the accessibility of a given molecular probe, e.g., Fab or IgG, may be quite different even though the specific combining sites are identical. The nature of such surface effects will depend on the conditions and type of cell and may be responsible for the differences in binding constants seen in Table I. In contrast to these results experiments with soluble antigens suggest W6/32 interacts very similarly if not identically with all HLA gene products.

A major problem in using cells for immunochemical analysis of HLA antigens has been highlighted by these studies. Do the binding constants measured reflect only the interaction between antigen and antibody or do other cell-associated factors contribute? The B cell lines are heterogeneous in morphology, growth characteristics and cell surface properties. Fresh peripheral blood lymphocytes might provide a more uniform source of target cells.

2.2. Structural studies

2.2.1. Amino acid sequencing of W6/32 precipitates. Amino-acid sequences of HLA antigens have largely been obtained by conventional methods using milligram quantities of purified proteins (Orr et al., 1979). A more versatile approach, radioactive sequencing, was possible for H-2K,D antigens as mouse alloantisera could be used for simple efficient purification. The use of monoclonal HLA antibodies for this purpose has been investigated.

For preliminary studies W6/32 was used as it was known to be of high affinity and bind Staphylococcus Protein A. The experiments were designed to investigate the variability of position 9 in the sequence of HLA antigens. Allison et al. (1978) postulated that it may represent an alloantigenically important residue and our analysis of the MB40.1 antibody (Parham and McLean, 1980) suggested its polymorphic reaction patterns might correlate with substitutions at this position. HLA-A,B homozygous B cell lines were biosynthetically radiolabelled with ^{35}S-methionine, ^{3}H-tyrosine and ^{3}H-phenylalanine as described in Table II. Extracts were precleared by immunoprecipitation (Kessler 1975) with 50 µl of normal rabbit serum and 500 µl of a 10% suspension of formalin fixed, heat-killed Staphylococcus aureus Cowan 1. HLA antigens were then precipitated by incubation of the extract with W6/32 (10 µg/ml) for 15 minutes on ice followed by 100 µl of Staph A, also for 15 minutes on ice. The precipitates were washed (Parham and Ploegh, 1980), frozen and thawed to break up the pellet, and then solubilized by boiling for 2 minutes with 100 µl of 0.05% SDS containing 2 mg/ml sperm whale myoglobin (Beckman). The bacteria were pelleted (5 minutes in Eppendorf centrifuge) and the supernatant removed. The pellet was reextracted with 50 µl of 0.05% SDS, 2 mg/ml sperm male myoglobulin and the two supernatant pooled for use in amino acid sequence analysis as described in Table III.

Electrophoretic analysis of the W6/32 immunoprecipitates obtained from a number of B cell lines is shown in Fig. 6. Each precipitate shows HLA-A,B (≈42,000 MW) and β_2-microglobulin (≈12,000 MW) as the major bands. Variable quantities of actin, the major protein component of these cells, was observed. This actin contamination was of little consequence

TABLE II
Biosynthetic Radiolabeling of LB cells

Number of successive labelling	Total Acid Precipitable Radioactivity (cpm x 10^{-6})	
	^{3}H-Tyr	^{3}H-Phe
1	133	102
2	40	23
3	3	4
4	1	2

10^{7} LB cells were incubated for 8 hours at 37°C in 6 ml of either Tyr, Phe or Met free RPMI 1640 medium containing 10% dialyzed fetal bovine serum and 1 mCi of the appropriate isotope. (^{3}H-Tyr 37.4 Ci/mmol, ^{3}H-Phe 100.0 Ci/mmol ^{35}S-Met 500 Ci/mmol.) Cells were harvested by centrifugation and the supernatant taken for labelling of a fresh aliquot of cells. The labelling was repeated four times. Cells were solubilized in 1 ml of 2% NP40, 0.01 M Tris-HCl pH 8.0, 0.1 mM PMSF, 0.1 M DTT, 0.02% azide for 45 minutes at 4°C and then centrifuged in the Eppendorf centrifuge for 5 minutes. The supernatant was removed, 10 μl aliquots taken for assay of TCA precipitable radioactivity, and the remainder frozen at -20°C. As shown above only the first two labelings resulted in incorporation of significant radioactivity and were routinely done. Radiolabelling with ^{35}S-Met has been previously characterized (Parham and Ploegh, 1980) and was not analyzed here.

TABLE III

NH_2 terminal sequencing of W6/32 immunoprecipitates biosynthetically labelled with ^{35}S-Met, ^{3}H-Tyr and ^{3}H-Phe

	A	B	C	D	E	
Step	^{35}S	^{3}H	Tyr	Met	Phe	Assignment
1	31	112				
2	17	43				
3	21	98				
4	19	78				
5	225	151	138	607	156	Met
6	27	105				
7	25	805	2892	26	208	Tyr
8	21	847	522	27	2700	Phe
9	23	886	2053	20	1568	Tyr, Phe
10	18	626	2053	15	375	Tyr
11	18	321	1630	26	293	
12	85	231	887	323	273	Met
13	22	168				

The Beckman spinning cup automatic sequencer was used as described (Orr et al. 1979). One fifth of the sample obtained at each step was assayed for ^{3}H and ^{35}S radioactivity (columns A,B) the remainder was converted to PTH derivatives and analyzed by reverse phase HPLC. Fractions were collected and assayed for radioactivity. Columns C,D,E give the amounts of radioactivity eluting in the peaks corresponding to the Tyr, Met, Phe derivatives.

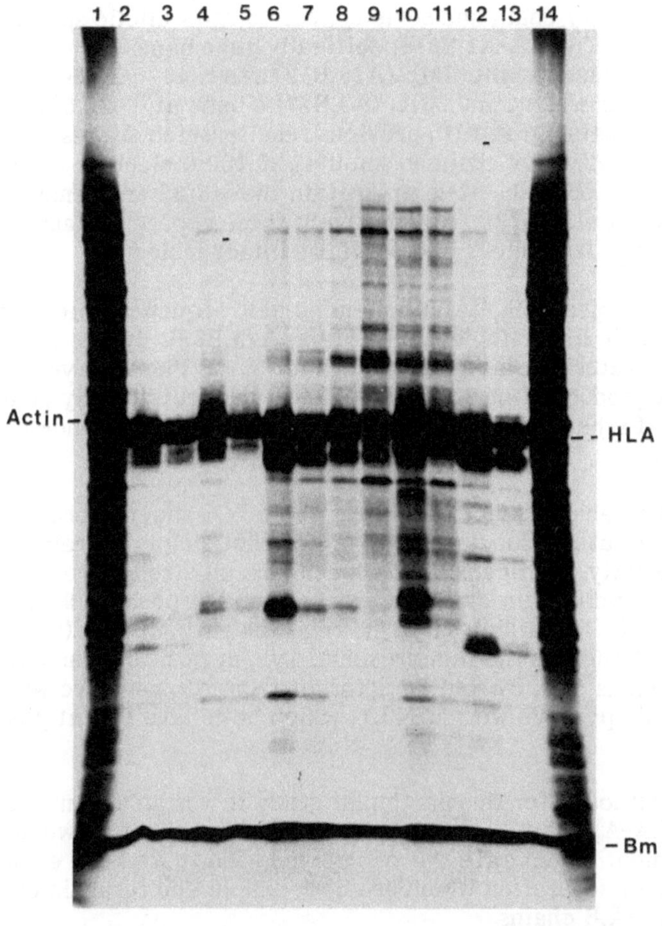

Figure 6. SDS-PAGE of W6/32 immunoprecipates from ^{3}H-Tyr (lanes 2,4,6,8,10,12) and ^{3}H-Phe (lanes 5,7,9,11,13) labelled B cell lines. The cell lines used and their HLA types are PALA (A9,-B8) 1,2; WT20 (A30, B18) 3,4; 23.1(A2, B27) 5,6; WT46 (A32, B13) 7,8; JMF (A25,B12) 9,10; LKT (A1,B8). Lanes 1 and 14 contain the total ^{3}H-TYR extract of PALA cells. Electrophoresis was on 25 cm, 7.5-15% gradient gels using the method of Laemmli. Fluorography and autoradiography was by the method of Bonner and Laskey (1974). The autoradiograph was over exposed in order to reveal minor bands.

for NH_2-terminal amino-acid sequence analysis as actin has a blocked NH_2-terminal amino-acid (Elizinga et al., 1973). Various bands of molecular weight < 42,000 were specifically precipitated by W6/32 from different cell lines. These may be proteolytic degradation products of the HLA-A,B heavy chain and some appear to show allelic specificity. For example, LKT (HLA-A1,B8) specifically gave bands at ≃ 26,000 and ≃ 19,000 molecular weight; JMF (A25,B12) bands at ≃ 23,000 and 21,000 molecular weight and 23.1, (A2,B27) a band at ≃ 22,000 molecular weight. These data agree with previous results which suggest that fragments of the HLA heavy chain as small as 25,000 molecular weight can be found associated with β2-m and retain the W6/32 antigenic determinant (Brodsky et al., 1979). Alternatively these apparent cleavage products may result from nicks in essentially intact molecules.

A typical NH_2-terminal amino acid sequence determination is shown in Table III for the cell line Maja (A23,B35). It was apparent from the characteristic assignments of Met 5, 12, Phe 8,9 Tyr 7,9,10 that an HLA product of purity equivalent to that obtained by conventional purification (Parham et al., 1977) was being sequenced (Terhorst et al., 1976).

Our analysis of 7 cell lines (Table IV) shows there is considerable variability at residue 9 and that it does not simply correlate with MB40.1 reactivity. Antigens B7, B40, B28 are all strongly reactive with MB40.1 yet two different amino acids are found at position 9; His in B40, Tyr for A28, B7. B35 which shows no reaction with MB40.1 (Parham and McLean, 1980) has the same residue, Tyr, at position 9 as two strongly reactive antigens A28, B7, but is different from A3, negative with MB40.1, and A2, weakly positive with MB40.1, which both have Phe at position 9.

Although for this particular study it was an advantage to sequence HLA-A,B and β_2-m together; for many studies it would be more convenient to separate the two polypeptides. We have explored two methods of separation which are both simple and result in efficient yields of HLA-A,B chains.

2.2.2. Separation by exchange with an excess of non-radioactive "cold" β_2-microglobulin. Hyafil and Strominger (1979) demonstrated that free β_2-m could be exchanged with HLA associated β_2-m. Therefore if radioactive HLA antigens could be exhaustively exchanged with non-radioactive β_2-m the material as isolated by W6/32 precipitation would sequence, by radioactive methods, as though it were pure HLA heavy chain. LKT cells were biosynthetically radiolablelled with ^{35}S-Met and then solubilized with either papain (Cresswell et al., 1973) or detergent (Kessler, 1975). Each extract was divided in two and to one aliquot was added non-radioactive β_2-m to a final concentration of 5 mg/ml. The extracts were incubated at 37°C for 1 hour, immunoprecipitated with W6/32, and then analyzed by SDS-PAGE. With papain

TABLE IV

Variability at position 9 in the amino acid sequence of HLA-A,B heavy chains

Residues identified by conventional sequencing (adapted from Strominger et al. 1981)

	1	2	3	4	5	6	7	8	9	10	11	12	13	14	15
β_2-m	Ile	Gln	Arg	Thr	Pro	Lys	Ile	Gln	Val	<u>Tyr</u>	Ser	Arg	His	Pro	Ala
B7	Gly	Ser	His	Ser	<u>Met</u>	Arg	<u>Tyr</u>	<u>Phe</u>	<u>Tyr</u>	Thr	Ser	Val	Ser	Arg	Pro
B40	Gly	Ser	His	Ser	<u>Met</u>	Arg	<u>Thr</u>	<u>Phe</u>	His	Thr	Ala	Met	Ser	Arg	Pro
A28	Gly	Ser	His	Ser	<u>Met</u>	Arg	<u>Tyr</u>	<u>Phe</u>	<u>Tyr</u>	Thr	Ser	Val	Ser	Arg	Pro
A2	Gly	Ser	His	Ser	<u>Met</u>	Arg	<u>Tyr</u>	<u>Phe</u>	<u>Phe</u>	Thr	Ser	Val	Ser	Arg	Pro

(continued)

TABLE IV (continued)

Residues of HLA-A,B heavy chains identified by radioactive sequencing of W6/32 immunoprecipitates

Cell Line	HLA A	HLA B	1	2	3	4	5	6	7	8	9	10	11	12
LB	28	40					Met		Tyr	Phe	Tyr -			Met
JY	2	7					Met		Tyr	Phe	Tye Phe			
23.1	2	27					Met		Tyr	Phe	Phe -			
HOM-2	3	27					Met		Tyr	Phe	Phe -			
MAJA	2	35					Met		Tyr	Phe	Phe Tyr			Met
WT20	30	18					Met		Tyr	Phe	Tyr -			
WT46	32	13					Met		Tyr	Phe	Tyr Phe			Met

Cell lines were biosynthetically radiolabelled with ^{35}S-Methionine, ^{3}H-Phenylalanine and ^{3}H-Tyrosine, solubilized with detergent and HLA-A,B antigens precipitated with W6/32. Immune precipitates containing equal amounts of ^{3}H-Phe and ^{3}H-Tyr and one tenth the amount of ^{35}S-Met were combined with 0.5 mg of sperm whale myoglobin and subjected to twelve cycles of Edman degradation. Radioactivity released at each step was assayed and the composition of the PTH derivatives determined by HPLC. Although mixtures of HLA-A,B heavy chains were present it was possible to estimate whether one or both chains had an amino acid at each position. This required the assumptions that: 1) equimolar quantities of HLA-A,B and β_2-m were sequenced and 2) the Tyrosine signal at position 10 was solely due to β_2-m. Thus (-) indicates that a residue other than Phe or Tyr is present at position 9. For LB cells we know this is a Histidine in B40. All our results indicated that Met 5, Tyr 7, and Phe 8 were conserved in all HLA-A,B gene products examined. The low amounts of HLA-C present in the precipitates was assumed to contribute negligibly to the results obtained.

Figure 7. Exchange of β_2-m. LKT (A1, B8) cells were synthetically labelled with ^{35}S-Met and treated with papain. The cells were removed by centrifugation and the supernatant divided in two. One half was incubated for 4 hour an excess (1 mg) of non-radioactive β_2-m (lane 1), the other similarly treated with addition of β_2-m (lane 2). W6/32 immunoprecipitates were made and analyzed by SDS-PAGE as described in the legend to Figure 6.

solubilized HLA antigens, a clear result was obtained indicating complete exchange of β_2-m (Fig. 7). Characteristic radioactive bands at $\simeq$ 34,000 molecular weight of HLA papain heavy chains were precipitated in the presence and absence of cold β_2-m. A radioactive band of β_2-m was only observed in the absence of cold β_2-m. For detergent solubilized HLA-A,B antigens some exchange was observed but it was by no means complete. In addition, the incubation at 37 °C increased the background precipitation, as assessed by SDS-PAGE, to an unacceptable degree and also resulted in considerable proteolysis of the HLA-A,B heavy chain. Similar differences in the exchange of β_2-m were also observed by Hyafil and Strominger (1979).

2.2.3. Separation by preparative SDS-PAGE. A simple, small-scale preparative electrophoresis apparatus (now made by BRL) proved useful in the preparation of HLA mRNA (Ploegh et al., 1980). We have adapted it for separation of proteins. Proteins are separated on a cylindrical acrylamide gel (Laemmli, 1979) with a scintered glass frit at the base which provides significant mechanical but little electrical resistance. Buffer is pumped from a port on the underside of the frit to a fraction collector and this flow efficiently collects all material that electrophoresis out of the gel. Figure 8 shows the separation of polypeptides from a combined W6/32 immunoprecipitate from a ^{35}S-Met, ^{3}H-Tyr and ^{3}H-Phe. The HLA labelled extract of Maja cells heavy chain (peak II) was completely separated from β_2-m (peak I) and significant separation of two HLA components and actin was seen within peak II. Yields in excess of 50% have usually been obtained and preliminary experiments show that material prepared in this fashion can be used for NH_2-terminal sequencing or peptide mapping.

Acknowledgements

This work was supported by two grants from the National Science Foundation, PCM78-14224 at Harvard University and PCM80-17834 at Stanford University, and a grant from the National Institutes of Health, IRO1-A1 17892. We thank Dr. Jack Strominger for his support and encourage-ment at Harvard, and Kathy Callahan for preparing the manuscript at Stanford. To Drs. Bob Rouse, Hiddle Ploegh, and Alan Williams, we are indebted for help with some of the various experiments described here.

REFERENCES

Allison, J.P., Ferrone, S., Walker, L.E., Pellegrino, M.A., Silver, J. and Reisfeld, R.A. 1978. Partial amino acid sequence of HLA-A9 antigen purified with a specific xenoantiserum. Transplantation 26:451-454.

Barnstable, C.J., Bodmer, W.F., Brown, G., Galfre, G., Milstein, C., Williams, A.F. and Ziegler, A. 1978. Production of monoclonal antibodies to Group A erythrocytes, HLA and other human cell surface antigens - new tools for genetic analysis. Cell 14:9-20.

Bonner, W.J. and Laskey, R.A. 1974. A film detection method for tritium-labelled proteins and nucleic acids in polyacrylamide gels. Eur. J. Biochem. 46:83-88.

Brodsky, F.M., Bodmer, W.F. and Parham, P. 1979. Characterization of a monoclonal anti- β_2-microglobulin and its use in the genetic and biochemical analysis of major histocompatibility antigens. Eur. J. Immunol. 9:536-545.

Brodsky, F.M., Parham, P., Barnstable, C.J., Crumpton, M.J., and Bodmer, W.F. 1979. Monoclonal antibodies for analysis of the HLA system. Immunol. Rev. 47:3-56.

Brodsky, F.M. and Parham, P. 1982. Monomorphic anti-HLA-A,B,C monoclonal antibodies detecting molecular subunits and combinatorial determinants. J. Immunol. 128:129-135.

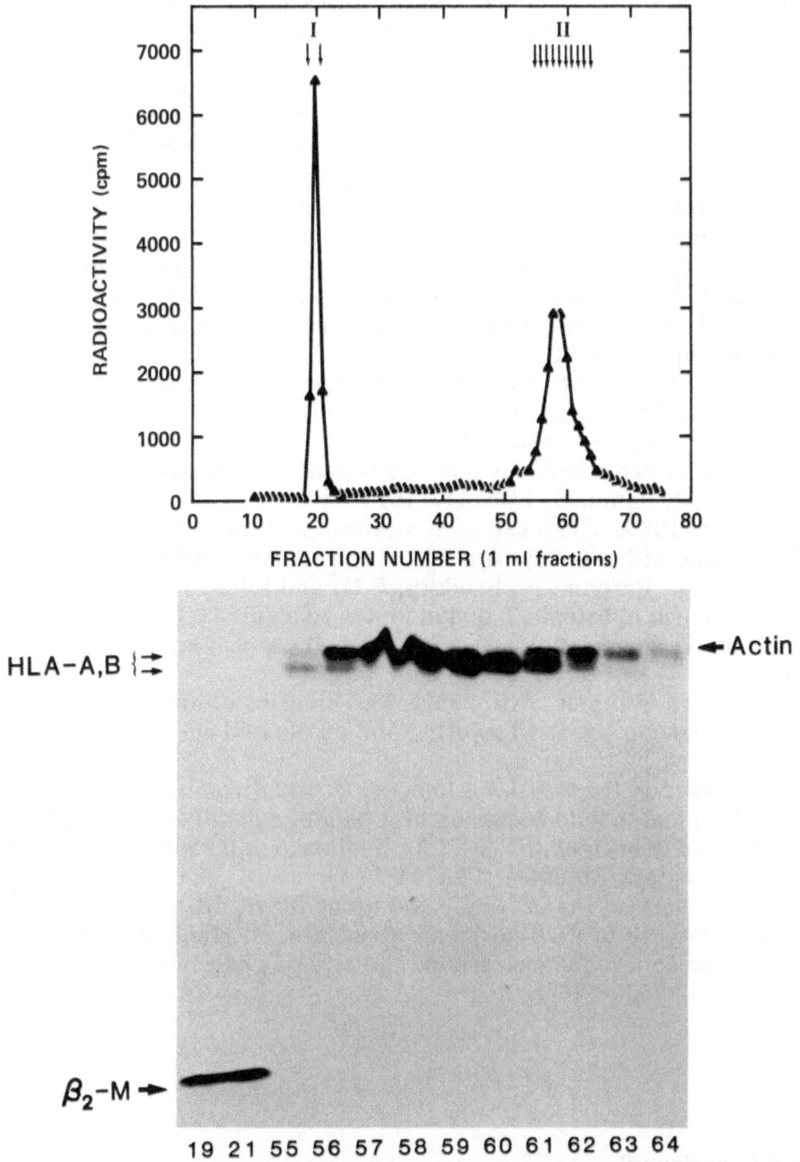

Figure 8. Separation of HLA-A,B, β_2-m and actin by SDS-PAGE. W6/32 immune precipitates from ^{3}H-Phe, ^{3}H-Tyr and ^{35}S-Met labelled extracts of Maja cells (A2,B35) were solubilized with sample buffer, combined and electrophoresed using the buffer systems of Laemmli (1970). A cylindrical gel 1 cm x 5 cm with a 1.5 cm stacking gel was used and electrophoresed at a constant potential difference of 120 volts. Reservoir buffer was pumped from the outlet port in the frit at the anode end of the gel at a rate of 20 mls per hour. One ml fractions were collected.

Coligan, J.E., Kindt, T.J., Uehara, H., Martinko,, J. and Nathenson, S.G., 1981. Primary structure of a murine transplantation antigen. Nature, 291:35-39.

Cresswell, P., Turner, M.J., Strominger, J.L. 1973. Papain-solubilized HL-A antigens from cultured human lymphocytes contain two peptide fragments. Proc. Natl. Acad. Sci. USA 70:1603-1607.

Elzinga, M., Collins, J.H., Kuehl, W.M. and Adelstein, R.S. 1973. Complete amino acid sequence of actin of rabbit skeletal muscle. Proc. Natl. Acad. Sci. USA 70:2687-2691.

Frankel, M.E. and Gerhard, W. 1979. The rapid determination of binding constants for antiviral antibodies by a radioimmunoassay. An analysis of the interaction between hybridoma proteins and influenza virus. Mol. Immunol. 16:101-106.

Hyafil, F. and Strominger, J.L. 1979. Dissociation and exchange of the β_2-microglobulin subunit of HLA-A and HLA-B antigens. Proc. Natl. Acad. Science, USA 5834-5838.

Kessler, S. 1975. Protein-A antibody adsorbent for isolation of cellular antigens. J. Immunol. 115:1617-1624.

Laemmli, U.K. 1970. Cleavage of structural protein during the assembly of the head of bacteriophage T_4. Nature 225:680-685.

McMichael, A.J., Parham, P., Brodsky, F.M. and Pilch, J.. 1980. Influenza virus specific cytotoxic T lymphocytes recognize HLA molecules, blocking by monoclonal anti-HLA antibodies. J. Exp. Med. 152:195s-203s.

Mason, D.W. and Williams, A.F. 1980. The kinetics of antibody binding to membrane antigens in solution and at the cell surface. Biochem. 187:1-20.

Orr, H.T., Lopez de Castro, J.A., Lancet, D. and Strominger, J.L. 1979. Complete amino acid sequence of a papain-solubilized human histocompatibility antigen, HLA-B7.2. Sequence determination and search for homologies. Biochem. 18:5711-5720.

Parham, P., Alpert, B.N., Orr, H.T. and Strominger, J.L. 1977. The carbohydrate moiety of HLA antigens: structure, antigenic properties and amino acid sequences around the site of glycosylation. J. Biol. Chem. 252:7555-7567.

Figure 8 (Continued)

The time from the start of electrophoresis to the point at which fractions were first assayed for radioactivity (fraction 1 was 3 hours). β_2-m eluted at a time 4 hours after the start and the HLA-A,B, actin peak had eluted by 6.5 hours (top panel). The arrows above the elution profile show fractions that were analyzed by SDS-PAGE on a 25 cm, 7.5 - 15% slab gel as shown in the lower panel. The fraction numbers are indicated below.

Parham, P. 1979. Isolation of immunologically active HLA-A and B antigens by a series of monoclonal antibody columns. J. Biol. Chem. 254:8709-8712.

Parham, P., Bodmer, W.F. 1978. Monoclonal antibody to a human histocompatibility alloantigen, HLA-A2. Nature 276:397-399.

Parham, P., Barnstable, C.J. and Bodmer, W.F. 1979. Use of a monoclonal antibody (W6/32) in structural studies of HLA-A,B,C antigens. J. Immunol. 123:342-349.

Parham, P., and McLean, J. 1980. Characterization, evolution and molecular basis of a polymorphic antigenic determinant shared by HLA-A and B products. Human Immunol. 1:131-139.

Parham, P. and Ploegh, H.L. 1980. Molecular characterization of HLA-A,B homologues in owl monkeys and other non-human primates. Immunogenetics 11:131-143.

Parham, P. 1981. Monoclonal antibodies against two separate alloantigenic sites of HLA-B40. Immunogenetics 13:509-527.

Parham, P. and Brodsky, F.M. 1981. Partial purification and some properties of BB7.2; a cytotoxic monoclonal antibody with specificity for HLA-A2 and a variant of HLA-A28. Human Immunol., 3:277-299.

Ploegh, H.L., Orr, H.T. and Strominger, J.L. 1980. Molecular cloning of a human histocompatibility antigen cDNA fragment. Proc. Nat'l. Acad. Sci. USA 77:6081-6085.

Rouse, R.M., Parham, P., Grumet, F.C. and Weissman, I.L. 1981. Expression of HLA antigens by human thymic ephithelial cells. Human Immunol. 5:21-34.

Strominger, J.L., Engelhard, V.H., Fuks, A., Guild, B.C., Hyafil, F., Kaufman, J.F., Korman, A.J., Kostyk, T.G., Krangel, M.S., Lancet, D. Lopez de Castro, J.A., Mann, D.L., Orr, H.T., Parham, P., Parker, K.C., Ploegh, H.L., Pober, J.S., Robb, R.J. and Shackelford, D.A. 1981. Structure of MHC products. In: The Role of the Major Histocompatibility Complex in Immunobiology, Ed. M.E. Dorf, Garland Press Publishing, Inc. pp. 115-172.

Terhorst, C., Parham, P., Mann, D.L., Strominger, J.L. 1976. Structure of HLA antigens: Amino-acid and carbohydrate composition and NH_2-terminal sequences of four antigen preparations. Proc. Nat'l. Acad. Sci. USA 73:910-914.

Trucco, M. de Petris, S. Garotta, G. and Cepellini, R. 1980. Quantitative analysis of cell surface HLA structure by means of monoclonal antibodies. Human Immunol. 1:233-243.

PART IV.

MONOCLONAL ANTIBODIES IN DIAGNOSIS AND THERAPY

8

MONOCLONAL ANTIBODIES TO HEPATITIS B SURFACE ANTIGEN

Vincent R. Zurawski, Jr. and Nancy T. Chang

Centocor, 244 Great Valley Parkway, Malvern, PA 19355

1. SUMMARY

We have established stable clones of murine somatic cell hybrids that produce high affinity monoclonal antibodies to hepatitis B surface antigen (HBsAg)[1]. Characterization of several of these antibodies has indicated that they bind to a 49000 dalton protein, likely encoded in the hepatitis B virus genome. One of these antibodies, designated 5D3 is an IgM antibody which binds to HBsAg with a $K_{ASN} = 4 \times 10^{11}$ liters/mole. This antibody, which has broad reactivity among HBsAg serotypes, can be used in an immunoradiometric assay. This assay has proven useful in evaluating clinical samples immunodiagnostically. Further studies with these unique monoclonal reagents should provide clinicians and research investigators with important new information about the hepatitis B virus.

2. INTRODUCTION

It is estimated that hepatitis B virus (HBV) persistently infects 200 million people in the world. Infection with HBV can cause an acute disease in humans commonly characterized by parenchymal cell necrosis and histocytic periportal inflammation in the liver (Dienstag, et al., 1980). Additionally, HBV infection can lead to a chronic persistent form and a chronic active form of the illness characterized by continuing hepatic necrosis and active inflammation often leading to fibrosis, cirrhosis, and eventual parenchymal liver cell failure resulting in death (Wands, et al., 1980a). Further, it has been suggested that persistent infection with HBV may play some role

[1]Abbreviations: HBsAg, Hepatitis B surface antigen; Anti-HBs, Antibody to hepatitis B surface angen; Anti-HBc, Antibody to hepatitis B core antigen; HBV, Hepatitis B virus; HBcAg, Hepatitis B core antigen; HBeAg, Hepatitis B e antigen.

in the development of primary hepatocellular carcinoma (Blumberg and London, 1981). Moreover, the perinatal vertical transmission of the virus from persistently infected mothers to infants and occult infection among close contacts of so-called carriers mark HBV as a particularly troublesome pathologic agent.

There are several recognizable and serologically distinct antigens associated with HBV (W.H.O., 1977). Among these are hepatitis B core antigen (HBcAg) (Barker, et al., 1974), hepatitis B e antigen (HBeAg) (Magnius and Epsmark, 1972), and hepatitis B surface antigen (HBsAg), originally designated Australia antigen (Blumberg, et al., 1965). The presence of HBsAg in patient serum has been determined to be a marker of HBV infection (Prince, 1968). HBsAG is, in fact, the first viral marker to appear in patient serum following infection with HBV. During the course of acute illness serum HBsAG levels decline and antibodies to HBcAg (anti-HBc) and to HBsAg (anti-HBs) appear (Dienstag, et al., 1980).

HBsAg can be located on the surface of 42 nm particles (Dane, et al., 1970) which represent intact HBV. HBsAg also appears in the serum of infected individuals in the form of 22 nm spheres and filamentous forms (Dane, et al., 1970; Dienstag, et al., 1980). HBsAg, whether as a part of a viral envelope or in the 22 nm forms, contains lipids and virally coded peptides designated PI (p23) and PII (gp28) (Tiollais, et al., 1981). In addition, host proteins have been found to be associated with the 22 nm forms (Neurath, et al., 1974; Skelly, et al., 1979). Mishiro et al. (1980) have identified an approximately 49,000 dalton protein isolated from HBsAg preparations, likely a disulfide-linked dimer of PI and PII, which was highly immunogenic in contrast to the free peptides. A structural model of the HBsAg in the viral envelope including this dimer has been proposed by Tiollais et al (1981) in their review of the molecular biology of HBV.

The application of monoclonal antibody technology has been successfully applied to the study of other viruses (Gerhard, et al., 1980; Koprowski and Wiktor, 1980). Therefore, we decided to produce monoclonal antibodies to purified HBsAg (Wands and Zurawski, 1981) in an effort to identify reagents capable of augmenting previous studies of HBV. Additional reports of monoclonal antibodies to HBsAg (anti-HBs) have also appeared (Shih, et al., 1980, Goodall, et al., 1980; Kalil, et al., in press).

3. RESULTS

3.1. Fusions and characteristics of monoclonal anti-HBs

We utilized a modified AUSAB assay (Abbott Laboratories, No. Chicago, IL) to detect the presence of monoclonal anti-HBs in cell culture supernatants (Wands and Zurawski, 1981). Figure 1 illustrates the assay results of one successful fusion. Extraordinarily high binding of $[^{125}I]$-HBsAg was noted for supernatants of several hybrid cells using this assay. Up to 95 percent of the labeled antigen was bound by monoclonal anti-HBs attached to HBsAg coated beads. In this and subsequent fusions, hybrids produced both IgG and IgM anti-HBs. The properties of a group of 5 monoclonal antibodies synthesized by twice-cloned hybrid cells are presented in Table I. Of particular interest are the extremely high hemagglutination titers and affinity binding constants of these monoclonal anti-HBs. The antibody designated 5D3 which binds well to both ad and ay serotypes of HBsAg has been the subject of continued study (Wands and Zurawski, 1981; Wands et al., 1981).

We examined the binding of 5D3 and other monoclonal anti-HBs to components of HBsAg. In preliminary studies Wands et al., (1980b) found that monoclonal anti-HBs 1F8, 2F11, 1C7, and 4E8 as well as 5D3 all bound to a 49,000 dalton protein. We have confirmed that result for the 5D3 antibody using electrophoretic transfer of protein to nitrocellulose sheets (Towbin, et al., 1979) following overlay with [^{125}I] - 5D3. Figure 2 illustrates the result. Moreover, we have utilized the immunoradiometric assay (see below) described by Wands et al (1981) to examine the binding of 5D3 anti-HBs to material in cell culture supernatants of the $b2g_{21}$ cell line described by Dubois, et al (1980). This cell line secretes HBsAg-like particles produced from cloned HBV DNA. It can be seen in Figure 3 that supernatants of this line were reactive in the 5D3 assay. Additionally we have shown (Shouval et al., in press) that the 5D3 antibody binds to HBsAg synthesized and secreted by the PLC/PRF/5 human hepatoma cell line (Macnab, et al, 1976). These data all support the conclusion that the 5D3 antibody binds to the 49,000 dalton protein described by Mishiro et al. (1980), the peptides of which are encoded in the HBV genome (Tiollais, et al., 1981).

3.2. Immunoradiometric assay for HBsAg

We have described the design of an immunoradiometric assay (IRMA) which utilizes the 5D3 IgM monoclonal anti-HBs (Wands, et al., 1981). Figure 4 provides a schematic representation of this assay. One hundred µl each of patient sample and tracer solution containing [^{125}I] - 5D3 (150,000 cpm) are added to a plastic well containing a one-quarter inch polystyrene bead coated with 5D3 anti-HBs. After an incubation period of 4 hr at 45°C the beads are washed 3-5 times and counted in a gamma well counter. The assay may also be run for 1 hour or 20 hours at 45°C.

Table II presents data indicating that the 5D3 assay is capable of detecting all relevant HBV serotypes. Shorey et al (1981) have also found that this monoclonal antibody binds all subtypes of HBsAg tested.

We have also found that this assay functions effectively in distinguishing reactive and non-reactive clinical samples. In Table III it can be seen that those samples from individuals diagnosed as having HBV infections were clearly reactive in the 5D3 assay whereas a number of other control samples were non-reactive. In Table IV, data is shown comparing the 5D3 assay to an existing commercial assay (AUSRIA II, Abbott Laboratories, No. Chicago, IL) in a set of samples collected longitudinally from HBV infected donors. Reactivity was similar in each assay.

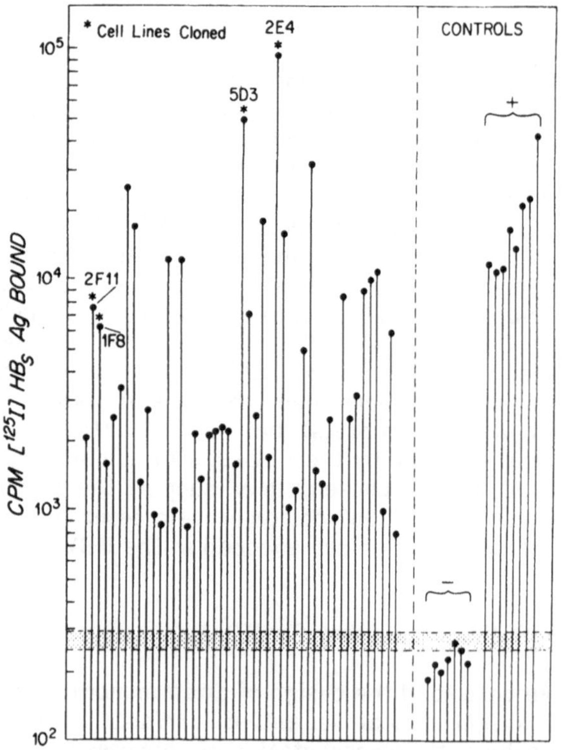

Figure 1. Anti-HBs production of hybrid cells resulting from the fusion of immunized splenocytes with NS1 myeloma cells. Of the 297 microtiter wells seeded, 216 were positive for hybrid growth. Of these, 47 (22%) were producing anti-HBs as determined in a solid-phase radioimmunoassay. [^{125}I] HBsAg binding greater than twice background (stippled area) was taken as an indication of antibody production. The negative controls in this case are sera from normal individuals with no prior exposure to hepatitis B. The positive controls represent binding activity of eight different hemophiliac sera. Cell lines H22F11, H21F8, H25D3, and H22E4 were cloned initially. Clones of H22E4 did not maintain stable anti-HBs production, whereas the other clones did.

TABLE I

Properties of monoclonal anti-HBs antibodies

Clone	$[^{125}I]$-HBsAg[a] cpm	Immunoglobulin Class	Immunoglobulin Subclass	Ascites[b] fluid concentration	Kasn[c]	Hemagglutination titers[d] adw	Hemagglutination titers[d] ayw
5D3	96.3	Igm	-	6.3	4.0×10^{11}	2.2×10^{11}	2.7×10^{12}
1F8	54.4	IgM	-	24.7	1.4×10^{10}	1.7×10^{10}	2.6×10^{4}
2F11	60.8	IgM	-	20.1	4.8×10^{9}	1.3×10^{11}	1.0×10^{6}
1C7	46.4	IgG	IgG_1	18.2	3.2×10^{10}	7.0×10^{10}	3.2×10^{4}
4E8	50.2	IgG	IgG_1	15.2	9.1×10^{9}	4.3×10^{9}	2.1×10^{6}
Controls[e]	0.2	-	-	-	-	4	4

[a] $[^{125}I]$ =HBsAg bound (cpm x 10^{-3}) in IRMA assays.
Initial sample volume was 200 μl of ascites fluid.

[b] Expressed as mg/ml of monoclonal anti-HBs in ascites fluid.

[c] Expressed as liters/mole per molecule of IgM or IgG.

[d] Reciprocal of highest dilution giving agglutination of HBsAg coated 0 negative red blood cells.

[e] Controls consist of ascites fluid containing IgG_1 monoclonal antibodies to cardiac myosin (Ehrlich, et al, 1979) reacting in the RIA and with HBsAg coated cells. Additional controls consist of the above 5 cloned cell lines reacting with human 0 negative red blood cells (HBsAg negative) and mouse myeloma IgM (Bionetics, Kensington, MD) 1 mg/ml in phosphate buffered saline reacting with HBsAg coated red blood cells and control red blood cells.

Reprinted from Proceedings of the National Academy of Science in modified form, with permission.

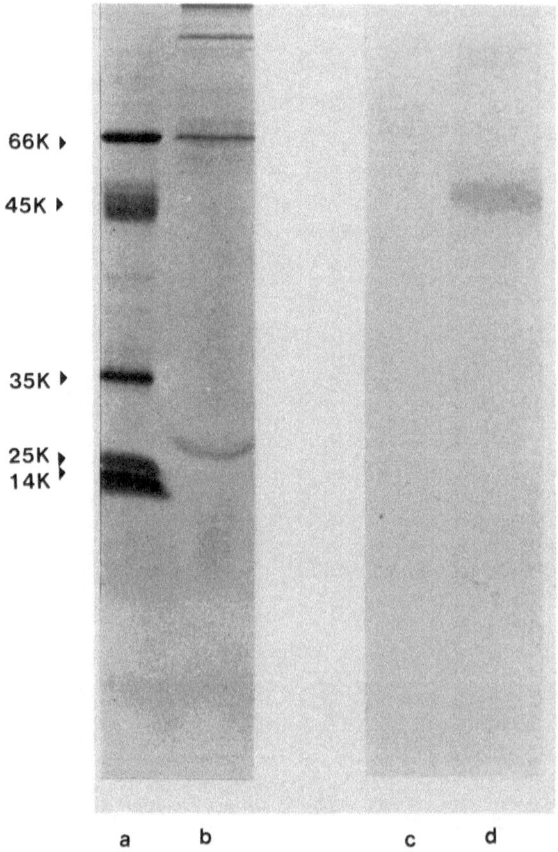

Figure 2. Binding of [^{125}I]-5D3 to electrophoretically resolved peptides (unreduced) of HBsAg (see Experimental Procedure). Lane a, molecular weight markers of 66000, 45000, 35000, 25000 and 14000 daltons stained with Commassie Brillant Blue. Lane b, purified HBsAg stained with Coomassie Blue. Lane c, transblotted molecular weight markers on nitrocellulose incubated with [^{125}I]-5D3 followed by autoradiography (no binding of labelled 5D3 was detected). Lane d, transblotted HBsAg on nitrocellulose incubated with [^{125}I]-5D3 followed by autoradiography [^{125}I]-5D3 was detected bound to a protein migrating at approximately 49000 daltons).

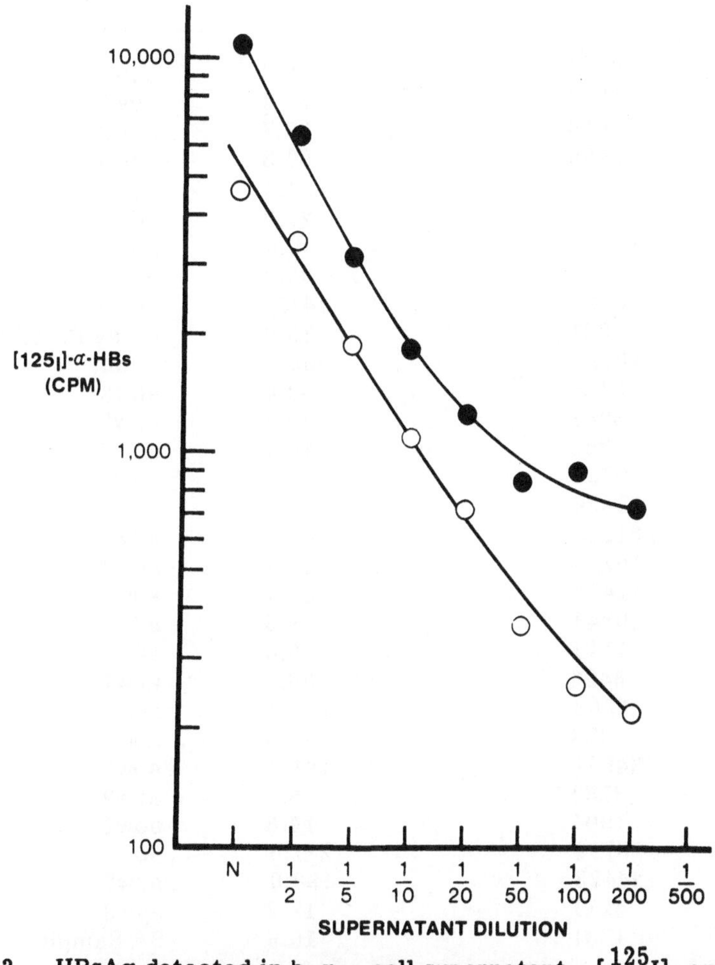

Figure 3. HBsAg detected in b_2g_{21} cell supernatant. [^{125}I] -anti-HBs bound versus supernatant dilution: ●-●, 5D3 assay; o-o, AUSRIA II assay (Abbott Laboratories). It should be noted that this is a semi-logrithmic plot.

TABLE II

5D3 assay: Detection of various serotypes of HBV

Sample	$[^{125}I]$-5D3 Bound (cpm)[a]	S/N[b]	Subtype
1	7350	56.5	aywl
2	8367	64.4	ayw2
3	7445	30.7	ayw2
4	3996	57.3	ayw3
5	25625	197.0	ayr
6	2974	22.9	adw2
7	13758	105.0	adw4
8	4386	33.7	adr
9	18329	141.0	adyw
10	9881	76.0	ad/ay Positive
11	40138	309.0	adw4
12	5126	39.4	adw2
13	3947	30.4	adw2
14	1358	10.4	adw3
15	15758	121.0	adw3
16	25480	196.0	adw2
17	31325	241.0	adw2
18	16244	125.0	adw2
19	11452	88.1	adr
20	10343	79.6	adr
21	11149	85.8	adw2
22	8480	66.2	ayw2
23	13369	103.0	adr
24	2473	19.0	adr
25	24801	191.0	adw2
26	2182	16.8	adw2
27	1904	14.6	adw2
28	28791	221.0	adw2
29	24498	188.0	adw2
30	2211	17.0	ayw3
31	1381	10.6	SA Sample[c]

[a]Counts bound represent the average of two runs. The negative control mean was 130 cpm. Samples with S/N greater than 3.5 are considered to be negative.

[b]S/N is signal to noise ratio.

[c]A rare South African serotype not currently subtyped by existing reagents.

We also compared another set of samples collected serially from patients known to have hepatitis A virus infection as confirmed immunodiagnostically using an assay which detects IgM anti-hepatitis A virus (Abbott Laboratories). In these samples the 5D3 assay and the AUSRIA II assay were negative in all cases except for patient 220 (Table V). In that case a serum sample collected on 14 May 1981 was clearly reactive in both assays. Subsequent samples were reactive only in the 5D3 assay. The samples collected on 21 May and 4 June 1981 also contained anti-HBc as determined by the CORAB assay (Abbott Laboratories). Consequently, patient 220 was likely a carrier of HBsAg who had been infected with hepatitis A virus. The lack of reactivity of the 21 May and 4 June samples in AUSRIA II suggests that the 5D3 assay may be capable of

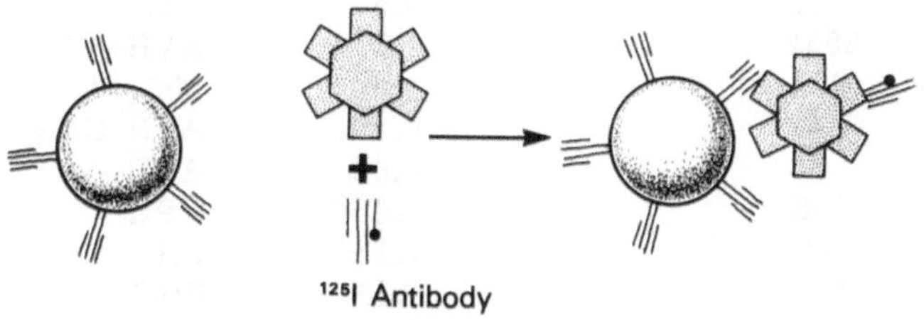

Figure 4. Schematic representation of the 5D3 assay. Monoclonal antibody (5D3) bound to the polystyrene bead is incubated with antigen (designated by the hexagonal symbol) and radiolabelled 5D3 simultaneously. Following a wash step beads are counted in a gamma well scintillation counter. Only samples containing antigen will retain radiolabelled antibody. This is a simultaneous sandwich immunoradiometric assay (see text for further details).

detecting HBsAg in samples where existing commercial assays were not effective.

4. DISCUSSION

We have prepared monoclonal antibodies to HBsAg (Wands and Zurawski, 1981; Zurawski, et al., 1981). Characterization of these antibodies has indicated that a number bind to HBsAg with particularly high avidity (Wands and Zurawski, 1981; Wands, et al., 1981; Zurawski, et al., 1981). These antibodies apparently bind to a protein that is coded in the HBV genome. Data presented here are particularly supportive of this conclusion for the IgM anti-HBs designated 5D3 (Figures 2 and 3).

TABLE III

5D3 Assay performance on a clinical panel

Sample	[^{125}I]-5D3 Bound[a] cmp	S/N	Diagnosis[b]
1	9451	73.40	CP
2	39	0.30	ALD
3	85	0.65	PBC
4	68	0.52	AVH-CMV
5	140	1.08	PBC
6	53	0.41	PBC
7	70	0.54	PBC
8	6532	50.20	AVH-B
9	208	1.60	MC
10	81	0.62	AVH-CMV
11	88	0.68	ALD
12	138	1.06	CPH
13	149	1.15	LH
14	150	1.15	PBC
15	121	0.93	AL
16	1230	9.46	CH-B
17	1346	10.40	AVH-B
18	186	1.43	ALD
19	1341	10.30	AVH
20	3529	27.10	CARRIER-B
21	1658	12.80	AVH-B
22	65	0.50	UC
23	5560	42.80	CAH-B
24	7192	55.30	CAH-B
25	98	0.75	BB
26	50	0.38	BB

[a]Approximately 150,000 cpm [^{125}I]-5D3 were added in the assay. The average of the negative controls was 130 cpm. Samples above S/N = 3.5 are considered to be reactive.

[b]Abbreviations: CP, positive control from a commercial assay; ALD, alcoholic liver disease; PBC, primary biliary cirrhosis; AVH-CMV, acute viral hepatitis with serologic evidence for cytomegalo virus infection; AVH-B, acute viral hepatitis B; CPH, chronic persistent hepatitis, cryptogenic; LH, lupoid hepatitis; AL, acute leukemia; CH-B, chronic hepatitis B; Carrier-B, HBsAg carrier; UC, ulcerative colitis; CAH-B, chronic active hepatitis B; BB, non-reactive sample from blood bank.

TABLE IV

5D3 Assay performance on selected HBV cases compared with an existing commercial assay

			5D3 Assay		CR Assay[b]	
Patient	Diagnosis[a]	Date of Bleeding	[^{125}I] -α-HBs Bound (cpm	Reactivity	[^{125}I] -α-HBs Bound (cpm)	Reactivity
17	AVH-B	5 MAR 81	15813	+	20002	+
		9 APR 81	185	–	66	–
44	CH-B	23 OCT 80	7785	+	4176	+
		20 JAN 81	7447	+	5338	+
83	CARRIER-B	9 OCT 80	12698	+	15293	+
		29 JAN 81	9483	+	18257	+
84	AVH-B	21 MAY 81	10201	+	7161	+
		11 JUN 81	9989	+	12163	+
86	CARRIER-B	22 JAN 81	21213	+	15663	+
		19 FEB 81	19428	+	13957	+
192	AVH-B	23 OCT 80	12353	+	22537	+
		30 OCT 80	15610	+	30209	+
		6 DEC 80	31362	+	31199	+

[a]Abbreviations: AVH-B, acute viral hepatitis B; CH-B, chronic hepatitis b; Carrier-B, HBsAG carrier.

[b]CR assay is AUSRIA II (Abbott Laboratories, No. Chicago, IL).

TABLE V

5D3 assay: Performance on selected HAV cases compared with an existing commercial assay

Patient	Diagnosis[c]	Date of Bleeding	CR Assay[a] $[^{125}I]$-α-HBs Bound (cpm)	Reactivity	5D3 Assay $[^{125}I]$-α-HBs Bound (cpm)	Reactivity
9	AVH-A	9 OCT 80	252	-	34	-
15	AVH-A	9 OCT 80	115	-	20	-
		25 MAR 81	131	-	12	-
		16 APR 81	159	-	28	-
		16 JUL 81	180	-	48	-
26	AVH-A	15 MAY 80	142	-	188	-
28	AVH-A	9 APR 80	139	-	28	-
32	AVH-A	23 MAY 81	99	-	34	-
75	AVH-A	25 JUN 81	154	-	89	-
220[b]	AVH-A	14 MAY 81	541	+	1169	+
		21 MAY 81	129	-	632	+
		4 JUN 81	115	-	1172	+

[a]CR assay is AUSRIA II (Abbott Laboratories, No. Chicago, IL).

[b]Patient serum contained anti-HBc.

[c]Abbreviations: AVH-A, acute viral hepatitis A.

This 5D3 antibody was found to be broadly reactive with HBV serotypes (Table II). Shorey et al (1981) have embarked upon a detailed study of the serotypic specificities of the monoclonal antibodies we have described (Wands and Zurawski, 1981; Zurawski, et al, 1981). They have confirmed the broad reactivity of the 5D3 antibody and demonstrated selective binding of some other of these monoclonal anti-HBs to certain HBV subtypes.

We have shown that this 5D3 antibody can be utilized in an immunodiagnostic assay for hepatitis B (Wands, et al., 1981). We have confirmed this initial study with several extensive follow-up investigations (Wands, unpublished data; Zurawski and Rakela, unpublished data). Data presented here are indicative of the performance of this assay (Tables II-V). Moreover, we have presented preliminary evidence (Zurawski, et al., 1981) that the 5D3 assay is capable of detecting HBsAg in samples where other asays which rely on the use of conventional heterogenous anti-HBs do not indicate reactivity. This finding may have pertinent implications for the diagnosis of HBV infection.

We have also found that monoclonal anti-HBs are useful in the purification of HBsAg from serum and plasma samples (Zurawski, et al., 1981). Additionally, the demonstration that these antibodies may have immunotherapeutic applications (Shouval, et al,m 1981; Shouval, et al., in press; Shouval, et al., submitted) is of great potential interest.

Continuing investigations utilizing monoclonal anti-HBs will undoubtedly provide information that will extend our knowledge concerning immunological analysis of HBsAg and provide us with a better understanding of the biology of HBV. We believe that such an understanding will ultimately prove to be clinically as well as scientifically rewarding.

5. EXPERIMENTAL PROCEDURES

5.1. Production of monoclonal antibodies to HBsAg

We have described the antigen preparation immunization of mice, the fusion and selection procedures that were utilized in establishing the somatic cell hybrids producing monoclonal antibodies to HBsAg. In some recent experiments, HBsAg, purified by affinity chromatography, was used as an immunogen. Additionally, the non-producing myeloma variants Sp 2/0-Ag. 14 (Shulman, et al, 1979) and X63-Ag 8.653 (Kearney, et al., 1979) were also used as parent cell lines for fusions.

Screening assays to identify the clones synthesizing and secreting monoclonal anti-HBs have also been described (Wands and Zurawski, 1981).

5.2. Characterization of monoclonal antibodies to HBsAG

Preliminary characterization of the monoclonal anti-HBs was accomplished as reported (Wands and Zurawski, 1981). Antibodies have been further characterized (Shorey, et al., 1981; Wands and Zurawski, 1981; Wands, et al., 1981; Zurawski, et al., 1981).

Binding of 5D3 antibody to the 49,000 dalton protein was accomplished using the method of Towbin, et al (1979). Briefly, purified HBsAg was pretreated but not reduced and run on 12.5% sodium dodecyl sulfate-polyacrylamide gels. Following this electrophoresis, the protein was transferred to nitrocellulose filters. [^{125}I]-5D3 was then overlaid on to the filter, washed off, and autoradiograms were done. Positive autoradiograms were taken to be evidence that a 5D3 immunoreactive peptide had been detected.

5.3. Tissue culture

The hybrid cells were cultured as reported (Wands and Zurawski, 1981). The b_2g_{21} cell line (Dubois, et al., 1980) was grown in tissue culture in Dulbecco's modified Eagle medium supplemented with 15% fetal calf serum and hypoxanthine-aminopterin-thymidine in a humid CO_2 incubator containing 95% air-5% CO_2.

5.4. Assay procedure

The 5D3 assay design and details of the assay protocol have been described (Wands et al., 1981; Zurawski, et al., submitted).

Acknowledgement

The authors wish to thank their colleague Dr. J.R. Wands whose continuing cooperation has been instrumental in furthering progress in these investigations. We wish also to thank Dr. James Shorey for the generous gift of serotypically well characterized HBsAg samples, Dr. Jorge Rakela who provided assays of some of the clinical samples described in this study, and Dr. Pierre Tiollais for the generous gift of the b_2g_{21} cell line. The continued support of Dr. K.J. Isselbacher is also deeply appreciated. The excellent technical assistance of Paul Brock, Micheal L. McClure, Gwen A. Melincoff and Adrienne Quinn is also appreciated.

REFERENCES

Barker, L.R., Almeida, J.D., Hoofnagle, J.H., Gerety, R.J., Jackson, D.D., and McGrath, P.P., 1974, Hepatitis B core antigen. Immunology and electron microscopy. J. Virol 14:1552-1558.

Blumberg, B.S., Alter, H.J., and Visnick, S., 1965, A "new" antigen in leukemia sera. J. Am. Med. Assoc. 191:541-546.

Blumberg, B.S. and London, W.T., 1981, Hepatitis B virus and the prevention of primary hepatocellular carcinoma. N. Engl. J. Med. 304:782-784.

Dane, D.S., Cameron, C.H., and Briggs, M., 1970, Virus-like particles in serum of patients with Australia-antigen-associated hepatitis. Lancet 1:694-700.

Dienstag, J.L., Wands, J.R., and Kaff, R.S., 1980, Acute hepatitis, in: Harrison's Principles of Internal Medicine (K.J. Isselbacher, eds), McGraw-Hill Book Co., New York, pp. 1459-1470.

Dubois, M.-F., Pourcel, C., Rousset, S., Chany, C., and Tiollais, 1980, Excretion of hepatitis B surface antigen particles from mouse cells transformed with cloned viral DNA. Proc. Natl. Acad. Sci. U.S.A. 77:4549-4553.

Ehrlich, P.H., Zurawski, V.R., Jr., Kha, B.A., and Haber, E., 1979, Hybridoma antibodies against human cardiac myosin. Circulation, 60, 139.

Gerhard, W., Yewdall, J., Frankel, M., Lopes, A.D., and Staudt, L., 1980, Monoclonal antibodies against influenza virus, in: Monoclonal Antibodies Hybridomas: A New Dimension in Biological Analyses. (R.H. Kenneth, T.J. McKearns, and K.B. Bechtol, eds.), pp. 317-332.

Goodall, A.H., Janossy, G., Shipton, U., and Thomas, H.C., 1980, Abstracts of the Fourth International Congress of Immunology (19.2.09).

Kalil, L., Dronet, J., Couroree, A.-M., Proceedings of the Third International Symposium on Viral Hepatitis (in press).

Kearney, J.F., Radbruch, A., Liesegang, B., and Rajewsky, K., 1979, A new mouse myeloma cell line that has lost immunoglobulin expression but permits the construction of antibody-secreting hybrid cell lines. J. Immunol. 123:1548-1550.

Koprowski, H. and Wiktor, T., 1980, Monoclonal antibodies against rabies virus, in: Monoclonal Antiodies, Hybridomas: A New Dimension in Biological Analyses, (R.H. Kennett, T.J. McKearn, K.B. Bechtol, eds.) New York, pp. 335-351.

Macnab, G.M., Alexander, J.K., Lecatsas, G., Bey, E.M., and Urbanwich, J.M., 1976, Hepatitis B surface antigen produced by human hepatoma cell line. Br. J. Cancer 34:509-515.

Magnius, L.O. and Epsmark, J.A., 1972, New specifications in Australia antigen positive sera distinct from Le Bouvier determinants. J. Immunol. 124:1589-1593.

Mishiro, S., Imai, M., Takahashi, K., Machida, A., Gotanda, T., Miyakawa, Y., and Mayumi, M., 1980, A 49000-Dalton polypeptide bearing all antigenic determinants and full immunogenicity of 22-nm hepatitis B surface antigen particles. J. Immunol. 124:1589-1593.

Neurath, A.R., Prince, A.M., and Lippin, A., 1974, Hepatitis B antigen: antigenic sites related to human serum proteins revealed by affinity chromatography. Proc. Natl. Acad. Sci., U.S.A. 71:2663-2667.

Prince, A.M., 1968, An antigen detected in the blood during the incubation period of serum hepatitis. Proc. Natl. Acad. Sci., U.S.A. 60:814-821.

Shulman, M., Wilde, C.D., and Kohler, G., 1978, A better cell line for making hybridomas secreting specific antibodies. Nature 276:269-270.

Shih, J.W.-K., Cote, P.J., Dapolito, G.M., and Gerin, J.L., 1980, Production of monoclonal antibodies against hepatitis B surface antigen (HBsAg) by somatic cell hybrids. J. Virol. Meth. 1:257-273.

Shorey, J., Brown, R.D., and Wands, J.R., 1981, A new analysis of HBV subtypes by epitopes identification using monoclonal anti-HBs radioimmunoassays. Hepatology 1:545.

Shouval, D., Wands, J.R., Zurawski, Jr., V.R., and Shafritz, D.A., 1981, Immunotherapy of human hepatoma using specific monoclonal antibodies against HBsAg viral determinants. Hepatology 1:547.

Shouval, D., Wands, J.R., Zurawski, Jr., V.R., Isselbacher, K.J., and Shafritz, D.A., (in press), Monoclonal antibodies to hepatitis B virus surface antigen bind to and lyse human hepatoma PLC/PRF/S cells in culture. Proc. Natl. Acad. Sci., U.S.A.

Shouval, D., Shafritz, D.A., Zurawski, Jr., V.R., Isselbacher, K.J., and Wands, J.R. (submitted), Immunotherapy of human hepatoma using monoclonal antibodies against hepatitis B virus surface antigen viral determinants.

Skelly, J., Howard, C.R., and Zuckerman, A.J., 1979, Analysis of heptitis B surface antigen components solubilized with Triton X-100. J. Gen. Virol. 44:679-689.

Tiollais, P., Charnay, P., and Vyas, G.N., 1981, Biology of hepatitis B virus. Science 213, 406-411.

Towbin, H., Staehelin, T., and Gordon, J., 1979, Electrophoretic transfer of proteins from polyacrylamide gels to nitrocellulose sheets: procedure and some applications. Proc. Natl. Acad. Sci., U.S.A., 76:4350-4354.

Wands, J.R. and Zurawski, Jr., V.R., 1981, High affinity monoclonal antibodies to hepatitis B surface antigen (HBsAg) produced by somatic cell hybrids. Gastroenterology 80:225-232.

Wands, J.R., Koff, R.S., and Isselbacher, K.J., 1980a, Chronic active hepatitis, in: Harrison's Priniciples of Internal Medicine (9th edition) (K.J. Isselbacher, R.D. Adams, E. Braunwald, R.G. Petersdorf, and J.D. Wilson, eds.) pp. 1470-1484.

Wands, J.R., Carlson, R.I., and Zurawski, Jr., V.R., 1980b, Identification of epitopes on HBsAg polypeptides by analysis with monoclonal anti-HBs antibodies. Gastroenterology 79:1063.

Wands, J.R., Carlson, R.I., Schoemaker, H.J.P., Isselbacher, J., and Zurawski, Jr., V.R., 1981, Immunodiagnosis of hepatitis B with high affinity monoclonal antiodies. Proc. Natl. Acad. Sci., U.S.A., 78:1214-1218.

Wands, J.R. (unpublished data).

World Health Organization, 1977, Report of the Expert Committee on Viral Hepatitis. Techn. Rep. Ser. Bo. 602. Geneva.

Zurawski, Jr., V.R. and Rakela, J., unpublished data.
Zurawski, Jr., V.R., Del Villano, B.C., Wands, J.R., 1981, Monoclonal antibodies to hepatitis B surface antigen, in: Monoclonal Antibodies and T Cell Hybridomas. (U. Hammerling, G.J. Hammerling, J.J. Kearney, eds.) Elsevier - N. Holland, New York, pp. 273-282.
Zurawski, Jr., V.R., Wands, J.R., Isselbacher, K.J., and Schoemaker, H.J.P., (submitted) A novel radioimmunoassay, utilizing monoclonal antibodies, for detection of hepatitis B surface antigen in serum.

9

WANDERING AROUND THE CELL SURFACE - MONOCLONAL ANTIBODIES AGAINST HUMAN NEUROBLASTOMA AND LEUKEMIA CELL SURFACE ANTIGENS

Madelyn Feder, Zdenka L. Jonak, Arthur A. Smith, Mary Catherine Glick, and Roger H. Kennett

Department of Human Genetics and Department of Pediatrics
University of Pennsylvania School of Medicine
Philadelphia, Pennsylvania 19104

1. SUMMARY

Monoclonal antibodies have been made against cell surface antigens on human neuroblastoma cell lines and leukemia cell lines. These antibodies were tested against a panel of human cell types to define those antibodies having a restricted specificity. We describe the use of monoclonal antibodies for detection of metastasis of neuroblastoma cells to bone marrow. Using a combination of two monoclonal antibodies, we are able to detect and identify neuroblastoma cells and to distinguish them from normal and malignant lymphoid cells. We have also used these antibodies to isolate the cell surface glycoproteins by affinity chromatography. The resultant glycoprotein antigens from different tumor cell lines have been compared.

One of the most promising applications of anti-tumor monoclonal antibodies may be the detection of cell surface molecules involved in the processes of cell growth and cell division. We have identified three monoclonal antibodies which inhibit the growth of the leukemia cell, designated Reh. The detection of these antibodies and their use in characterization of cell surface receptors are discussed.

2. INTRODUCTION

We began several years ago to make antisera against human tumor cells with the long term goal of determining whether tumor specific antigens are present on human tumors. If detected, our intention was to then analyze the possible relationship between their expression and the malignant phenotype. Concentrating on human neuroblastoma cells and non-T, non-B acute lymphoblastic leukemia (ALL)[1], we postulated that if there were "new" antigens on these tumors cells, they would be the result of structural mutations, viral infection, or the aberrant expression of normal gene products such as fetal or differentiation antigens. Detection of such new antigens was seen as a first step toward understanding the relationship of their expression and the molecular genetic events responsible for the cells growing out of control.

When Kohler and Milstein (1975) described production of monoclonal antibodies, we saw hybridoma technology as the way out of the jungle of classical serology with its endless "cross reactions," absorptions, and limited supplies of antisera. We began to make monoclonal antibodies against human neuroblastomas and ALL cells using a variety of immunization protocols and antigen preparations (Kennett et al., 1978; Kennett and Gilbert, 1979). The general procedure was to immunize with the tumor cells and to screen the monoclonal antibodies against the immunizing cell and at least one other type of human cell such as fibroblasts, peripheral blood cells or B-cell lymphoblastoid lines. Hydridomas producing antibodies reacting with the immunizing tumor cells but not the other types of human cells were cloned and the antibodies tested against a more extensive panel of human cell types. We outline here our experiences in using this procedure to search through the cell surface antigens of human ALL cells and neuroblastoma cells for the thus far elusive "tumor specific antigens."

3. RESULTS

3.1. An Anti-neuroblastoma antibody

One of the first anti-neuroblastoma antibodies that we reported, PI153/3 (Kennett and Gilbert, 1979), was shown to react with human tumors derived from neuroectoderm (neuroblastoma, glioblastoma, retinoblastoma). It reacts with fetal brain but not with adult brain, other tumors, or with other normal human cells that were tested in the original panel. Using immunoperoxidase staining, it was possible to use this antibody to detect neuroblastomas metastasized to bone marrow (Kenneth et al., 1980). The antibody could detect the neuroblastoma cells in all cases in which they were detectable by standard clinical procedures (clumps of cells with neuroblastoma morphology in the marrow) and in two cases where these procedures did not permit this detection.

---

[1]Abbreviations: ALL, Acute Lymphoblastic Leukemia; DMSO, Dimethylsulfoxid TPA 12-0-tetraclearocyl phorbol-13-acetate; SDS-PAGE, Sodium Dodecylsulpha Polyacrylamide Gel Electrophoresis; PBS, Phosphate Buffered Saline.

3.2. The neuroblastoma determinant is also detectable on leukemia cells

Further analysis using marrow from ALL patients as a control indicated that the antibody also reacted with the same or a similar determinant on leukemia cells (Kennett et al., 1980). Tests on a panel of leukemia cells demonstrate that the antibody reacts with B-cell leukemias, null cell leukemias but not with T-cell leukemias (Kennett et al., 1980b; Greaves et al., 1980) and is thus helpful in classification of leukemia cells.

3.3. Using a combination of monoclonal antibodies to detect neuroblastoma cells in marrow

Being concerned that the anti-neuroblastoma antibody might react with a small subpopulation of normal lymphocytes corresponding to the stage of differentiation of the leukemia cells expressing the antigen, we chose a second monoclonal antibody which reacts with lymphocytes but not neuroblastomas. Using this combination of two monoclonal antibodies, we were able to detect neuroblastomas and distinguish them from lymphocytes expressing the neuroblastoma-leukemia antigen (Kennett et al., 1981).

3.4. Other monoclonal antibodies against neuroblastomas or leukemias

Using the above screening procedures, we have isolated more than 300 monoclonal antibodies against leukemia cells (non-B, non-T ALL) and more than 200 against neuroblastomas. Each of these did not react with the other cell type used in the initial screening. They were produced after immunization using whole cells, KCl extracts (Reisfeld et al., 1971), membrane vesicles (Scott, 1976), or solubilized membrane proteins. In each case, it was found that the antigenic determinant could be detected on other types of human tumor and normal cells. Although we identified no antigens specific for tumor cells, the variety of reaction patterns against panels of human cells indicated that a large number of determinants was being detected.

Our results are consistent with those of other investigators who have reported monoclonal antibodies reacting with human tumor cells. The general theme presented over the past few years has been identification of an antigen that appears to be tumor specific whereupon further work shows that the antigen is present on some other type(s) of human cells. Another anti-neuroblastoma monoclonal antibody purported to be specific for neuroblastomas (Kemshead et al., 1981) was later found to react with pancreatic islet cells (J. Starkie, personal communication).

3.5. "Operationally tumor specific" antibodies

Antibodies that react with tumor antigens also present on normal cells may, in fact, be useful for detection or even therapy of tumors. It is possible that they can be used for detection, in combination with other monoclonal antibodies as shown above, or for therapy in situations where toxicity to the normal cells bearing the antigen would not delete essential cells such as irreplaceable stem cells. On the other hand, if the motivation for identifying tumor specific cell surface alterations is to define molecular changes responsible for malignancy rather than tumor detection or therapy, some other approach than simply screening through more and more monoclonal antibodies must be devised. With this end in mind, we have started to use two new approaches which have given promising preliminary results. The first is further development of the in vitro immunization system described by Luben and Mohler (1980), and the second is detection of monoclonal antibodies that inhibit the proliferation of tumor cells.

3.6. In vitro immunization

The ability to produce hybridomas from spleen cells sensitized in vitro presents the possibility of immunizing with smaller amounts of antigen and also the possibility of manipulating the immune response in ways not possible by in vivo stimulation. We began by hypothesizing that if non-tumor cells were used to stimulate spleen cells in vitro, the responding and dividing cells could be selectively killed by incorporation of bromodeoxyuridine treatment with the dye Hoechst 33258 and exposure to light (Stetten et al., 1976). Restimulation with tumor cells may then enrich for cells responding to antigens present on the tumor cell but not on the original non-tumor cell used for stimulation.

After confirming that Luben and Mohler's (1980) in vitro immunization system worked as described, we modified it in two ways: (1) we found that the thymus conditioned medium could be replaced by addition of peritoneal exudate cells to the culture, and (2) that the in vitro stimulation could be done in the serum-free medium recently described by Murakami et al., 1981. Tables I and II show the results obtained by in vitro immunization with whole cells (Reh, a non-B non-T ALL cell line, Rosenfeld et al., 1977) and with a soluble protein.

Our initial results with BrdU and Hoechst 33258 indicate that hydridomas can be obtained after two rounds of stimulation, but whether the antibodies show increased specificity for the cell used in the second stimulation remains to be determined.

In any case, the consistent results now obtained with these modifications of Luben and Mohler's system does not make the use of the in vitro system a practical procedure that can be applied to this and other types of potentially selective immunization protocols.

TABLE I

In vitro immunization with human complement component, C_2

Fusion #	Immunization With C_2	# of Clones	# of Clones With Anti-C_2 Activity
30	in vitro	93	27
31	in vitro	13	1
39	in vitro	7	2
46	in vitro	21	5
50	in vitro	27	7
72	in vivo]] in vitro]	40	6
89	in vivo]] (serum-free in vitro] medium)	29	

3.7. Detection of antibodies inhibiting leukemia cell proliferation

Considering the possibility that neuroblastoma cells or leukemia cells may not in fact express a new tumor specific antigen, we considered the possibility that some of the cell surface molecules we had already detected with monoclonal antibodies might be altered in some way in comparison to similar antigens on normal cells. With the large number of antibodies available, it came to the question of which ones should be used to look at the cell surface molecules in more detail. We decided that since one of the basic properties that distinguishes a tumor from a normal cell is the lack of growth control, we would attempt to detect cell surface molecules that are involved in the regulation or mechanisms of cell growth division or differentiation (Kennett et al., 1981). Considering the fact that antibodies against insulin receptors can react with cell surface insulin receptors and mimic the effects of insulin (Kahn et al., 1977), we decided to screen for monoclonal antibodies reacting with the ALL cell line Reh that would stop the proliferation of the tumor cells as measured by tritrated thymidine incorporation. McGrath et al. (1980) have also recently reported an anti-T-cell monoclonal antibody that inhibits the proliferation of a cultured

TABLE II

In vitro immunization with ALL cell line, Reh

Fusion #	Immunization With Reh Cells	# of Clones (300 Wells Total)	#of Clones With Anti-Reh Activit;
63	in vitro (BrdU treatment)	42	6
64	in vitro (BrdU treatment)	41	8
65	in vitro	45	7
71	in vivo a/	75	19
	in vitro b/(BrdU treatment)	32	5
77	in vivo	21	6
78	in vitro (serum-free medium)	0	

mouse T-lymphoma. By screening more than 80 antibodies, we obtained three that inhibited the proliferation of Reh (Kennett et al., 1981). See Figures 1 and 2. Two of these antibodies also inhibit the proliferation of the leukemia cell line HL60 (Collins et al., 1977). With this cell line, which after treatment with reagents such as DMSO and TPA exhibits in vitro differentiation, we can ask whether the antibodies also induce differentiation. If so, the antibodies provide a way to isolate the cell surface molecules involved in the inhibition of proliferation and induction of differentiation. Although at this point there are several possible explanations for the effects of these antibodies, isolating the cell surface molecules involved and comparing them to those on normal cells provides what could be an important way of detecting molecules which could be involved in transformation to the malignant phenotype.

3.8. Isolation of cell surface molecules using monoclonal antibodies

One question that arises is if such cell surface molecules are detected whether they can actually be isolated with monoclonal antibodies. To provide an affirmative example to this question, we will describe the work we have done with our first anti-neuroblastoma antibody PI153/3. As

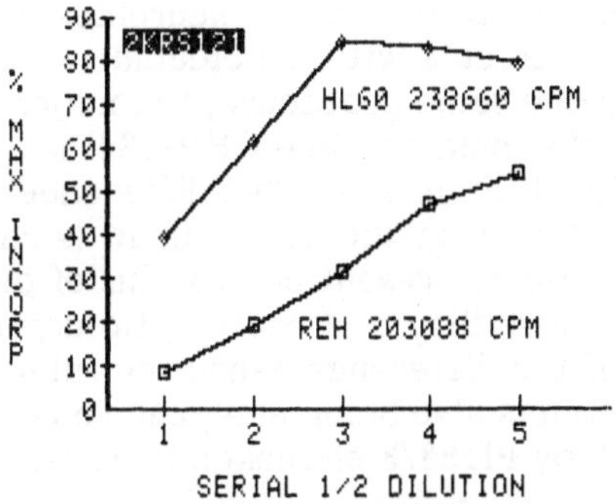

Figure 1. Inhibition of proliferation of leukemia cell lines Reh and HL60 by dilutions of monoclonal antibody 2KRS121. Dilutions of hybridoma supernatant (100 μl) were added to 1.6 x 10^4 cells per well in 100 μl of fresh medium. After incubation for 3 days, 1.25 μC of [^{3}H] thymidine was added to each well. After overnight incubation, the cells were harvested and thymidine incorporation measured. Samples were done in duplicate. The maximum CPM listed on the graph for each cell line are those incorporated when the hybridoma supernatant is replaced by spent culture medium from the parental plasmacytoma SP2/0-Ag-14.

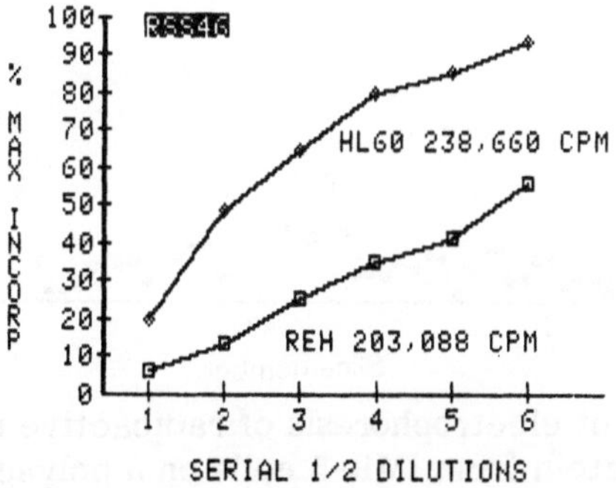

Figure 2. Inhibition of proliferation of leukemia cell lines Reh and HL60 by dilution of monoclonal antibody RSS46. The procedure was identical to that described in Figure 1.

with many monoclonal antibodies, we were unable to precipitate an antigen from neuroblastoma extracts using indirect immunoprecipitation procedures. Inhibition of PI153/3 cytotoxicity with cell fractions showed that the antigen is a glycoprotein and that the antigenic determinant is probably in the carbohydrate portion of the molecule (Momoi et al., 1980). Analysis of the molecule isolated from the neuroblastoma cell line IMR-5 by SDS-PAGE showed a molecule with a molecular weight of approximately 20,000 (Figure 3). Using the same procedure, the major antigen isolated by PI153/3 from the neuroblastoma cell line CHP-134 has an apparent molecular weight of 70,000 (Figure 4). This difference is consistent with the data indicating that the antigenic determinant is on a carbohydrate chain which could of course be present on proteins of different molecular weights. However, the possibility that proteolytic degradation in the IMR-5 extract is the cause of the difference cannot be ruled out at this time. The leukemia cell line Reh is also being analyzed to determine the nature of the antigen detected by PI153/3 on leukemia cells.

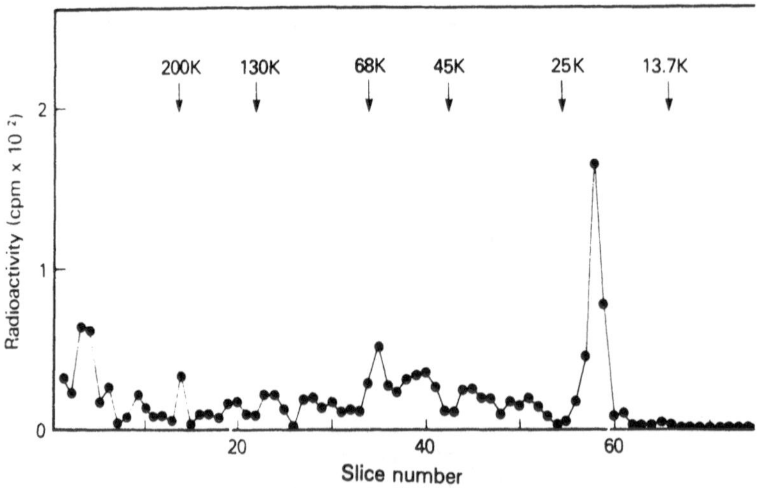

Figure 3. Profile of electrophoresis of radioactive affinity purified glycoprotein from IMR-5 cells on a polyacrylamide slab gel under denaturing conditions. IMR-5 cells were labeled metabolically with D-[^{3}H] glucosamine and extracted with NP-40 in the presence of alpha-toluene/sulfonyl fluoride, included also in all subsequent procedures. The NP-40 extract was adsorbed to an immunoaffinity column of monoclonal antibody PI153/3. Eluate from the column was run on polyacrylmamide gels in 1% SDS. Reproduced from Momoi et al. (1980) with permission.

3.9. Monoclonal antibodies can be used to isolate minor components of the cell surface

The antigen on neuroblastomas detected by PI153/3 reacts with an antigen making up less than 1% of the total cell surface glycoprotein (Momoi et al., 1980). Successful isolation of this antigen shows that even monoclonal antibodies that do not readily precipitate the molecule can be used to isolate and characterize the antigen. Molecules such as hormone receptors or other molecules that might be detected by the growth inhibition test above can thus be isolated with monoclonal antibodies. Certainly enough material could be isolated using the original monoclonal antibody to do an in vitro immunization so that other antibodies against other determinants in the molecule could be isolated. These other antibodies could then be used in combination to further characterize and more readily isolate more of the material of interest.

3.10. Concluding summary

1. Monoclonal antibodies against antigenic determinants on human neuroblastoma and leukemia cells that have been reported are not "tumor specific."

2. Even those that appear to have restricted patterns of reactivity often react with: other tumors, human brain, lymphocyte subsets, and fibroblasts.

3. This may represent "cross reacting" antigenic determinants on different molecules or differentiation antigens shared by nerve cells and lymphoid cells and sometimes other cells (Thy-1).

4. Even when not tumor specific, monoclonal antibodies can be used in combinations to detect tumor metastasis and to classify tumor cells.

5. These same antibodies may be useful for immunotherapy if they do not delete essential "stem cell" clones.

6. Monoclonal antibodies can detect differentiation antigens that are minor (< 1%) components of the total antigen mixture of a cell.

7. Even monoclonal antibodies that do not readily immunoprecipitate these minor components can be used to isolate and characterize these molecules.

8. Monoclonal antibodies that react with both tumor and normal cells will be useful for characterizing differences between tumor and normal cell surface and for identifying molecules influencing cell growth and differentiation.

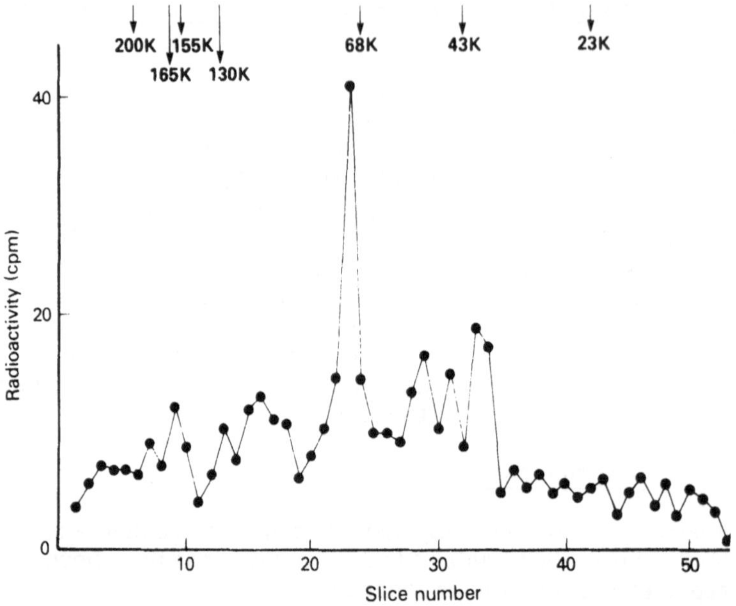

Figure 4: Profile of electrophoresis of affinity purified radioactive glycoprotein from CHP-134 cells on a polyacrylamide gel under denaturing conditions. This [^{3}H] glucosamine labeled glycoprotein was isolated following the same procedure as that described for the IMR-5 glycoprotein shown in Figure 3.

4. DISCUSSION

Attempts to make monoclonal antibodies against human tumor specific antigens have not led to detection of antigens that have been shown to be truly tumor specific (also see other chapters in this volume). On the other hand, many of the antibodies detect antigens which are "operationally tumor specific" and can thus be used for tumor detection and potentially for therapy.

The reason for the lack of detection of a human tumor specific antigen may of course be that they do not exist. This, of course, may depend on the tumor type and its etiology. Another possibility that should be considered is that because monoclonal antibodies are so exquisitely specific and react with a single or group of closely related antigenic determinants or epitopes, it will be highly unlikely that a tumor specific epitope will be detected. Even if a monoclonal antibody against a tumor specific antigen were obtained, what would be the possibility that it would not react with the same or a similar determinant on at least one other molecule on normal cells? If one considers that the haemagglutinin molecule of influenza virus (M.W. = 220,000) can be recognized at least

1000 different ways by monoclonal antibodies (Klinman et al., 1981), the possibility of having a tumor specific epitope seems extremely low. This seemingly pessimistic argument is meant only as a caution and an indication of the need to use monoclonal antibodies to isolate the molecules from the cell surface of tumor cells and normal cells so that these molecules can be compared by other complementary biochemical methods. This may make it possible to detect subtle but significant differences between tumor cells and their normal counterparts. Among the large number of cell surface molecules, it would seem that antigens that play a role in cell growth, division, or differentiation may be prime candidates for such an analysis. As indicated above, monoclonal antibodies may be useful in detecting such molecules and in their initial isolation.

In spite of the apparent current complexity involved in the use of monoclonal antibodies, one cannot help but be optimistic in light of the increased sensitivity and discrimination that these reagents provide for the analysis of gene products at the same time that recombinant DNA technology provides the means to analyze gene structure. We can expect to see these complementary technologies shed light on many aspects of normal and abnormal cell growth, regulation and differentiation during the next few years.

5. EXPERIMENTAL PROCEDURES

5.1. Production and characterization of monoclonal antibodies

Monoclonal antibodies were produced, screened and characterized as reported previously (Kennett et al., 1978; Kennett et al., 1981c).

5.2. In vitro immunization

The in vitro immunization was a modification of the procedure described by Luben and Mohler (1980). Their thymus conditioned medium was replaced by incorporating peritoneal exudate cells (PEC) into the cultures in which the spleen cells were exposed to antigen. Four days prior to the stimulation, BALB/c mice were injected with 0.5 ml of thioglycolate broth. The cells were removed from the peritoneum by infusion of PBS or HY medium and irradiated at 4500 R. Spleen cells were perfused from the spleen as described previously (Kennett et al., 1978) and suspended at a concentration of 10^7 spleenic lymphocytes/ml. PEC were added at a concentration of 10^5 cells/ml. Human C_2 was added at a concentration of 10 μg/$5x10^7$ spleen cells and Reh cells at a concentration of 10^5 Reh/ml of spleen cells.

The culture medium used was RPMI 1640 plus the concentrations of insulin, transferrin, selenium and ethanolamine used by Murakami et al. (1981) and 5 x 10^{-5}M 2-mercaptoethanol. An additional 1 ml/100 ml of medium of a stock solution of 25 g/100 ml glucose was added to the RPMI 1640 to raise the glucose concentration. When serum was used in the medium, fetal bovine serum tested to confirm efficient cloning of mouse plasmacytoma cells was used at a concentration of 10%.

In vitro immunization was continued for four days. The stimulated cells were then treated as previously reported (Kennett et al., 1978) for hybridoma production including the treatment with an ammonium chloride solution to lyse red blood cells and used for fusion to SP2/0-Ag-14 (Shulman et al., 1978). We routinely use this cell line because it gives on the order of 200 - 300 hybrids when fused with spleen cells from an in vivo immunized animal. In our experience, the non-producing myeloma cell lines S194/5.XXO.BU.1 (Trowbridge, 1978) or P3X63-Ag-8653 (Kearney et al., 1979) give more hybrids under the same conditions and in fact may be an advantage when fusing with in vitro immunized spleen cells. We are currently testing this in parallel fusion experiments.

5.3. Detection of monoclonal antibodies inhibiting Reh cell proliferation

Detection of inhibiting antibodies was carried out as described previously (Kennett et al., 1981).

5.4. Affinity chromatography of neuroblastoma antigens

These procedures are described in detail in Momoi et al. (1980).

Acknowledgements

This work was supported by NIH grants CA 24263, CA 14489, and GM 20138 and NSF grant PCM 79-26757. We would like to thank Sue Newberry, Virginia Braman and Barbara Meyer for their expert technical assistance.

REFERENCES

Collins, S.J., Gallo, R.C., and Gallagher, R.E.; 1977; Continuous growth and differentiation of human myeloid leukemic cells in suspension culture. Nature 270: 347-349.

Greaves, M.F., Verbi, W., Kemshead, J., and Kennett, R., 1980, Monoclonal antibody identifying a cell surface antigen shared by common acute lymphoblastic leukemias and B lineage cells. Blood 56, 1141-1144.

Kahn, C.R., Barid, K., Flier, J.S., and Jarrett, D.B., 1977, Effects of autoantibodies to insulin receptor on isolated adipocytes. J. Clin. Invest. 60, 1094-1106.

Kearney, J.F., Radbirch, A., Liesagen, B., and Rajewsky, K., 1979, New mouse myeloma cell-line that has lost immunoglobulin expression but permits the construction of antibody-secreting hybrid cell lines. J. Immunol. 123, 1548-1550.

Kemshead, J.T., Walsh, F., Pritchard J., and Greaves, M., 1981, Monoclonal antibody to ganglioside CQ discriminates between haemopoietic cells and infiltrating neuroblastoma cells in bone marrow. Int. J. Canc. 27, 477-452.

Kennett, R.H., Denis, K.A., Tung, A.S., and Klinman, N.R., 1978, Hybrid plasmacytoma production: Fusions with adult spleen cells, monoclonal spleen fragments, neonatal spleen cells, and human spleen cells. Curr. Topics in Microbiol. and Immunol. 81, 77-94.

Kennett, R.H., and Gilbert, F., 1979, Hybrid myelomas producing antibodies against a human neuroblastoma antigen present on fetal brain. Science 203, 1120-1121.

Kennett, R.H., Jonak, Z.L., Bechtol, K.B., and Byrd, R., 1981, Monoclonal antibodies as probes for cell surface changes in human malignancy, in: Fundamental Mechanisms in Human Cancer Immunology, (J.P. Saunders, J.C. Daniels, B. Serrou, C. Rosenfeld, and C.B. Denney, eds), Elsevier/North-Holland, New York, pp. 331-348.

Kennett, R.H., Jonak, Z.L., and Bechtol, K.B., 1980, Monoclonal antibodies to tumor associated antigens, in: Monoclonal Antibodies: A New Dimension in Biological Analyses (R.H. Kennett, T.J. McKearn, and K.B. Bechtol, eds.), Plenum Press, New York, pp. 155-168.

Kennett, R.H., Jonak, Z.L., Bechtol, K.B., 1980b, Characterization of antigens with monoclonal antibodies, in: Advances in Neuroblastoma Research: Progress in Cancer Research and Therapy, Volume 12, (A.E. Evans, ed.), Raven Press, New York, pp. 209-226.

Kennett, R.H., McKearn, T.J., and Bechtol, K.B., ed., 1981c, Appendix: Method for production and characterization of monoclonal antibodies, in: Monoclonal Antibodies: A New Dimension in Biological Analyses, Plenum Press, New York, pp. 361-419.

Klinman, N.R., Denis, K.A., and Sherman, L.A., 1981, Defining the immune mechanism with monoclonal antibodies. In Vitro, 17:200.

Kohler, G., and Milstein, C., 1975, Continuous cultures of fused cells secreting antibody of predefined specificity. Nature 271, 461-462.

Luben, R.A., and Mohler, M.A., 1980, In vitro Immunization as an adjunct to the production of hybridomas producing antibodies against the lymphokine osteoclast activating factor, Mol. Immunol. 17, 635-639.

McGrath, M.S., Pillemer, E., and Weissman, I.L., 1980, Mouse leukemogenesis: monoclonal antibodies to T cell determinants arrest T lymphoma proliferation, Nature 285, 259-261.

Momoi, M., Kennett, R.H., and Glick, M.C., 1980, A membrane glycoprotein from human neuroblastoma cells isolated with the use of a monoclonal antibody. J. Biol. Chem. 255, 11914-11921.

Murakami, H., Masui, H., Sato, G.H., Sueoka, N., Chow, T.P., and Kano—Sueoka, T., 1982. Growth of hybridoma cells in serum-free medium: identification of ethanolamine as an essential component. Proc. Natl. Acad. Sci. USA, 79(4):1158-1162.

Reisfeld, R.A., Pelligrino, M.A., and Kahan, B.D., 1971, Salt extraction of soluble HLA antigens. Science 172, 1134-1136.

Rosenfeld, C., Goutner, A., Choquet, C., Venaut, A.M., Kayibanda, B., Pico, J.L., and Greaves, M.F., 1977, Phenotypic characterization of a unique non-T, non-B acute lymphoblastic leukemic cell line. Nature 267, 841-843.

Scott, R.E., 1976, Plasma membrane vesiculation: a new technique for isolation of plasma membranes. Science 194, 743-745.

Shulman, M., Wilde, C.D., and Kohler, G., 1978, A better cell line for making hybridomas secreting specific antibodies. Nature 276, 269-270.
Stetten, G., Latt, S.A., and Davidson, R.L., 1976, 33258 Hoechst enhancement of the photo sensitivity of bromodeoxyuridine-substituted cells. Somatic Cell Genetics 2, 285-290.
Trowbridge, I., 1978, Interspecies spleen-myeloma hybrid producing monoclonal antibodies against mouse lymphocyte surface glycoprotein, T200. J. Exp. Med. 148, 313-323.

10

MONOCLONAL ANTIBODIES TO HUMAN MELANOMA-ASSOCIATED ANTIGEN p97

Joseph P. Brown, Karl Erik Hellström and Ingegerd Hellstrom

Division of Tumor Immunology, Fred Hutchinson Cancer Research Center, 1124 Columbia Street, Seattle, Washington 98104, and Departments of Pathology and Microbiology/Immunology, University of Washington Medical School, Seattle, Washington 98195

1. SUMMARY

Monoclonal antibodies obtained by fusing myeloma cells with spleen cells from mice that had been immunized with human melanoma were used to identify and characterize human melanoma-assocated antigen p97, a 97,000 molecular weight cell surface glycoprotein. Sequential immunoprecipitation experiments and competition binding assays showed that seven IgG antibodies recognized three distinct epitopes on the p97 molecule. Cultured melanoma cells expressed up to 400,000 molecules of p97 per cell, whereas other cell types, such as fibroblasts, carcinomas, and lymphoid cells expressed fewer than 10,000 molecules per cell. Tissues were assayed for p97 by an immunoradiometric assay. Melanomas contained the highest levels of p97, and benign nevi and certain fetal tissues, particularly intestine, contained significant amounts, whereas normal adult tissues contained only traces. Amino acid sequence studies showed that p97 is structurally related to transferrin and lactotransferrin, and a functional relationship of p97 to transferrin and lactotransferrin was established by the observation that p97 binds iron.

2. INTRODUCTION

Monoclonal antibodies produced by somatic cell hybrids (Kohler and Milstein, 1975) are now being used to identify cell surface antigens of human tumors. Many studies have concentrated on melanoma, primarily because of the evidence that these tumors express antigens that are recognized by the melanoma patient's immune system (Hellström et al., 1968; Shiku et al., 1976, Hellstrom and Brown, 1979). The studies with monoclonal antibodies have confirmed the existence of melanoma-associated antigens, several groups having described antigens present in larger amounts on melanoma cells than on other cell types (Koprowski et al., 1978; Yeh et al., 1979; Carrel et al., Woodbury et al., 1980; Dippold et al., 1980; Loop et al., 1981; Wilson et al., 1981; Morgan et al., 1981).

We describe here some aspects of our studies of melanoma-associated antigen p97 (Woodbury et al., 1980; 1981; Brown et al., 1980, 1981a, 1981b, 1981c). We describe the isolation and characterization of monoclonal antibodies to p97, summarize results of assays of cultured cells and of tissues for p97, and review recent structural studies of p97. Finally, we discuss the relationship of p97 to other proteins and speculate on its function.

3. RESULTS

3.1. Monoclonal antibodies to p97

We now have a total of seven hybridomas producing antibodies specific for p97. They were obtained by fusing mouse myeloma cells with spleen cells from mice that had been immunized with human melanoma cells (Woodbury et al., 1980; Brown et al., 1981; 1981b, 1981c). Initially we used an autoradiographic protein A binding assay (Brown et al., 1979) to screen for antibodies that bound to melanoma cells but not to autologous fibroblasts. Subsequent fusions, once we knew the molecular weight (MW) of p97, were screened by immunoprecipitation and sodium dodecyl sulfate-polyacrylamide gel electrophoresis (SDS-PAGE) (Brown et al., 1980, 1981b). The antibodies were purified from ascites fluid by affinity chromatography on protein A-sepharose (Ey et al., 1978), and their immunoglobulin isotypes were determined by double immunodiffusion. Antibodies 4.1 and 8.2 were IgG1, antibodies 96.5, 118.1, 133.1, and 133.2 were IgG2a, and antibody 133.3 was IgG2b. Each of these antibodies immunoprecipitated from lysates of surface-radioiodinated melanoma cells a protein with a MW on SDS-PAGE of approximately 97,000. Sequential immunoprecipitation experiments, in which the cell lysates were depleted of antigen by pretreatment with one antibody and then tested with other antibodies, showed that all seven antibodies bound the same molecule (Brown et al., 1981b).

Binding assays with radioiodinated antibody using unlabeled antibodies as competitors showed that the seven antibodies recognized three distinct epitopes on the p97 molecule. Antibodies 96.5 and 4.1 recognized epitope $p97^{a}$, antibodies 118.1, 133.1, and 133.3 recognized epitope $p97^{b}$, and antibodies 8.2 and 133.2 recognized epitope $p97^{c}$.

Antibodies to epitopes $p97^{a}$ and $p97^{b}$ have been used in an immunoradiometric assay, in which one antibody is coupled to a solid phase to which antigen binds, and the other antibody is radiolabelled and used to detect the bound antigen (Brown et al., 1981a, 1981c). Furthermore, antibodies to these two epitopes have been shown to interact synergistically in complement-dependent cytotoxicity, melanoma cells being killed by a mixture of the two antibodies at concentrations as low as 10^{-10} M, whereas either antibody alone was ineffective even at ten-fold higher concentrations (Hellström et al., 1981).

3.2. Tissue distribution of p97

Cultured cells were assayed for p97 by binding assays with radioiodinated antibodies (Woodbury et al., 1980; Brown et al., 1981a, 1981b; Hellström et al., 1981). Melanoma cells were found to bind up to 400,000 molecules of antibody per cell, median binding being 50,000 molecules per cell. In contrast, carcinoma cells bound 2,000 to 10,000 molecules per cell, and fibroblasts and lymphoid cell lines bound approximately 2,000 molecules per cell. The immunological specificity of low levels of binding was established by using unlabelled antibodies as competitors.

Tissues were tested both as membrane suspensions by binding assays with radioiodinated antibodies (Brown et al., 1981c) and also as detergent-lysates by the immunoradiometric assay described above (Brown et al., 1981a). Melanomas contained the highest levels of p97, up to 1,200 nanograms per milligram of membrane protein, median 60 ng/mg. Benign nevi and fetal intestine contain significant amounts, 60 and 200 ng/mg. Immunoperoxidase staining of cryostat sections of tissues was used to examine p97 expression at the microscopic level. Most primary melanomas expressed the antigen, whereas a variety of normal cells examined in the same sections did not (Garrigues et al., submitted for publication).

3.3. Structural studies of p97

For initial studies of the structure of p97 we used monoclonal antibodies to immunoprecipitate the antigen from lysates of radiolabelled cells, followed by SDS-PAGE and autoradiography. Several labelling methods were used. Lactoperoxidase-catalysed radioiodination (Gudjonsson and Johnson et al., 1978) and periodate oxidation followed by reduction with tritiated borohydride (Gahmberg et al., 1976) showed that p97 is a cell surface sialoglycoprotein with a MW of approximately 97,000 (phosphorylase b, MW 97,400, was used as a marker). Digestion of radioiodinated p97 with neuraminidase resulted in a slight decrease in MW, consistent with the removal of terminal sialic acid residues. Metabolic labelling with radiolabelled amino acids confirmed that p97 was synthesized by the melanoma cells. Gel filtration of the native protein and SDS-PAGE of p97 under non-reducing conditions indicated that p97 is monomeric, probably with intrachain disulfide bridges, while digestion with trypsin produced a stable, antigenic fragment with a MW of 40,000 (Brown et al., 1981b and unpublished).

Amino acid sequence studies of p97 were undertaken in the hope of obtaining insight into the relationship of p97 to other proteins and into its function (Brown et al., submitted for publication). However, the milligram amounts of protein required for analysis by a conventional amino acid sequencer and the small amounts of p97 present in melanoma cells made this a difficult undertaking. A new, highly sensitive sequencer (Hewick et al., 1981) made the project more feasible, since as little as 50 micrograms of p97, which was obtained from 10 grams of melanoma cells (50 roller bottles), was enough for N-terminal amino acid sequence determination with this machine.

Antigen p97 was purified from cultured melanoma cells by affinity chromatography. Antibody 96.5 was added to a detergent-lysate of the cells, and the mixture was passed down a protein A-sepharose column, which efficiently retained the IgG2a antibody. The antibody and p97 were eluted with pH 5 citrate buffer. Analytical SDS-PAGE of the eluate revealed IgG heavy and light chains and p97, with virtually no contaminating proteins. The proteins were precipitated, and p97 was separated from IgG heavy and light chains by preparative SDS-PAGE. The purified p97 was eluted from the gel and sequenced in collaboration with Drs. W. Dreyer and R. Hewick at the California Institute of Technology. We obtained a sequence of 12 amino acids at the N-terminus.

A search of an amino acid sequence library (Doolittle, 1981) revealed that 7 of the 12 N-terminal amino acid residues of p97 were present at the N-terminus of human serum transferrin, and 6 of these 7 residues were conserved in lactotransferrin. Independent evidence for a structural relationship between p97 and transferrin and lactotransferrin was obtained by using an antiserum produced by immunizing mice with p97 that had been denatured by exposure to SDS and 2-mercaptoethanol at 100°C. This antiserum precipitated denatured p97, and cross-reacted with denatured transferrin and lactotransferrin. However, a 25-fold higher antiserum concentration was required to precipitate transferrin and lactotransferrin than to precipitate p97.

Since transferrin and lactotransferrin bind iron, p97 was examined for iron binding. Melanoma cells were incubated with ^{59}Fe (as ferric citrate), lysed with detergent, and p97 was purified by affinity chromatography. The purified p97 contained a significant proportion (4%) of the ^{59}Fe taken up by the cells, whereas uptake by other cell surface antigens, such as HLA, was negligible.

3.4. Relationship of p97 to other tumor-associated antigens

Several other tumor-associated antigens with MWs in the range 90,000 to 100,000 have been described (Bramwell and Harris, 1978; Dippold et al., 1980; Pesando et al., 1980; Omary, and Trowbridge, 1980; Judd et al., 1980). Of these, gp95 (Dippold et al, 1980) appeared very similar to p97 in its tissue distribution. In particular, SK-MEL 28, the melanoma cell line with the greatest amount of gp95, also expressed the greatest amount of p97. Sequential immunoprecipitation experiments subsequently established that p97 and gp95 were identical (Brown et al., 1981b).

Several groups have obtained antibodies to the human transferrin receptor, a protein with a subunit MW of 90,000 (Sutherland et al., 1980, Trowbridge and Omary, 1981), and on the basis of the similar MWs of the transferrin receptor and p97 the former authors have suggested that the two proteins are identical. However, the tissue distribution of the transferrin receptor is quite different from that of p97, and monoclonal antibodies to p97 did not bind the transferrin receptor and vice versa (Brown et al., unpublished). Also, when the transferrin receptor was analyzed by SDS-PAGE without prior reduction, it ran as a disulfide-crosslinked dimer with a MW of 180,000 (Omary and Trowbridge, 1980; Sutherland et al. 1980), whereas under the same conditions p97 ran as a monomer (Brown et al., 1981b).

An antigen of common acute lymphoblastic leukemia (CALLA)[1] has a MW of approximately 100,000 (Pesando et al., 1980). We have found that CALLA is also present on human melanoma cells (Brown et al., unpublished). However, when the two proteins were electrophoresed in adjacent tracks of the same gel it was clear that CALLA was higher in MW than p97. In addition, we observed no cross-reactions between CALLA and p97 using monoclonal antibodies specific for the two proteins (Brown et al., unpublished), and we conclude that they are unrelated.

4. DISCUSSION

The hybridoma technique provides a new tool for identifying and characterizing human tumor-associated antigens (Hellström et al., 1980). The procedure is simple. A mouse is immunized with human tumor cells, its spleen cells are fused with myeloma cells, and the antibodies that are produced by the resulting hybridomas are screened against neoplastic and normal cells.

At first it was surprising that antibodies of any degree of tumor specificity could be obtained by immunizing mice with human cells. Nevertheless, over the past few years a number of tumor-associated antigens have been identified in this way. We shall discuss some general aspects of

---

[1]CALLA, Common acute lymphocytic leukemia antigen; MW, molecular weight; SDS, sodium dodecyl sulfate; and PAGE, polyacrylamide gel electrophoresis.

the identification and characterization of human tumor antigens with monoclonal antibodies, restricting our discussion to glycoprotein antigens, with which we have most experience, although many of our arguments also apply to glycolipid antigens.

4.1. Screening hybridomas

In order to identify hybridomas producing antibodies to biologically interesting tumor cell surface antigens, large numbers of hybridomas must be screened. This is commonly done by using binding assays such as the autoradiographic assay described by Brown et al. (1979) or ELISA assays. Other screening methods are also of value and may be more informative, though slower. For example, we have used immunoprecipitation to screen hybridomas for antibodies specific for melanoma cell surface proteins, and we obtained 5 hybridomas producing antibodies specific for p97 in this way (Brown et al., 1980, 1981b). The major advantage of this approach is that proteins are identified directly by their MWs on SDS-PAGE, rather than by the pattern of antibody binding to a number of cell lines. Also, since many tumors are difficult to grow in culture, we are now using immunohistological techniques, particularly the immunoperoxidase method on cryostat sections, to screen for antibodies that bind to primary tumors (Garrigues et al., submitted for publication). The primary advantage of this approach is that a much wider range of tumors and normal cells can be tested, and one can determine immediately whether an antigen is expressed in vivo.

4.2. Antigen specificity

Once a hybridoma has been selected and cloned, the next step is to examine in more detail the specificity of the antigen identified by the antibody. We wish to emphasize three points. First, the assays commonly used for screening are not necessarily optimal for determinating antigen distribution on a wide range of cell types, since they are designed for speed rather than for sensitivity and versatility. Second, the amounts of antigen in various cells and tissues should be expressed quantitatively, for example as molecules per cell or nanograms of antigen per milligram of protein, rather than as "positive" or "negative". Assays vary in sensitivity, and an amount of antigen that may be neglible for some purposes may be important for others. Third, one must test tissues and not rely on the use of cultured cells alone. For example, an antigen, 3.1, which is present in vitro on only very few melanoma cell lines (Yeh et al., 1979), is present in vivo on a number of normal tissues (Brown et al., 1981c).

The advent of monoclonal antibodies has led to the development of new serological procedures. These procedures offer advantages in sensitivity, convenience, and quantitation over procedures designed for use with antisera, which contain antibodies varying in concentration, isotype, affinity, and specificity. For assays of cultured cells we have found that the most convenient method is to use radioiodinated antibody in a binding assay. In this way one can detect fewer than 100 molecules per cell. To test tissues we used both a binding assay with radioiodinated antibody and membrane fragments, and an immunoradiometric assay. Both methods are quantitative and can detect antigen at concentrations as low as 1 ng per mg protein. For the immunoradiometric assay, however, antibodies to two epitopes of the same antigen are required. Absorption techniques are also useful and are preferred by some investigators (Dippold et al., 1980; Ueda et al., 1981). Immunohistological methods should also be included, since they allow detection of minor antigen-positive cell populations, even though they do not provide quantitative information.

Of the considerable number of human tumor cell surface antigens that have been identified by monoclonal antibodies, few if any have been convincingly demonstrated to be entirely specific for neoplastic cells. In many cases specificity has been claimed after only a few cell types, such as cultured fibroblasts and B cell lines, have been tested. In others, a range of cultured cells was tested, but tissues were not. In still other cases, the assays used were incapable of detecting small amounts of antigen. By using highly sensitive assays and by testing in vivo material, we have been able to detect all of the antigens that we have identified in certain normal cells and tissues, although often in much smaller amounts than in melanomas (Brown et al., 1981).

4.3. Comparison of tumor-associated antigens

It is clearly important to determine whether newly identified tumor-associated antigens are related to proteins that have been described previously. Similar tissue distributions and MWs on SDS-PAGE may suggest two antigens are related, but more definitive methods of comparison must also be used. Sequential immunoprecipitation allows one to determine whether two monoclonal antibodies recognize the same antigen molecule, and we have used this method to show that p97 and gp95 are identical (Brown et al., 1981b). Competition binding assays can establish that two antibodies bind to the same epitope (or closely adjacent epitopes). However, the latter method is inconclusive if two antibodies do not compete, since they could be binding to different epitopes of the same molecule. To detect more distant relationships antisera to denatured proteins can be used. For example, we found that an antiserum obtained by immunizing mice with denatured p97 cross-reacted with transferrin and lactotransferrin, even though these proteins differ in 50% of their amino acid residues.

Amino acid sequencing is the method of choice for detecting structural homology. Until recently, such studies required larger amounts of protein than could be readily purified from the plasma membranes of cultured cells. The development of a miniaturized protein sequencer (Hewick et al., 1981), which can work with picomole amounts of protein, now means that one can purify sufficient antigen by affinity chromatography and SDS-PAGE to obtain an N-terminal amino acid sequence. Related sequences can then be searched for systematically in an amino acid sequence library, as is exemplified by our work on p97.

4.4. Nature, function, and practical application of tumor-associated antigens defined by monoclonal antibodies

Three major questions arise with respect to tumor-associated antigens. First, what are they? Second, what are their functions? Third, how useful are they in cancer diagnosis or as targets for therapy? The first question can be answered in part by stating that most human tumor-associated antigens so far detected are differentiation antigens. This acknowledges the fact that the antigens are present on some cell types but not on others (or present in smaller amounts), and that they are not present exclusively on neoplastic cells, and also implies that they are subject to developmental regulation. The question as to their function is harder to answer. There is little evidence that their expression is a cause of malignant transformation, although they may be expressed as a result of uncontrolled growth. The transferrin receptor and p97, two cell surface proteins identified by monoclonal antibodies, are illustrative in this regard.

In 1978, Bramwell and Harris described a cell surface protein associated with malignancy, and in 1980 Omary et al. and Judd et al. described a cell surface glycoprotein present on dividing hematopoietic cells. In each case the proteins had MWs in the range 90,000 to 100,000. Subsequently, the distribution of these antigens was found to be similar to that of a cellular receptor for transferrin and it is now believed that they are identical (Trowbridge and Omary, 1981; Sutherland et al., 1981). Our work on p97 provides another example of how one can gain some insight into antigen function. Knowledge of the tissue distribution of p97 was not helpful, but amino acid sequence studies provided the key. An N-terminal amino acid sequence enabled us to search an amino acid sequence library for homologous sequences, revealing its homology to transferrin and lactotransferrin. Like transferrin and lactotransferrin, p97 is an iron-binding protein, although the details of its function have yet to be elucidated. In both of the above examples one might speculate, however, that the proteins are expressed in tumors as a consequence of the increased demand for iron that results from rapid growth.

With respect to the third question, relating to the practical usefulness of tumor-associated antigens and antibodies specific for them, generalizations are hard to make. Although a degree of tumor specificity is essential, the absolute amount of antigen at the tumor cell surface and the nature of any normal cells that express significant amounts of the antigen are also important. For example, an antigen expressed on a small but essential population of normal stem cells may be useless as a therapeutic target, even if most other adult cells express very little of it, whereas an antigen strongly expressed on certain differentiated adult cells may still prove to be an excellent therapeutic target. This is supported by the demonstration that monoclonal antibodies to the normal differentiation antigen Thy-1 had a therapeutic effect in mice with spontaneous lymphomas (Bernstein et al., 1980).

In conclusion, monoclonal antibodies have made it possible to identify and characterize tumor-associated antigens at a rate and with a precision that was unimaginable a few years ago. We hope that in coming years they will have a major impact on the diagnosis and treatment of cancer.

Acknowledgements

We thank all our colleagues and collaborators who have contributed to this work, particular Drs. M.-Y. Yeh, R.G. Woodbury, K. Nishiyama, R.M. Hewick, W.J. Dreyer and R.F. Doolittle. This work was supported by Grants CA 27841, CA 25558, and CA 19149 from the National Institutes of Health, and Grant IM 241 from the American Cancer Society.

REFERENCES

Bernstein, I.D., Tam, M.R. and Nowinski, R.. 1980. Mouse leukemia therapy with monoclonal antibodies against a thymus differentiation antigen. Science 207:68-71.

Bramwell, M.E., and Harris, H.. 1978. An abonormal membrane glycoprotein associated with malignancy in wide range of different tumors. Proc. Roy. Soc. Lond. B 201:87-106.

Brown, J.P., Tamerius, J.D., and Hellström, I.. 1979. Indirect ^{125}I -labelled protein A assay for monoclonal antibodies to cell surface antigens. J. Immunol. Methods 31: 201-209.

Brown, J.P., Wright, P.W., Hart, C.E., Woodbury, R.G., Hellström, K.E., and Hellström, I.. 1980. Protein antigens of normal and malignant human cells identified by immunoprecipitation with monoclonal antibodies. J. Biol. Chem. 255:4980-4983.

Brown, J.P., Woodbury, R.G., Hart, C.E., Hellström, I., and Hellström, K.E.. 1981a. Quantitative analysis of melanoma-associated antigen p97 in normal and neoplastic tissues. Proc. Natl. Acad. Sci. USA 78:539-543.

Brown, J.P., Nishiyama, K., Hellström, I., and Hellström, K.E.. 1981b. Structural characterization of human melanoma-associated antigen p97 using monoclonal antibodies. J. Immunol. 127:539-546.

Brown, J.P., Hellström, K.E., and Hellström, I.. 1981c. Use of Monoclonal antibodies for quantitative analysis of antigens in normal and neoplastic tissues. Clin. Chem. 27:1592-1596.

Carrel, S., Accolla, R.S., Carmagnola, A.L., and Mach, J.P.. 1980. Common human melanoma-associated antigens detected by monoclonal antibodies. Cancer Res. 40:2523-2528.

Dippold, W.G., Lloyd, K.O., Li, L.T., Ideda, H., Oettgen, H.F., and Old, L.J.. 1980. Cell surface antigens of human malignant melanoma; definition of six antigenic systems with monoclonal antibodies. Proc. Natl. Acad. Sci. USA 77: 6114-6118.

Doolittle, R.F.. 1981. Similar amino acid sequences: chance or common ancestry? Science 214:149-159.

Ey, P.L., Prowse, S.J., and Jenkin, C.R.. 1978. Isolation of pure IgG1, IgG2a, and IgG2b immunoglobulins from mouse serum using protein A-sepharose. Immunochem. 15:429-436.

Gahmberg, C.G., Hagvy, P., and Andersson, L.C.. 1976. Characterization of surface glycoproteins of mouse lymphoid cells. J. Cell Biol. 68:642-653.

Gudjonsson, H., and Johnsen, S.. 1978. HeLa cell plasma membranes IV. Iodination of membrane proteins in synchronized cells. Exp. Cell Res. 112:289-295.

Hellström, I., Brown, J.P., and Hellström, K.E.. 1981. Monoclonal antibodies to two determinants of melanoma antigen p97 act synergistically in complement-dependent cytotoxicity. J. Immunol. 127:157-160.

Hellström, I., Hellström, K.E., Pierce, G.E., and Yang, J.P.S.. 1968. Cellular and humoral immunity to different types of human neoplasms. Nature 220:1352-1354.

Hellström, K.E., and Brown, J.P.. 1979. Tumor antigens. The Antigens 5:1-82.

Hellström, K.E., Brown, J.P., and Hellström, I.. 1980. Monoclonal antibodies to tumor antigens. Contemporary Topics in Immunobiology 11:117-137.

Herlyn, M., Clark, W.H., Jr., Mastrangelo, M.J., Guerry, D., IV, Elder, D.E., LaRossa, D., Hamilton, R., Bondi, E., Tuthill, R., Steplewski, Z., and Koprowski, H.. 1980. Specific immunoreactivity of hybridoma-secreted monoclonal anti-melanoma antibodies to cultured cells and freshly derived human cells. Cancer Res. 40:3602-3609.

Hewick, R.M., Hunkapiller, M.W., Hood, L.E., and Dreyer, W.J.. 1981. A gas-liquid solid phase peptide and protein sequenator. J. Biol. Chem. 256:7990-7997.

Judd, W., Poodry, C.A., and Strominger, J.L.. 1980. Novel surface antigen expressed on dividing cells but absent from nondividing cells. J. Exp. Med. 152:1430-1435.

Kohler, G., and Milstein, C.. 1975. Continuous cultures of fused cells secreting antibody of predefined specificity. Nature 256:495-497.

Koprowski, H., Steplewski, Z., Herlyn, D., and Herlyn, M.. 1978. Study of antibodies against human melanoma produced by somatic cell hybrids. Proc. Natl. Acad. Sci. USA 75:3405-3409.

Loop, S.M., Nishiyama, K., Hellstrom, I., Woodbury, R.G., Brown, J.P., and Hellstrom, K.E.. 1981. Two human tumor-associated antigens, p155 and p210, detected by monoclonal antibodies. Int. J. Cancer 27:775-781.

Morgan, A.C., Jr., Galloway, D.R., and Reisfeld, R.A.. 1981. Production and characterization of monoclonal antibody to a melanoma-

specific glycoprotein. Hybridoma 1:27-36.
Omary, M.B., Trowbridge, I.S., and Minowada, J.. 1980. Human cell surface glycoprotein with unusual properties. Nature 286:888-891.
Pesando, J.M., Ritz, J., Levine, H., Terhorst, C., Lazarus, H., and Schlossman, S.. 1980. Human leukemia-associated antigen: relation to a family of surface glycoproteins. J. Immunol. 124:2794-2799.
Shiku, H., Takahashi, T., Oettgen, H.F., and Old, L.J.. 1976. Cell surface antigens of human malignant melanoma. II. Serological typing with immune adherence assays and definition of two new antigens. J. Exp. Med. 144:873-881.
Sutherland, R., Delia, D., Schneider, C., Newman, R., Kemshead, J., and Greaves, M.. 1981. Ubiquitous cell-surface glycoprotein on tumor cells is proliferation-associated receptor for transferrin. Proc. Natl. Acad. Sci. USA 78:4515-4519.
Trowbridge, I.S., and Omary, M.B.. 1981. Human cell surface glycoprotein related to cell proliferation is the receptor for transferrin. Proc. Natl. Acad. Sci. USA 78:3039-3043.
Ueda, R., Ogata, S., Morrissey, D.M., Finstad, C.L., Szudlarek, J., Whitmore, W.F., Oettgen, H.F., Lloyd, K. O., and Old, L.J.. 1981. Cell surface antigens of human renal cancer defined by mouse monoclonal antibodies: identification of tissue specific kidney glycoproteins. Proc. Natl. Acad. Sci. USA 78:5122-5126.
Wilson, B.S., Imai, K., Natali, P.G., and Ferrone, S.. 1981. Distribution and molecular characterization of a cell-surface and a cytoplasmic antigen detectable in human melanoma cells with monoclonal antibodies. Int. J. Cancer 28-293-300.
Woodbury, R.G., Brown, J.P., Yeh, M.-Y., Hellström, I., and Hellstrom, K.E.. 1980. Identification of a cell surface protein, p97, in human melanoma and certain other neoplasms. Proc. Natl. Acad. Sci. USA 77:2183-2187.
Woodbury, R.G., J.P., Brown, Loop, S.M., Hellström, K.E., and Hellström, I.. 1981. Analysis of normal and neoplastic human tissues for the tumor-associated protein p97. Int. J. Cancer 27:145-149.
Yeh, M.-Y., Hellström, I., Brown, J.P., Warner, G.A., Hansen, J.A., and Hellström, K.E.. 1979. Cell surface antigens of human melanoma identified by monoclonal antibody. Proc. Natl. Acad. Sci. USA 76:2927-2931.

11

A BIOCHEMICAL AND BIOSYNTHETIC ANALYSIS OF HUMAN MELANOMA-ASSOCIATED ANTIGENS WITH MONOCLONAL ANTIBODIES

Thomas F. Bumol, John R. Harper, Darwin O. Chee and Ralph A. Reisfeld

Scripps Clinic and Research Foundation, 10666 N. Torrey Pines Road La Jolla, California 92037

1. SUMMARY

The molecular profile and cellular topography of two melanoma associated antigens defined by monoclonal antibodies 9.2.27 and F11 were investigated. The 9.2.27 antibody recognizes a 250K glycoprotein which associates with a high molecular weight complex (HMW-C)[1] M_r >500K in indirect immunoprecipitation analysis of detergent extracts obtained from biosynthetically labeled melanoma cells. The HMW-C appears late in the biosynthesis of this antigenic complex as judged by pulse-chase studies and is sensitive to degradation by chondroitinase ABC lyase suggesting that it is a chondroitin sulfate proteoglycan. Direct binding analysis of iodinated 9.2.27 IgG to isolated membrane fractions of melanoma cells reveal strong antigenic activity of this antigen demonstrating its membrane association. In contrast to these results, the F11 antibody recognized three glycoproteins in detergent extracts of biosynthetically labeled melanoma cells at M_r 75, 77 and 100K. The primary antigenic activity recognized by F11, however, is in the spent media of melanoma cells. Indirect immunoprecipitation analysis of secreted spent media components revealed that the F11 antibody recognizes only the 100K component as a secreted melanoma associated antigen and a cross reactive lower molecular weight

---

[1]HMW-C, high molecular weight complex

antigen in spent media of two neuroblastoma cell lines. Thus, monoclonal antibodies can be used to dissect the molecular nature and cellular topography of melanoma associated antigens through the combination of biosynthetic, enzymatic and cell fractionation techniques allowing the identification of these potentially tumor associated gene products. This information will allow the design of experiments to investigate the functional roles of these molecules in malignant melanoma.

2. INTRODUCTION

The adaptation of the fluid mosaic model for membrane structure has provided a working hypothesis from which to explore the eukaryotic cell surface (Singer and Nicolson, 1972). This model proposed that proteins, glycoproteins and other glycoconjugated molecules reside in a fluid lipid bilayer in a variety of asymmetric relationships; some peripheral others transmembrane and still others on the cytosol side of the plasma membrane (Singer, 1974). In addition, a complicated network of proteins, glycoproteins and proteoglycan molecules exist loosely associated with the plasma membrane at the cell surface comprising the extracellular matrix or glycocalyx of the cell (Roden, 1980).

Many membrane and extracellular matrix components have been identified and implicated in a variety of cellular processes such as adhesion, motility, growth control and differentiation (Nicolson, 1979). These precise properties are often aberrant in the malignant or transformed cell and remain a focal point for investigations in tumor biology.

The advent of hybridoma technology has provided for the first time uniquely specific monoclonal antibody probes to identify potential tumor associated gene products at the cell surfaces and extracellular matrices of malignant cells. Our laboratory has focused on the development of monoclonal antibodies to cell surface and secreted antigens of human melanoma cells. In this regard, we have obtained several monoclonal antibody products with high specificity for melanoma associated antigens and have used these reagents as probes to study the cell surface of malignant melanoma (Morgan et al., 1981a; Morgan et al, 1981b; Bumol and Reisfeld, 1981).

Our initial approach focused on obtaining a molecular profile of these monoclonal antibody defined antigens followed by a combination of biosynthetic, enzymatic and biochemical characterizations of the antigenic components involved. The results of these approaches on two monoclonal antibody defined melanoma associated antigens will be the subject of the following report.

3. RESULTS

All initial experiments were performed exclusively with biosynthetically labeled detergent extracts and spent media of melanoma cell lines in order to eliminate potential artefacts often caused by cell surface labeling techniques because of contamination with tissue culture serum proteins. In addition, biosynthetic labeling allows the use of multiple glycoprotein precursor components to be used to biosynthetically establish the glycoconjugated nature of the antigens under study.

The 9.2.27 and F11 monoclonal antibodies used in our studies both recognize antigens associated with melanoma cell lines as determined by radioimmunometric binding analyses (Table I). An analysis of spent media antigens by solid phase radioimmunometric binding assay, however, demonstrates that the F11 antibody also recognizes an antigenic determinant on a molecule released by melanoma cells, whereas the 9.2.27 antigenic activity remains largely cell associated (Table I). Thus, the results of these analyses suggest a differential cellular topography of the two melanoma associated antigens.

A delineation of the molecular profile of these two antigens was achieved by indirect immunoprecipitation of detergent lysates obtained from M21 melanoma cells labeled with ^{3}H-glucosamine followed by SDS-PAGE and fluorography (Figure 1). The F11 antibody recognizes three glycoprotein components with apparent molecular weights (M_r) of 75,000 (75K), 77,000 (77K) and 100,000 (100K). The 9.2.27 antibody recognizes two glycoconjugated components, one with M_r 250,000 (250K) and the other, a high molecular weight component (HMW-C) with M_r >500,000 (500K). The complicated molecular profiles observed led us to investigate the biosynthesis of these antigens by pulse-chase studies utilizing ^{35}S-methionine.

An example of a pulse-chase analysis of the antigens recognized by antibody 9.2.27 is shown in Figure 2. Indirect immunoprecipitation analysis of detergent lysates of M21 cells at early time points in the chase (3, 10 and 20 minutes) reveal that three moieties are recognized with M_r 210, 220 and 240K. The component with lower M_r disappears as the chase continues with the appearance of a 250K entity and the HMW-C at 30 minutes. At 60 minutes into the chase and at subsequent time points the antigenic profile appears identical with the typical profile observed in melanoma cells which were biosynthetically labeled for long periods of time (Figure 1). Thus, the 9.2.27 monoclonal antibody recognizes an antigenic determinant on the 250K glycoprotein and on several apparent biosynthetic precursors of this component that show apparent M_r of 210, 220 and 240K. The appearance of the high molecular weight complex occurs later in the chase period indicating that the monoclonal antibody recognizes biosynthetically modified precursor components of this antigen complex.

Additional studies on this antigen complex revealed that the HMW-C could be exclusively labeled with $^{35}SO_4^=$ and was sensitive to β-elimination in dilute base (Bumol and Reisfeld, 1981). These data suggested to us that the HMW-C may be a sulfated proteoglycan component. The results of a chondroitinase ABC digestion experiment on the immunoprecipitated 9.2.27 antigens shown in Figure 3 supported this notion since the

TABLE I

Radioimmunometric Binding Analysis of F11 and 9.2.27

Monoclonal Antibodies

	Specific ^{125}I-Protein A Bound[a]	
Antibody	M14 CDM Cells[b]	CSM[c]
9.2.27	48,354	4,951
F11	13,418	33,110

[a] Specific cpm bound equals cpm bound experimental minus cpm bound control using X63-myeloma spent media as first antibody. Results are duplicate determinations.

[b] M14 CDM cells designates M14 melanoma cells adapted to chemically defined serum free media utlized as targets (2 x 10^5 cells) in this assay.

[c] CSM designates a 30X concentrated spent media target (10 μg protein) in this solid phase radioimmunometric binding assay.

HMW-C is sensitive to degradation by this enzyme, whereas the 250K glycoprotein is resistant, clearly indicating the chondroitin sulfate nature of the HMW-C.

Studies on the antigen recognized by monoclonal antibody F11 initially focused on the antigenic activity we observed in the spent media of human melanoma cells (Table I). Indirect immunoprecipitation analysis of biosynthetically labeled spent media from M21 cells revealed that a component of M_r 100,000 (100K) was secreted by these cells(Figure 4 A, B). In addition, we examined the spent media of two neuroblastoma cell lines and found a cross reacting species of apparent lower molecular weight in those cell types recognized by monoclonal antibody F11 (Figure 4 C, D). Thus, of the three molecular species revealed in the detergent extracts of melanoma cells by F11 immunoprecipitation analysis (Figure 1), only the 100K component is found in the extracellular spent media.

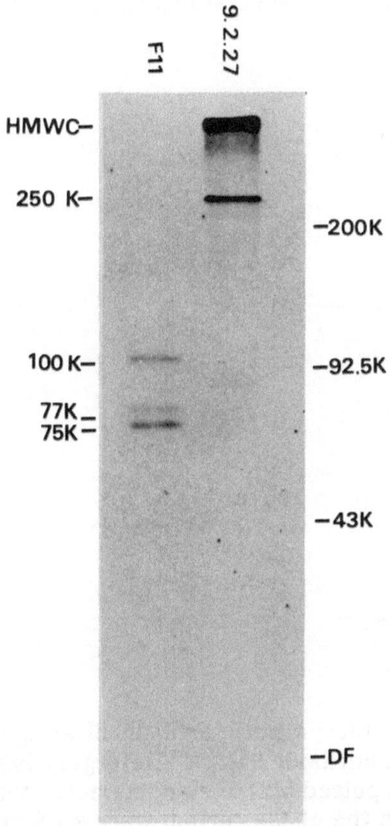

Figure 1. Indirect immunoprecipitation analysis of detergent lysates from ^{3}H-glucosamine labeled M21 melanoma cells with the F11 and 9.2.27 monoclonal antibodies. Antigens were eluted and analyzed on a 5-15% acrylamide gradient gel followed by fluorography, utilizing molecular weight standards of myosin (200K), phosphorylase B (92.5K) and ovalbumin (43K). DF designates the position of the dye front on this fluorograph exposed for six days.

While both the F11 and 9.2.27 monoclonal antibodies recognize melanoma associated glycoproteins, our results suggest that the 9.2.27 antigen is mainly cell associated while the F11 antibody recognizes both cell associated and secreted antigens. Since we have tentatively identified the HMW-C of the 9.2.27 defined antigenic complex as a chondroitin sulfate proteoglycan, a known component of cell surfaces and extracellular matrix, we focused our efforts on establishing the membrane association

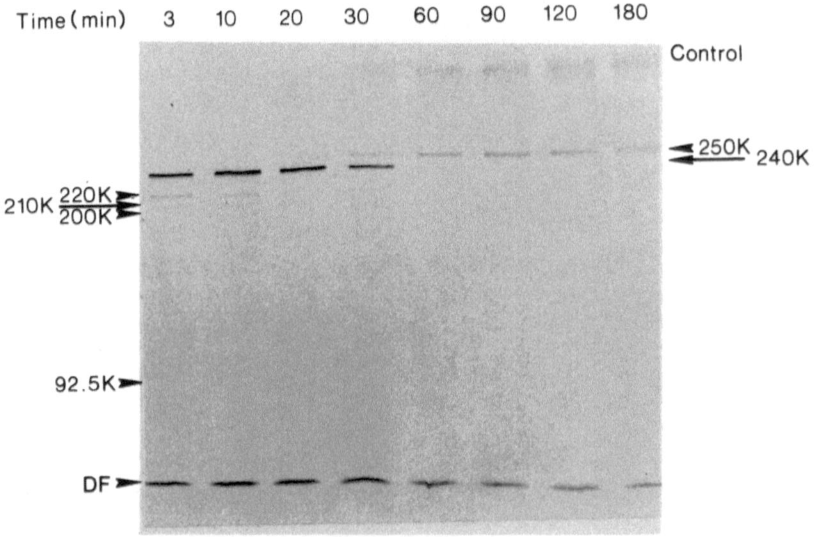

Figure 2. Pulse-chase biosynthetic analysis of antigens recognized by monoclonal antibody 9.2.27. Detergent lysates from ^{35}S-methionine pulsed M21 melanoma cells, from the time points indicated in the chase period, were subjected to indirect immunoprecipitation, SDS-PAGE on 5% acrylamide gels and fluorography. DF designates the dye front on this 4 day exposure.

of this particular antigen complex. To this end, a crude membrane preparation was obtained from M14 melanoma cells, grown in synthetic serum free media by a combination of hypotonic lysis, dounce homogenization and differential centrifugation. These membranes were then plated direct onto soft vinyl microtiter wells at concentrations ranging from 1-10 μg protein/ well and assayed for binding of ^{125}I-IgG isolated from 9.2.27 hybridoma ascites in a solid phase direct binding assay. The

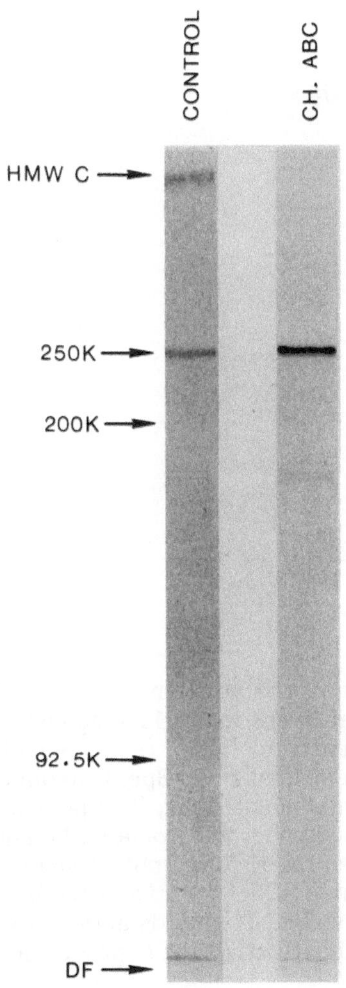

Figure 3. Chondroitinase ABC lyase digestion of antigens defined by monoclonal antibody 9.2.27. Immunoprecipitates of ^{3}H-glucosamine-labeled M21 detergent lysates were digested with 0.01 units of chondroitinase ABC lyase for 1 hour at 37°C prior to analysis on 5% acrylamide SDS-PAGE and fluorography. Migration of molecular weight standards are indicated and DF designates the dye front on this fluorograph exposed for six days.

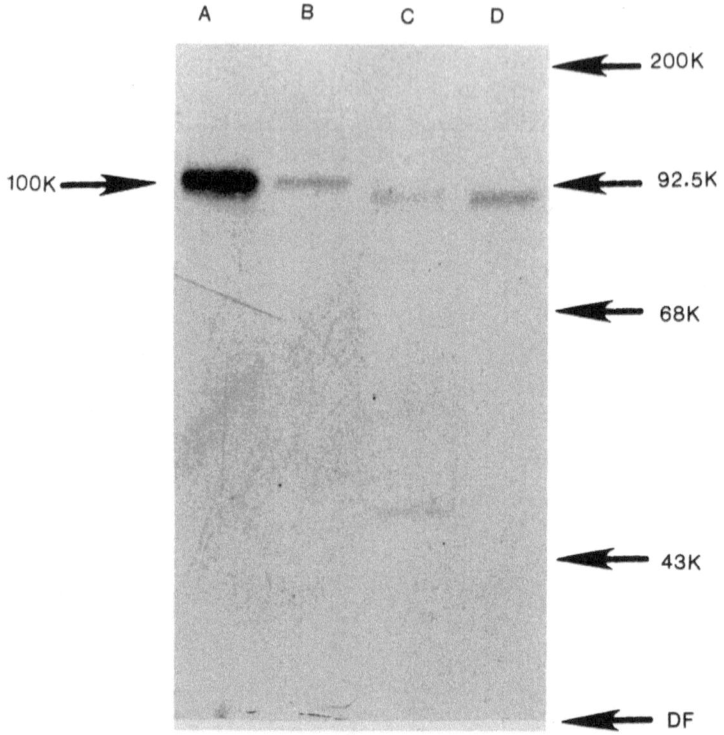

Figure 4. Indirect immunoprecipitation analysis of biosynthetically labeled spent media antigens with F11 monoclonal antibody. Immunoprecipitated antigens from: (A) ^{35}S-methionine labeled M21 melanoma spent media; (B) Spent media of ^{3}H-glucosamine labeled M21 melanoma cells; (C) Spent media from ^{3}H-leucine labeled CHP-100 neuroblastoma cells, and (D) Spent media of ^{3}H-leucine labeled LA-N-5 neuroblastoma cells were analyzed on 7.5% acrylamide SDS-PAGE followed by fluorography. Migration of molecular weight standards is indicated and DF designates dye front in this fluorograph exposed for 7 days.

results of a typical binding curve obtained with multiple inputs of ^{125}I-9.2.27 IgG are shown in Figure 5.

As can be observed, the binding is saturable and demonstrates increased levels of binding as the protein concentration of the target membranes increases. This binding is specific as judged from the complete displacement of iodinated IgG by the addition of 3 μg cold 9.2.27 IgG in parallel experiments (data not shown). These data demonstrate that the 9.2.27 antigens either as the 250K glycoprotein, the HMW-C, or both are tightly associated with membrane fractions isolated from melanoma cells. The precise nature of this association with these membranes is currently under investigation.

4. DISCUSSION

In this study we have examined the molecular profile, cellular topography and biosynthesis of two melanoma associated antigens defined by monoclonal antibodies F11 and 9.2.27.

The F11 antibody has a broad range of specificity among human tumors including reactivity with melanoma, carcinoma and neuroblastoma cell lines (Chee et al., 1981; T.F. Bumol, D.O. Chee and R.A. Reisfeld, in preparation). In this study we focused on delineating the molecular profile of both the cell associated and secreted antigens of melanoma cells and discovered that the F11 antibody recognizes three glycoproteins of M_r 75, 77 and 100 K in detergent extracts of whole cells with only the 100K antigen being secreted into the spent media. A priori, these data would suggest the existence of several cross reactive antigenic determinants on melanoma cells, but current data obtained by pulse-chase analysis of detergent extracts of biosynthetically labeled M21 melanoma cells with the F11 antibody suggest that the 75 and 77K antigens are biosynthetic precursors of the 100K molecule. While these observations present an interesting problem of understanding melanoma cell glycoprotein biosynthesis, an even more intriguing problem is provided by the finding of the lower molecular weight antigen secreted by neuroblastoma cells which also is recognized by the F11 monoclonal antibody. Previous studies from this laboratory described alterations in this 100K component secreted from fetal melanocytes, as it displayed a faster migrating, lower molecular weight glycoprotein recognized by a monoclonal antibody, 250/165, that also recognizes the 100K component secreted by melanoma cells (Morgan et al., 1981a). Thus, the particular antigen appears to exist as several secreted molecules with different molecular weights originating from different fetal and malignant tissue types. Current studies are designed to examine the biochemical basis for the structural changes of this secretory glycoprotein which appears to be an oncofetal antigen (Morgan et al., 1981a).

A combination of biosynthetic and enzymatic digestion experiments with the 9.2.27 monoclonal antibody has allowed us to establish that this reagent recognizes an antigenic complex consisting of a 250K glycoprotein which either associates with or is a precursor to a high molecular weight component (M_r >500K) which has all the characteristics of a chondroitin sulfate proteoglycan. The 9.2.27 antibody does recognize an antigenic determinant on the 250K glycoprotein and several of its biosynthetic precursors ranging in M_r from 210-240K in the absence of the HMW-C. This clearly demonstrates that the antigenic site recognized by this antibody is formed early in the biosynthesis of this glycoprotein molecule. Tryptic peptide map experiments by HPLC, comparing the 250K component and the HMW-C, indicate many identical peptides suggesting that the 250K component may be contained within the HMW-C proteoglycan (T.F. Bumol, L.E. Walker and R.A. Reisfeld, in preparation). Taken together, these data suggest that the 250K component is a precursor glycoprotein of a chondroitin sulfate proteoglycan in human melanoma cells.

Our data on the localization of this antigen in isolated membrane fractions of melanoma cells (Figure 5) provide further evidence for the tight membrane association of this antigenic complex in melanoma cells. The fact that proteoglycans are important cell surface constituents in many cell types (Roden, 1980) allows us currently the unique possibility to study these monoclonal antibody-defined molecules on the cell surface of a highly malignant tumor. Thus far, the 9.2.27 monoclonal antibody recognizes antigens on many melanoma and on a single neuroblastoma cell line among a large variety of normal and malignant cell types examined thus far, suggesting a limited distribution of this antigen. The association of this glycoprotein antigen with a common proteoglycan suggests that tumors may have alterations in their cell surface constituents due to the expression of unique gene products capable of accepting biosynthetic modifications to form tumor analogues of normal cell surface constituents. Further studies are necessary to explore this hypothesis.

In summary, our studies demonstrate that monoclonal antibodies can be used as probes to study glycoprotein/glycoconjugate biosynthesis as well as cell surface expression and secretion of human tumor cells. A combination of approaches utilizing intrinsically labeled cells, pulse-chase and enzymatic digestion experiments with specific immunoprecipitation analysis has allowed us to identify and obtain biosynthetic and structural information on two melanoma associated antigens. These data have permitted us to pinpoint the cellular topography of these antigens to better design future studies that focus on the functional role of these glycoproteins at the surfaces of human melanoma cells.

5. MATERIALS AND METHODS

5.1. Cells

The M21 and M14 melanoma cell lines were originally derived from metastatic lesions and obtained from Dr. Donald Morton, UCLA. The M21 cell line was maintained in RPMI 1640 media supplemented with 10% calf serum (GIBCO), 2 mM glutamine and 50 μg/ml gentamycin sulfate while the M14 cell line was adapted to grow in serum free, Synmed media (Centaurus). The LA-N-5 and CHP-100 neuroblastoma cell lines were provided by Dr. Seeger at UCLA.

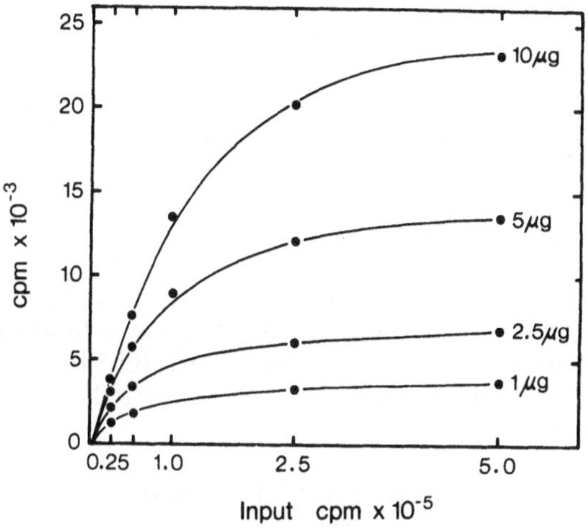

Figure 5. Direct binding analysis of iodinated 9.2.27 monoclonal antibody IgG to isolated membrane fractions of the M14 melanoma cell line. Results indicate the specific cpm bound to solid phase target membranes ranging from 1-10 μg with varied input counts of iodinated 9.2.27 IgG. Points represent triplicate determinations.

5.2. Monoclonal antibodies

The 9.2.27 hybridoma was developed in this laboratory (Morgan et al., 1981B) and the F11 hybridoma was originally produced by Chee et al. (1981). Both hybridoma products were screened extensively and demonstrate serological activity highly associated with melanoma associated antigens. The F11 antibody also shows serological reactivity with several other tumor tissues as well, including several carcinoma cell lines (Chee et al., 1981). Spent tissue culture fluid from hybridoma cultures served as a source of antibody for these studies. In some cases, the 9.2.27 hybridoma clone was grown in ascites to isolate larger quantities of IgG by protein-A Sepharose affinity chromatography.

5.3. Serological assays

A solid phase radioimmunometric binding assay previously developed in this laboratory was used to examine antigenic activity in spent media (Morgan et al., 1980). A modification of this assay using whole cells to examine cell surface antigenic expression involved maintaining the target cells in PBS supplemented with 5% chicken serum (PCS).

A direct binding analysis of 9.2.27 IgG iodinated by the chloramine T method was achieved by incubation of solid phase target crude membrane fractions in microtiter wells with varied input amounts of iodinated 9.2.27 IgG for 3 hours at 4 °C. After extensive washing of the solid phase wells with PCS, the plates were dried, cut and counted in a Rackgamma scintillation spectrometer (LKB).

5.4. Biosynthetic labelling procedures

All cultures were intrinsically labelled with glycoprotein precursors for pulse-chase studies and subjected to long term labelling by incubation in Selectamine media kits (GIBCO) made deficient in the appropriate amino acid or oligosaccharide precursor as previously described (T.F. Bumol and R.A. Reisfeld, 1981).

5.5. Indirect immunoprecipitation/SDS-PAGE

Indirect immunoprecipitation on protein-A Sepharose immunoadsorbents (IADS) was carried out as previously described (Galloway et al., 1980; T.F. Bumol and R.A. Reisfeld, 1981), except that in the analysis of whole cell extracts, RIPA lysis buffer (0.01 M Tris-HCl pH 7.2, 0.15 M NaCl, 1% Triton X-100, 1% deoxycholate and 0.1% sodium lauryl sulfate, 1% Trasylol) was used to extract cell associated antigens under study (Gilead et al., 1976).

Antigens were eluted from IADS in Laemmli sample buffer (Laemmli, 1970) and subjected to electrophoresis on polyacrylamide slab gels as previously described (Galloway et al., 1980). Fluorography of slab gels to visualize immunoprecipitation antigens was performed according to previously published procedure (Bonner and Laskey, 1974).

In some experiments, washed IADS were first digested with 0.01 units chondroitinase ABC (Sigma) for 1 hour at 37 °C prior to analysis by SDS-PAGE.

5.6. Cell fractionation procedure

M14 melanoma cells grown in serum free, chemically defined media (Synmed) were harvested and exposed to hypotonic swelling buffer (0.01 M Tris-HCl pH 1.4, 1.5 mM $MgCl_2$, 0.1 M NaCl, 2% PMSF) for 30 minutes on ice prior to homogenization in a ground glass Dounce homogenizer. The homogenate was then centrifuged at 1000 x g to sediment nuclei and cell debris, followed by a 10,000 x g centrifugation step to remove sediment and mitochondria. The 10,000 g supernatant was then subjected to a 100,000 x g sedimentation run to pellet all subcellular and plasma membranes. This fraction, designated as "crude membrane fraction" was then resuspended, its protein content determined by the method of Lowry (Lowry et al., 1951) and subsequently used as target in solid phase serological assays.

5.7. Materials

D-[2-^{3}H]-glucosamine (15 Ci/mmole), L-[4,5-^{3}H (N)]-leucine (42 Ci/mmole) and L-[^{35}S]-methionine (1035 Ci/mmole) were obtained from New England Nuclear. All other materials were reagent grade or better.

Acknowledgment

The authors acknowledge the expert technical assistance of Ms. Vicky McCarthy and Ms. Kathy Rocker and thank Ms. Dee Davidson for typing of this manuscript. Research mentioned in this report was supported by NIH grant CA 28420 and grant IM 218 from the American Cancer Society. T.F. Bumol is supported by fellowship DRG-HHH-F-2 from the Damon Runyon-Walter Winchell Cancer Fund. This is publication number 2602 from Scripps Clinical and Research Foundation.

REFERENCES

Bonner, W.M. and Laskey, R.A., 1974, A film detection method for tritium labelled proteins and nucleic acids in polyacrylamide gels, Eur. J. Biochem 46:83.

Bumol, T.F. and Reisfeld, R.A., 1981, A unique glycoprotein-proteoglycan complex defined by monoclonal antibody on human melanoma cells, Proc. Natl. Acad. Sci. U.S.A., in press.

Chee, D.O., Yonemoto, R.H., Leong, S.B.L., Richards, G.S., Smith, V.R., Klotz, J.L., Goto, R.M., Gascan, R.L. and Drushella, M.M., 1981, Mouse monoclonal antibody to a melanoma/carcinoma antigen synthesized by a human melanoma cell line propagated in serum free media, Cancer Res., in press.

Galloway, D.R., McCabe, R.P., Pellegrino, M.A., Ferrone, S. and Reisfeld, R.A., 1981, Tumor associated antigens in spent medium of human melanoma cells: immunochemical characterization with xenoantisera. J. Immunol 126:62.

Gilead, Z., Jeng, Y., Wold, W.S.M., Sugawara, K., Rho, H.M., Harter, M.L. and Green, M., 1976, Immunological identification of two adenovirus Z-induced early proteins possibly involved in cell transformation. Nature 264:263.

Laemmli, U.K., 1970, Cleavage of structural proteins during the assembly of the head of bacteriophage T4. Nature 227:680.

Lowry, O.H., Rosebrough, N.J., Farr, A.L. and Randall, R.J., 1951, Protein measurement with the Folin phenol reagent. J. Biol. Chem. 193:265.

Morgan, A.C., Galloway, D.R., Jensen, F.C., Giovanella, B.C. and Reisfeld, R.A., 1981a, Human melanoma associated antigens, presence on cultured normal fetal melanocytes. Proc. Natl. Acad. Sci. U.S.A. 78:3834.

Morgan, A.C., Galloway, D.R. and Reisfeld, R.A., 1981b, Production and characterization of a monoclonal antibody to a melanoma associated glycoprotein, Hybridoma 1:27.
Morgan, A.C., Galloway, D.R., Wilson, B.S. and Reisfeld, R.A., 1980, Human melanoma associated antigens: a solid phase assay for detection of specific antibody. J. Immunol Methods 39:233.
Nicolson, G.L., 1979, Topographic display of cell surface components and their role in transmembrane signalling in Current Topics in Developmental Biology, Vol. 13 (M. Friedlander, ed.), Academic Press, New York, pp. 305-338.
Rodén, L., 1980, Structure and metabolism of connective tissue proteoglycans in The Biochemistry of Glycoproteins and Proteoglycans (W.J. Lennarz, ed.), Plenum Press, New York, pp. 267-355.
Singer, S.J., 1974, The molecular organization of membranes, Ann. Rev. Biochem. 43:805.
Singer, S.J. and Nicolson, G.L., 1972, The fluid mosaic model of the structure of cell membranes, Science 175:720.

12

MONOCLONAL ANTIBODIES TO A TUMOR SPECIFIC ANTIGEN ON RAT MAMMARY CARCINOMA Sp4 AND THEIR USE IN DRUG DELIVERY SYSTEMS

R. W. Baldwin, M. J. Embleton, G. R. Flannery, B. Gunn, J. A. Jones, J. G. Middle, A. C. Perkins*, M. V. Pimm, M. R. Price and R. A. Robins

Cancer Research Campaign Laboratories, University of Nottingham and *Department of Medical Physics, University Hospital Nottingham, United Kingdom

1. INTRODUCTION

With the development of techniques for producing monoclonal antibodies to cell surface antigens, the identification and typing of tumor-associated antigens is entering a new and sophisticated phase (Baldwin et al, 1981). There are, for example, a range of monoclonal antibodies which react with antigens associated exclusively, or more frequently at greatly increased levels, on human tumors including malignant melanoma, osteogenic sarcoma, neuroblastoma and colon carcinoma (see chapters 9-11, this volume). These monoclonal antibodies may have several applications in addition to their use for isolating and identifying tumor-associated antigens. This includes the use of radioisotopically labelled antibodies for identification of tumor deposits by γ-scintigraphy. Additionally monoclonal antibodies may be used as carriers for anti-tumor agents, thus providing an approach for selective attack upon malignant cells. There are several options available in the choice of anti-tumor agents but currently, attention is being given to plant and bacterial toxins, cytotoxic drugs and biological response modifiers including interferon (Baldwin et al, 1981: Baldwin and Byers, 1982). These approaches are considered in this paper which reviews the current status of research on the use of monoclonal antibodies specifying antigens associated with a naturally arising rat mammary carcinoma (Gunn et al, 1980) both for tumor localization and as carriers for anti-tumor agents.

2. RESULTS AND DISCUSSION

2.1. Monoclonal antibody specifying tumor specific antigen associated with rat mammary carcinoma Sp4

Immunization of syngeneic rats against tumor Sp4, a naturally arising mammary carcinoma in WAB/Not strain of rats, induces resistance to rechallenge with this tumor, but not other mammary carcinomas (Baldwin and Embleton, 1979). Sera from syngeneic rats immunized against this tumor also contain antibody reacting specifically in membrane immunofluorescence tests with Sp4 tumor but not other mammary carcinomas (Baldwin and Embleton, 1970). Developing from these studies, showing that mammary carcinoma Sp4 expresses a highly characteristic tumor associated antigen, interspecies hybridomas have been prepared following fusion of spleen cells from a syngeneic Sp4 tumor-immune rat and mouse myeloma P3NS1 (Gunn et al, 1980). Table I summarizes the specificity of binding of antibody (IgG2b) produced by one hybridoma Sp4/A4 with a range of target cells, both normal and neoplastic. This binding was detected most extensively by a second reaction with ^{125}I-labelled sheep F(ab')$_2$ anti-rat IgG, and the results expressed as a binding ratio in comparison with target cells treated with P3NS1 spent medium. As indicated in Table I, Sp4/A4 antibody reacted with Sp4 target cells, but not other rat tumors or normal tissues. Binding of Sp4/A4 antibody has also been examined using a fluorescence activated cell sorter (FACS IV) following reaction of cells with fluorescein isothiocyanate-labelled sheep F(ab')$_2$ anti-rat IgG (Fig. 1). For comparative purposes, the mean fluorescence intensity (FACS channel number) was calculated following reaction of tumor cells with standard amounts of antibody. By this criterion, it was again shown that cells derived from tumor Sp4 but not other tumors or normal tissues, including mammary epithelial cells reacted with Sp4/A4 antibody (Table I).

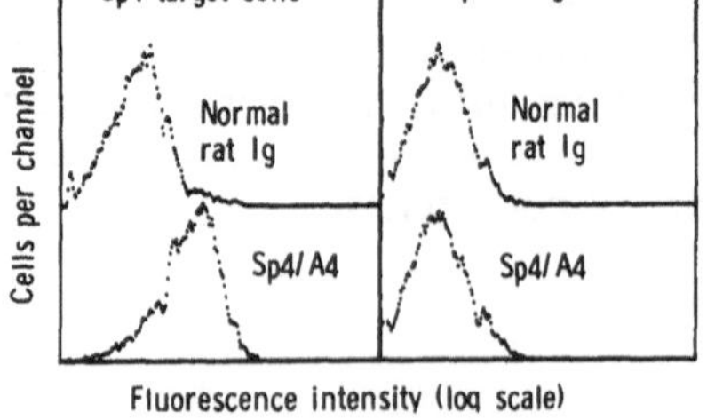

Figure 1. FACS analysis of Sp4/A4 monoclonal antibody binding to mammary carcinoma Sp4 and Sp15 target cells. Tumor cells were incubated with Sp4/A4 monoclonal antibody or normal rat Ig, in controls washe and bound Ig detected by reaction with fluorescein isothiocyanate-labelled sheep F(ab')$_2$ antirat IgG.

TABLE I

Reactions of monoclonal antibody Sp4/A4 to spontaneous rat mammary carcinoma Sp4 against various rat target cells

Target cells		Cell binding of antibody	
		Radio-immunoassay	FACS Assay
Mammary carcinoma	Sp4	5.61[a]	7.53[b]
	Sp15	1.26	0.98
	Sp22	1.21	1.11
Hepatoma	D23	1.32	NT
	D192A	1.28	0.83
Sarcoma	Mc7	1.07	1.08
	Mc57	1.08	NT
	Mc97A	1.21	NT
	Mc98	1.48	NT
	Mc100	1.33	NT
Normal cells	Spleen	0.66	NT
	Lymph node	0.96	NT
	Erythrocytes	0.77	NT
	Peritoneal exudate	0.85	NT
	Areolar tissue	1.39	NT
	Liver	0.96	NT
	Embryo	1.07	NT
	Normal breast	NT	0.90

[a]Binding ratio = mean cpm monoclonal antibody ÷ mean cpm with P3NS1 spent medium

[b]Fluorescence activated cell sorter (FACS) analysis

$$\text{Binding ratio} = \frac{\text{mean fluorescence/cell: cells treated with Sp4/A4}}{\text{mean fluorescence/cell: cells treated with P3NS1 spent medium}}$$

2.2. Anti-tumor activities of Sp4/A4 monoclonal antibody

The capacity of Sp4/A4 monoclonal antibody to modulate growth of tumor Sp4 was examined in a number of ways (Table II). In one series of experiments Sp4/A4 antibody (hybridoma supernatant or affinity column purified preparations) were incubated with Sp4 tumor cells and the mixture injected into normal WAB/Not rats. This treatment had no influence on tumor growth, as assessed by tumor incidence or mean tumor diameter when expriments were terminated, in comparison with control rats receiving tumor cells treated with medium or P3NS1 supernatants. Similarly repeated injection of Sp4/A4 antibody into rats challenged subcutaneously with tumor cells did not modulate tumor growth.

Consistent with the lack of response against tumors growing in challenged rats, Sp4/A4 antibody did not elicit complement- or lymphocyte-mediated reactions against Sp4 tumor cells. Table III summarizes comparative tests on the lytic response of Sp4 tumor cells treated either with Sp4/A4 monoclonal antibody or an alloantiserum (KX anti-WAB; $RT1^{y}$ anti-$RT1^{e}$), using either guinea pig or rabbit serum as a source of complement. The alloantiserum elicited significant cytotoxicity as assayed by release of radioactivity from ^{51}Cr-labelled target cells. In comparison Sp4/A4 culture supernatant or purified antibody preparations were devoid of cytotoxic activity. Table IV records similar studies comparing the ability of Sp4/A4 antibody and $RT1^{y}$ anti$RT1^{e}$ alloantiserum to mediate antibody dependent cell mediated (ADCC) lysis. Alloantibody, even at high dilution, sensitized Sp4 tumor cells to ADCC mediated lysis by KX-rat spleen cells, resulting in significant and titratable cytotoxicity whereas Sp4/A4 antibody showed no evidence of significant tumor cells lysis.

2.3. In vivo localization of Sp4/A4 antibodies

One of the objectives in developing monoclonal antibodies to tumor associated antigens is to provide drug delivery systems for localizing anti-tumor agents in tumor deposits (Baldwin et al, 1981). For this purpose, it is necessary to show that Sp4/A4 monoclonal antibody does localize in tumors developing in vivo. In these studies, antibody from hybridoma culture supernatants was isolated and purified by binding to Sepharose 4B-linked goat anti-rat IgG followed by dissociation with elution with 3M NaSCN (Pimm et al, 1982). Antibody preparations, labelled with ^{125}I using Iodogen reagent (Fraker and Speck, 1978), were then injected into WAB/Not rats bearing subcutaneous grafts of tumor Sp4 and/or other other sygeneically transplanted rat tumors. Typical organ distribution studies, summarized in Fig. 2, show that two to three days after intravenous injection of ^{125}I-Sp4/A4 antibody there is preferential localization

TABLE II

Effect of Sp4/A4 Monoclonal antibody on subcutaneous growth of Sp4 tumor cells in syngeneic rats

Antibody[a] Treatment Protocol	Antibody Reagent	Tumor Incidence[b]	Mean Tumor Diameter ± SD (cm)	Statistical Significance[c]
Admixture with cells	Sp4/A4 supernatant	6/7	2.50 ± 0.55	NS
" "	P3NS1 supernatant	7/7	2.40 ± 0.28	-
" "	Purified Sp4/A4	6/6	2.30 ± 0.56	NS
" "	HBSS	5/6	2.35 ± 0.37	NS
	-	5/5	2.42 ± 0.42	-
Intraperitoneal injection	Sp4/A4 supernatant	7/7	2.80 ± 0.60	NS
" "	P3NS1 supernatant	7/7	2.77 ± 0.39	-

[a] Details of treatment and the reagents used are shown in Materials and Methods

[b] Subcutaneous challenge with 5×10^3 SP4 cells

[c] Data as analyzed for statistical significance using the Wilcoxon rank test for unpaired data. NS: not significant

TABLE III

Complement-dependent cytotoxicity of

Sp4/A4 monoclonal antibody against Sp4 tumor cells

ANTIBODY DILUTION[a]		%CYTOTOXICITY MEDIATED BY[b]	
		RABBIT C'[c]	GUINEA PIG C'[c]
Sp4/A4 supernatant	1	4.1	0.6
	1/10	4.3	3.1
	1/100	5.5	3.5
Purified Sp4/A4	1	6.0	7.0
	1/10	6.4	7.4
	1/100	10.9	7.0
P3NS1 supernatant	1	4.0	3.2
	1/10	4.2	3.5
	1/100	3.8	3.8
KX anti-WAB serum	1/10	28.4	16.0

[a]Sp4/A4 supernatant contained, on average, approximately 30 μg/ml antibody protein. Purified Sp4/A4 was used at 35.5 μg/ml. KX anti-WAB was an $RT1^y$ anti-$RT1^e$ antiserum.

[b]% cytotoxicity = % release of ^{51}Cr, corrected for spontaneous release in medium controls by subtraction.

[c]Rabbit serum and purified guinea pig complement were added at a final dilution of 1/20.

TABLE IV

Antibody-dependent cell-mediated cytotoxicity of Sp4/A4 monoclonal antibody against Sp4 tumor cells

ANTIBODY[a] DILUTION	EFFECTOR[b]: TUMOR CELL RATIO	% CYTOTOXICITY[c] AGAINST CELLS TREATED WITH:	
		Sp4/A4	KX anti-WAB Serum
1/10	200:1	7.0	33.0
	100:1	9.0	18.5
	50:1	3.5	11.5
1/100	200:1	6.5	27.0
	100:1	7.0	16.0
	50:1	1.0	10.0
1/1000	200:1	-2.0	19.0
	100:1	-1.5	9.5
	50:1	0.5	6.5

[a]Sp4/A4 culture supernatant or KX WAB (RTL^y anti- RTL^e) antiserum

[b]KX rat spleen cells

[c]$\% \text{ cytotoxicity} = 100 \times \frac{\%\text{release with antibody} - \%\text{ release medium control}}{100 - \%\text{ release in medium control}}$

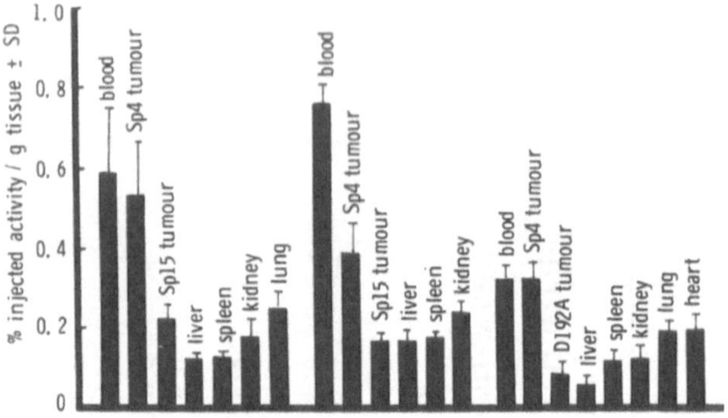

Figure 2. In vivo distribution of ^{125}I labelled Sp4/A4 monoclonal antibody three days after intravenous injection into rats with growths of Sp4 and a second tumor at contralateral subcutaneous sites. Three to four rats per group.

of radioactivity in Sp4 tumors compared to other organs. Moreover, as shown in Fig. 2, ^{125}I-Sp4/A4 antibody localized in Sp4 tumor but not other tumors, including an immunologically unrelated mammary carcinoma Sp15, and a rat hepatocellular carcinoma D192A.

The preferential localization of Sp4/A4 antibody in subcutaneous deposits of Sp4 was also demonstrated by γ-camera scintigraphy of rats which had received ^{131}I labelled antibody. After acquisition of the ^{131}I image, rats were injected intravenously with ^{113m}In chloride, to label the blood pool, a second image acquired, and a computerised subtraction used to visualise the tumor localization of ^{131}I (Fig. 3).

2.4. Anti-tumor activity of Sp4/A4 antibody-adriamycin conjugates

In vivo organ distribution studies of Sp4/A4 antibody indicated its preferential localization in tumor compared to normal organs in tumor-bearing rats (Figs 2,3). Developing from these experiments, the potential of Sp4/A4 antibody as a carrier for anti-tumor agents has been evaluated in the treatment of Sp4 tumor-bearing rats. Adriamycin was selected for these studies since the tumor is sensitive to this compound when tested by suppression of subcutaneous tumor growth (Pimm et al, 1982) or by inhibition of cellular DNA synthesis in cultured tumor cells.

Conjugation of adriamycin (Doxorubicin) to Sp4/A4 antibody or normal rat IgG was effected through a dextran bridge using essentially the procedure develped by Hurwitz and colleagues (1975). Briefly, T-

40 dextran (Sigma) of mean molecular weight 40K was oxidized (25%) to polyaldehyde (PAD) with sodium periodate and this product incubated with adriamycin (Montedison Pharmaceuticals Ltd.) at a molar ratio of 5:1. Protein was then added to the reaction mixture at an equimolar ratio to the PAD and after further incubation, the Schiff bases thus formed reduced with sodium borohydride. Conjugates were separated on Sephadex G25 columns to remove free adriamycin. This conjugation procedure

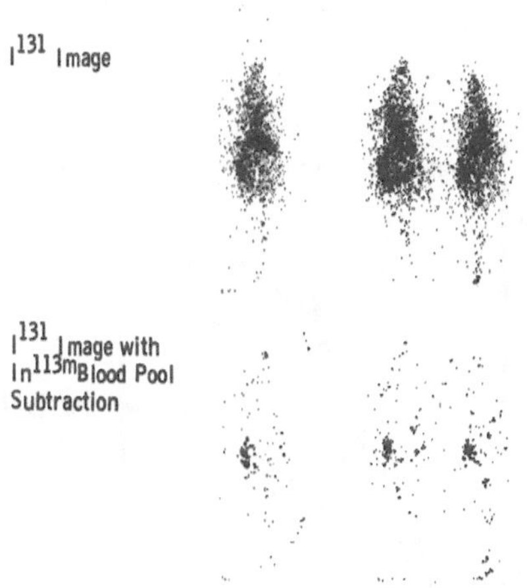

Figure 3. Gamma camera scintigraphy of tumor localization of ^{131}I-Sp4/A4 monoclonal antibody in rats with subcutaneous Sp4 growths. Rats were imaged two days after intravenous injection of ^{131}I-Sp4/A4 preparation (50 Ci ^{131}I/rat). A second image was acquired after intravenous injection of ^{113m}In chloride (50 Ci/rat) to label the blood pool, and computerized subtraction applied to show tumor localization of ^{131}I.

yielded products with adriamycin: protein molar ratios up to 28:1 (Pimm et al, 1982).

In order to establish that Sp4/A4 antibody-adriamycin conjugates retained anti-tumor activity, they were incubated (30 minutes at 37°C) with Sp4 tumor cells. After extensive washing, tumor cells were injected into normal rats and tumor growth profiles compared with Sp4 untreated or treated with free adriamycin or adriamycin conjugated to normal rat

IgG. As illustrated in Fig. 4, Sp4 cells incubated with adriamycin or adriamycin-normal IgG conjugates produced tumors developing at rates almost identical to those in untreated controls. In comparison, the adriamycin-Sp4/A4 antibody conjugate significantly suppressed tumor growth.

Systemic treatment of Sp4 tumor-bearing rats with adriamycin-Sp4/A4 antibody conjugates also produced a therapeutic response. This is illustrated in Fig. 5 which shows survival data of Sp4 tumor-injected rats treated with adriamycin or protein-adriamycin conjugates between days 7 and 28 following tumor implantation. Treatment with adriamycin

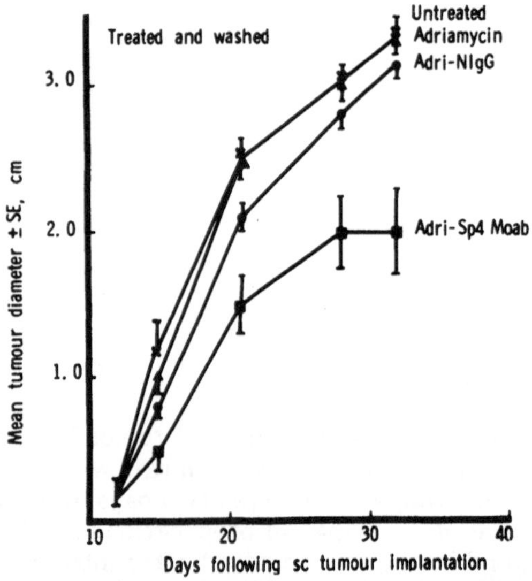

Figure 4. Tumor growth from subcutaneous inocula of 2×10^4 Sp4 cells following in vitro treatment with free adriamycin or conjugates. Cells were incubated at $37^{o}C$ for 30 minutes with adriamycin or adriamycin-immunoglobulin conjugates at 4 μg adriamycin/10^6 cells and washed three times in Hank's BSS before injection. ADR-normal IgG= adriamycin conjugated to normal rat IgG. ADR-Sp4MoAb =adriamycin conjugated to Sp4/A4 monoclonal antibody.

or adriamycin-normal IgG conjugates did not significantly improve survival, rats rapidly dying within 30-60 days following tumor implantation. Treatment with adriamycin-Sp4/A4 antibody conjugates significantly improved survival (Fig. 5). For example, at day 40 all of the rats receiving antibody conjugates were alive whereas there were no survivals in the untreated control group of rats. Moreover, this therapeutic response was substantially greater than that achieved with free adriamycin or adriamycin-normal

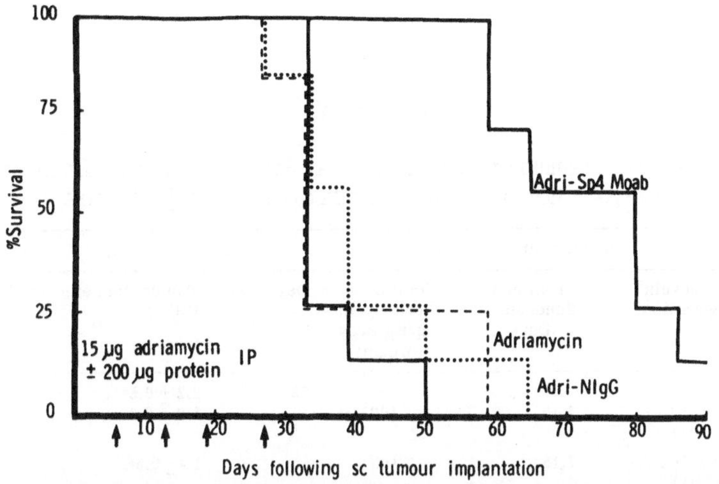

Figure 5. Survival of rats with subcutaneous Sp4 growth following repeated treatment with adriamycin or adriamycin-immunoglobulin conjugates. Each injection contained 15 g adriamycin either free, or conjugated to normal IgG (ADR-normal IgG) or Sp4/A4 monoclonal antibody (ADR-Sp4MoAb). Rats were killed when tumors reached 3 cm mean diameter. Seven rats per group.

IgG conjugates where the survivals were 25% and 10% respectively. Table V summarizes a series of experiments where the tumor-inhibitory activities of adriamycin-Sp4/A4 antibody conjugates were compared with free adriamycin and/or adriamycin-normal IgG conjugates. For comparative purposes in separate experiments tumor inhibition has been determined as the mean tumor diameter at the time after tumor implantation when control animals had to be killed. In all 5 experiments treatment with adriamycin-Sp4/A4 antibody conjugates produced a significant inhibition of tumor growth, T/c values ranging from 0.44 to 0.64, whereas treatment with free adriamycin at doses comparable to those in the conjugates was ineffective. The response to adriamycin-Sp4/A4 conjugates could not be ascribed to its retention nor uptake when attached to protein since conjugates prepared with normal rat IgG were ineffective.

3. CONCLUSIONS

These investigations with Sp4/A4 monoclonal antibody reacting with tumor associated antigen(s) on rat mammary carcinoma Sp4 clearly indicate that radioisotope labelled preparations can be used to identify tumor deposits using γ-scintigraphy when blood pool subtraction techniques are employed. Monoclonal antibodies reacting with a human osteogenic sarcoma (791T) have been used also to identify tumor developing as xenografts in immunodeprived mice (Pimm et al, 1981). These two examples point to the potential of monoclonal antibodies for developing new techniques for tumor localization.

TABLE V

Therapy of mammary carcinoma Sp4 with adriamycin Sp4/A4 monoclonal antibody conjugates

Expt.	Treatment			Tumor Growth		
	Adriamycin[a] Preparation	Treatment[b] Schedule Days	Total Dose Adriamycin μg/Kg Body Weight[c]	Day	Mean Tumor Diameter (cm)[d] ±SD	T/c[e]
1	None		-	28	2.2 ± 0.5	-
	Adria-NIgG (26:1)	7,15	100		1.9 ± 0.3	0.86
	Adria-Sp4/A4 (27:1)	7,15	100		1.4 ± 0.5*	0.64
2	None	-	-	38	2.6 ± 0.8	-
	Adriamycin	21,29	60		2.3 ± 0.5	0.88
	Adria-Sp4/A4 (28:1)	"	60		1.4 ± 0.8*	0.54
3	None	-	-	49	2.3 ± 0.7	-
	Adriamycin	8,15,22,29	200		1.7 ± 0.7	0.74
	Adria-NIgG (26:1)	"	200		1.9 ± 1.0	0.83
	Adria-Sp4/A4 (26:1)	"	200		1.0 ± 0.8**	0.44
4	None	-	-	33	3.0 ± 0.5	-
	Adriamycin	6,13,19,27	300		2.8 ± 0.5	0.93
	Adria-NIgG (27:1)	"	300		2.8 ± 0.5	0.93
	Adria MoAb	"	300		1.8 ± 0.6***	0.60
5	None	-	-	28	2.9 ± 0.3	-
	Adriamycin	7,15,21,28	300		2.9 ± 0.7	1.00
	Adria-NIgG (17:1)	"	300		2.6 ± 0.4	0.89
	Adria-Sp4/A4 (18:1)	"	300		2.1 ± 0.4*	0.63

[a] Figures in parentheses indicate drug:protein molar ratios

[b] Adriamycin or protein conjugates administered intraperitoneally.

[c] Total dose of adriamycin either in free form or as conjugates.

[d] Data analysed for statistical significance using the Wilcoxon rank test for unpaired data.

* $P < 0.025$

** $P < 0.01$

*** $P < 0.005$

[e] $T/c: \frac{\text{mean tumor diameter in treated rats}}{\text{mean tumor diameter in controls}}$

Monoclonal antibodies reacting exclusively, or preferentially, with tumor associated antigens also provide new approaches to therapy and the potential of this approach is emphasized by the therapeutic tests with Sp4/A4 monoclonal antibody-adriamycin conjugates (Pimm et al, 1982). This approach to tumor therapy is still very much in its infancy and there are many problems to be elucidated. The initial concept suggested that antibody binding to cell surface antigens leads to its internalization thus providing a mode of entry of cytotoxic agents into the internal milieu of the cell. This may occur, for example, by a process such as pinocytosis leading to internalization of antibody-drug conjugates. In many cases, however, drugs may be released at the cell surface following antibody binding or the antibody may even bind to extracellular tumor antigen in the milieu of the tumor before being released. In either case, the end result may be a localization of drug in tumor to allow an improved therapeutic response, or even a reduction in associated non-specific toxic effects at other sites.

4. MATERIALS AND METHODS

4.1. Tumors

Mammary carcinomas Sp4 and Sp15 arose naturally in breeding female rats of the syngeneic WAB/Not strain (Baldwin and Embleton, 1969). Hepatoma D192A was induced by oral feeding 4-dimethylaminoazobenzene. These tumors were maintained by subcutaneous transplantation into WAB/Not rats.

4.2. Mammary carcinoma Sp4 monoclonal antibody (Sp4/A4)

Sp4/A4 monoclonal antibody was derived from a hybridoma obtained following fusion of Sp4-immune rat spleen cells with the P3NS1 mouse myeloma (Gunn et al, 1980). Antibody was obtained either as culture supernatants or following purification by binding and elution from goat antirat IgG-Sepharose 4B (Pimm et al, 1982). Hybridoma supernatants were assayed for antibody activity using an indirect radioisotopic antiglobulin test with tissue culture derived Sp4 tumor target cells and ^{125}I-labelled sheep $F(ab')_2$ anti-rat IgG reagent (Gunn et al, 1980). Analysis of antibody binding to tumor cells was also determined in a fluorescence activated cell sorter (FACS IV) using fluorescein isothiocyanate labelled sheep $F(ab')_2$ anti-rat IgG as the anti-globulin agent.

4.3. Treatment of Sp4 with Sp4/A4 monoclonal antibody

In some tests, Sp4 cells were incubated for 20 minutes at 37 $^{\circ}C$ with Sp4/A4 hybridoma supernatant or affinity purified antibody at a concentration of 2.5×10^4 cells per ml. Syngeneic rats were then inoculated subcutaneously with 0.2 ml aliquots of the antibody, containing

5 x 10^3 Sp4 cells. Control rats were treated with 5 x 10^3 cells in 0.2 ml of P3NS1 myeloma supernatant or phosphate-buffered saline, similarly incubated.

As an alternative approach, rats were given 0.2 ml of Sp4/A4 hybridoma supernatant or P3NS1 supernatant by intraperitoneal injection at the following times with respect to a subcutaneous challenge with 5 x 10^3 Sp4 cells: days -3, -1, 0, 1 and 3 (day 0 being the day of challenge). The growth of the subcutaneous Sp4 inocula was monitored weekly, and compared with growth of 5 x 10^3 Sp4 cells in untreated rats.

4.4. Complement-dependent cytotoxicity test

Sp4 cells (10^6) were cultured overnight as a monolayer in 4 ml of Eagle's minimum essential medium containing 10% newborn calf serum and 15 µCi of ^{51}Cr sodium chromate. The monolayer was washed 5 times with Hank's balanced salt solution (HBSS) containing 0.1% bovine serum albumin and the cells were harvested with a mixture of trypsin (0.25%) and EDTA (0.02%) centrifuged and resuspended in RPMI 1640 + 10% newborn calf serum, and given one further wash.

A round-bottom microtiter plate (Sterilin M24A) was prepared with dilutions of Sp4/A4 supernatant (neat, 1/10 and 1/100) or a KX/Not anti-WAB/Not ($RT1^y$ anti-$RT1^e$) antiserum at 1/10 dilution in RPMI 1640, using 100 µl per well. 10^4 ^{51}Cr-labelled Sp4 cells were added in 50 µl of medium, followed by incubation for 20 minutes at 37°C. Complement was then added, in the form of 100 µl of 1/20 diluted rabbit serum previously absorbed with rat platelets, red cells and thymocytes, or 100 µl of 1/20 diluted guinea pig complement (Burroughs Wellcome, Ltd.).

The cultures were incubated for 6 hours at 37°C, then thoroughly mixed and centrifuged at 2000 xg. Supernatant samples (100 µl per well) were taken and placed in marked plastic tubes and the cell pellets and remaining supernatants were dried down. The plate was sealed with a plastic film spray and cut into separate wells. Wells and tubes of supernatant were counted separately in a gamma counter and % ^{51}Cr release was calculated by: $\frac{A \times 2.5 \times 100}{A + B}$, where A = counts per minute (cpm) in 100 µl supernatant and B = cpm in well. % cytotoxicity was expressed as % ^{51}Cr release in antibody and complement treated wells, minus % spontaneous release in medium controls. No increased ^{51}Cr release was stimulated by the addition of complement alone.

4.5. Antibody-dependent cell-mediated cytotoxicity

^{51}Cr-labelled Sp4 cells were incubated for 30 minutes on ice with Sp4/A4 monoclonal antibody at dilutions of 1/10, 1/100 and 1/1000, using 250 µl of cell suspension containing 2.5 x 10^5 cells and 100 µl of antibody. The mixture was made up to 5 ml with RPMI 1640 + 10% newborn calf serum and 100 μl aliquots of cells (5×10^3 per well) were plated in flat-bottom Microtiter plates (M29 ART). Spleen cells were prepared from KX/Not rats and these were added in 200 μl to give final overall ratios of 50, 100 or 200 effector cells per target cell. The cells were incubated 6 hours at 37 °C, then 100 μl aliquots of supernatant were removed from gamma-counting.

Negative controls were provided by normal rat serum, culture medium and P3NS1 supernatant, and a positive control was provided by KX/Not anti WAB/Not antiserum. P3NS1 supernatant and normal rat serum induced no cytotoxicity. % ^{51}Cr release was calculated by:

$\frac{A \times 3}{C} \times 100$, where A = cpm in 100 μl supernatant, and C = originally added cpm. % cytotoxicity was expressed:

$$100 \times \frac{\text{\% release with antibody - \%release in medium control}}{\text{100 - \% release in medium control}}$$

4.6. Radiolabelling of Sp4/A4 monoclonal antibody for in vivo localization

Anti-Sp4 antibody purified from hybridoma supernates was labelled with ^{125}I or ^{131}I using the Iodogen reagent (Pimm et al, 1982). Rats were injected intravenously through a lateral tail vein and organ distribution studies carried out with ^{125}I-labelled antibody by analysis of isolated organs. The levels of radioactivity are then expressed as % injected activity/g tissue.

Gamma scintigraphy was carried out 48 hours after injection of ^{131}I-labelled Sp4/A4 monoclonal antibody (50 microcuries/rat). Images were acquired using a gamma camera having a 40 cm diameter field of view with a parallel hole (400 KeV maximum) collimator. Views were stored and processed by computer in a matrix of 64 x 64 cells. To prevent movement artefacts rats were anaesthetized with ether prior to imaging. Image enhancement was carried out by computerized subtraction of a second image acquired after injection of a blood pool label. ^{113m}In chloride (50 microcuries/rat) was used to label the blood since its 393 KeV emission provides a blood background image with scatter information more representative of the data in images acquired with the 364 KeV emission of ^{131}I.

4.7. Anti-tumor activity of Sp4/A4 monoclonal antibody conjugated to adriamycin

Adriamycin (Doxorubicin) conjugated to Sp4/A4 monoclonal antibody or normal rat IgG through a dextran bridge (Pimm et al, 1982) was tested for anti-tumor activity by determining the inhibition of subcutaneous growth of tumor Sp4 cells. In these tests, Sp4 cells were incubated for 30 minutes at 37 °C with conjugates. Then after washing tumor

Sp4 cells were injected into normal WAB/Not rats and growth curves assessed over a period of 30 days. Conjugates were also evaluated for their capacity to inhibit growth of a developing tumor. In these tests conjugates were injected intraperitoneally into rats at weekly intervals commencing 7 to 10 days after subcutaneous challenge with Sp4 cells.

Acknowledgements

This work was supported by the Cancer Research Campaign.

REFERENCES

Baldwin, R.W. and Byers, V.S.. 1982. Prospects for Immunotherapy of Cancer. In: Growth Arrest and Cancer Treatment, Ed. B. Stoll, John Wiley, London.

Baldwin, R.W. and Embleton, M.J.. 1969. Immunology of spontaneously arising rat mammary adenocarcinomas. Int. J. Cancer, 4, 430-439.

Baldwin, R.W. and Embleton, M.J.. 1970. Detection and isolation of tumour specific antigens associated with a spontaneously arising rat mammary carcinoma. Int. J. Cancer, 6, 373-382.

Baldwin, R.W., Embleton, M.J. and Price, M.R.. 1981. Monoclonal antibodies specifying tumour-associated antigens and their potential for therapy, In: Molecular Aspects in Medicine, 4, 329-368, ed. H. Baum, Pergamon Press.

Fraker, P.J. and Speck, J.C.. 1978. Protein and cell membrane iodinations with a sparingly soluble chloromide, 1,3, 4,6-tetrachloro-3α,α-diphenylglycoluril. Biochem. Biophys. Res. Comm., 80, 849-857.

Gunn, B., Embleton, M.J., Middle, J.G. and Baldwin, R.W.. 1980. Monoclonal antibody against a naturally occurring rat mammary carcinoma. Int. J. Cancer, 26, 325-330.

Pimm, M.V., Dawood, F.A., Price, M.R., Perkins, A.C., and Baldwin, R.W.. 1981. Localisation of human osteogenic sarcoma xenografts with radiolabelled monoclonal antibody. Br. J. Cancer, 45(4):637.

Pimm, M.V., Jones, J.A., Price, M.R., Middle, J.G., Embleton, M.J. and Baldwin, R.W.. 1982. Tumour localization of monoclonal antibody against a rat mammary carcinoma and suppression of tumour growth with adriamycin-antibody conjugates. Cancer Immunol. Immunother. 12:125-134.

13

MONOCLONAL ANTIBODY TO TUMOR SPECIFIC EPITOPE OF MURINE LYMPHOMA CELLS: USE IN CHARACTERIZATION OF ANTIGEN AND IN IMMUNOTHERAPY

James P. Allison, Bradley W. McIntyre, James Irvin* David Bloch**, and Garrie B. Kitto**

The University of Texas System Cancer Center Science Park - Research Division, Smithville, Texas 78957 *Southwest Texas State University, San Marcos, Texas **The University of Texas at Austin, Austin, Texas

1. SUMMARY

A monoclonal antibody was produced that recognizes a cell surface antigen of an X-ray induced lymphoma, C6XL. The antibody is highly specific for C6XL cells and does not react with normal syngeneic lymphoid cells or with any of a panel of other murine lymphomas. The antigen reactive with MAb 124-40 was isolated and found to be a glycoprotein composed of disulfide-linked subunits of 39Kd and 41Kd. A molecule with similar structure, but not reactive with MAb 124-40, could be detected in extracts of thymocytes and T lymphocytes, but not B cells. These results suggest that the epitope reactive with MAb 124-40 might be a clonally expressed epitope carried by a T cell specific surface component.

Studies were also performed to determine whether the antigen reactive with MAb 124-40 might serve as a target for specific immunotherapy. Administration of antibody to lymphoma-bearing animals produced only a slight increase in survival time. However, administration of MAb 124-40 conjugated to pokeweed antiviral protein (PAP)[1] produced significant prolongation of survival.

---

[1] Abbreviations: PAP, Pokeweed antiviral protein; MULV, murine leukemia virus; TL, thymus leukemia; BA, binding assay; Mab, monoclonal antibody; SPDP, N-succinimidyl 3-(2-pyridyldithio) propionate; FMF, flow microfluorometry, RIP, radioimmunoprecipitation; SDS-PAGE, sodium dodecyl-polyacrylamide gel electrophoresis

These results demonstrate that T cell lymphomas may express tumor-specific antigens, and that these may be exploited in specific immunotherapy using immunotoxins containing pokeweed antiviral protein.

2. INTRODUCTION

One of the goals of tumor immunology has been the identification and characterization of cell surface antigens that might serve to distinguish tumor cells from normal tissues. Both theoretical and clinical considerations provide impetus for this work, since knowledge of the structure and genetic origin of tumor-associated antigens will not only contribute to our understanding of the mechanism of neoplastic transformation but will also aid in the development of immunological approaches to the diagnosis, prognosis, and therapy of cancer. The advent of the hybridoma technique for the production of monoclonal antibodies (Kohler and Milstein, 1975) has provided incisive probes with which to dissect the tumor cell surface mosaic. This report concerns the application of this approach to the study of surface antigen expression by a murine T cell lymphoma.

Serological studies employing conventional antisera have revealed that murine lymphoma cells often express antigens not found on normal lymphoid cells. These include antigens related to endogenous murine leukemia virus (MuLV) genomes (Old and Stockert, 1977, O'Donnell et al, 1980), and in some strains, products of the TL locus expressed as the result of derepression of normally silent genes (Old et al, 1963, Old and Stokert, 1977). Lymphoma cells do not appear to express antigens of the extreme individual specificity characteristic of the tumor-specific transplantation antigens of chemically-induced tumors (Price and Baldwin, 1975). On the other hand, lymphoid tumors appear to arise by clonal expansion from a single neoplastic cell (Nowell, 1976, Canaani and Aaronson. 1979) and may express antigens that are shared by all tumor cells, but that are undetectable in the normal animal. For example, the clonally expressed idiotypic markers of the surface immunoglobulin of B cell tumors are, in effect, tumor specific antigens and have been successfully used as specific targets for immunotherapy in both animal models and in human cancer (Beatty et al 1976, Haughton et al 1978, Krolick et al 1979, and Miller et al 1982). To date, however, no individually specific markers have been demonstrated on T cell tumors.

The T cell lymphoma C6XL, induced by X-irradiation of C57BL/Ka mice, expresses 4 cell surface glycoproteins related to MuLV; 2 *gag*-related glycoproteins, gp95 and gp110, and 2 *env*-related glycoproteins, gp90 and gp105 (Allison and Kendall 1980, McIntyre and Allison submitted). Sera from syngeneic mice which had been immunized with mitomycin-C inactivated C6XL cells invariably contained antibodies which reacted with the *env*-related gp105. No evidence was found for reactivity of the syngeneic antisera with any othe tumor cell surface antigens. This report describes the production and characterization of a monoclonal antibody that is highly specific for C6XL T lymphoma cells.

The antigen defined by this antibody is not structurally or immunologically related to MuLV, and may represent a clonally-expressed T cell surface component. Finally, the antigen can serve as a target for specific destruction of the tumor cells by conjugates of monoclonal antibody and pokeweed antiviral protein.

3. RESULTS AND DISCUSSION

3.1. Production of monoclonal antibody

Splenocytes from female BALB/c mice immunized by i.p. injection of viable C6XL lymphoma cells were fused with P3X63Ag8 myeloma cells and distributed among 960 microwells. Hybridoma colonies appeared in 785 of the wells and preliminary screening using the binding assay revealed that 43 of these wells contained antibody reactive with C6XL cells. These positive cultures were then screened for reactivity with splenocytes and thymocytes from normal C57BL mice, which allowed identification of two hybridomas capable of discriminating C6XL lymphoma cells from normal cells. One of these, designated 124-40, was cloned three times by limiting dilution, and once in soft agar. The antibody produced by this hybridoma was IgG1, did not directly bind to *Staphylococcus aureus* protein A, and was not active in complement-mediated cytotoxicity.

3.2. Reactivity of MAb 124-40 with C6XL lymphoma cells

Three independent assays were used to demonstrate the reactivity of MAb 124-40 with C6XL. The titration of 124-40 ascitic fluid in a binding assay is shown in Fig. 1. Titers in excess of 1:100,000 and specificity ratios of greater than 5:1 were regularly obtained. Flow microfluorometry was also used to assess the binding of MAb 124-40 to C6XL cells. Fig. 2 shows the fluorescence histograms of spleen cells and thymus cells from C6XL-bearing mice. Greater than 85% of the spleen cells from tumor-bearing mice were stained with MAb 124-40, while only 3% were stained by control reagents. In other experiments (not shown), it was established that the MAb 124-40 positive cells were larger than normal splenocytes and, like other lymphoma cells, expressed cell surface antigens related to the MuLV *gag* protein p30. The spleen cells staining with the control reagents were p30-negative and probably represent residual nonmalignant B lymphocytes.

In order to determine whether the antigen reactive with MAb 124-40 could be detected in extracts of C6XL, radioiodinated cells were solubilized with NP-40 and then immunoprecipitated. As shown in Fig. 3,

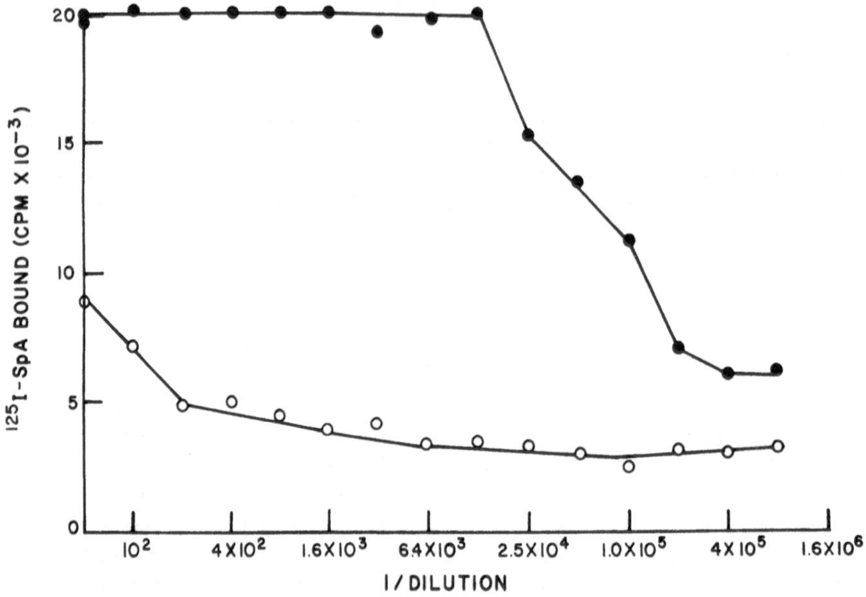

Figure 1. Reactivity of MAb 124-40 with C6XL lymphoma cells. Viable cells were incubated with, successively, dilutions of 124-40 (●-●) or P3 (o-o) culture supernatants, RAMIG, and ^{125}I-labeled protein A, with intervening washes. Each point is the mean of triplicate assays.

no radiolabeled antigen was precipitated by P3 IgG. MAb 124-40 precipitated material of approximately 40Kd under reducing conditions, and approximately 75Kd under non-reducing conditions.

3.3. Lack of reactivity of MAb 124-40 with other normal or malignant lymphoid cells

While the initial screening assays indicated that MAb 124-40 was not reactive with normal lymphocytes, it was possible that an alloantigen expressed at very low levels or by a small subpopulation of cells might have escaped detection. This does not appear to be the case, however, since an extensive analysis using more sensitive assay methods failed to detect Ag 124-40 on normal lymphoid cells. As shown in Table I, there was no significant binding of MAb 124-40 to normal adult or fetal lymphoid cells in a quantitative binding assay. In quantitative absorption experiments (not shown), reactivity of MAb 124-40 supernatants with C6XL cells was removed by as few as 2.5×10^6 C6XL cells, but absorption with as many as 4×10^7 normal splenocytes or thymocytes had no effect. An examination of normal lymphocytes by flow microfluorometry (Fig. 2, Table I), which is capable of detecting antigen on a very small fraction of the cells analyzed (<5%), failed to reveal any subpopula-

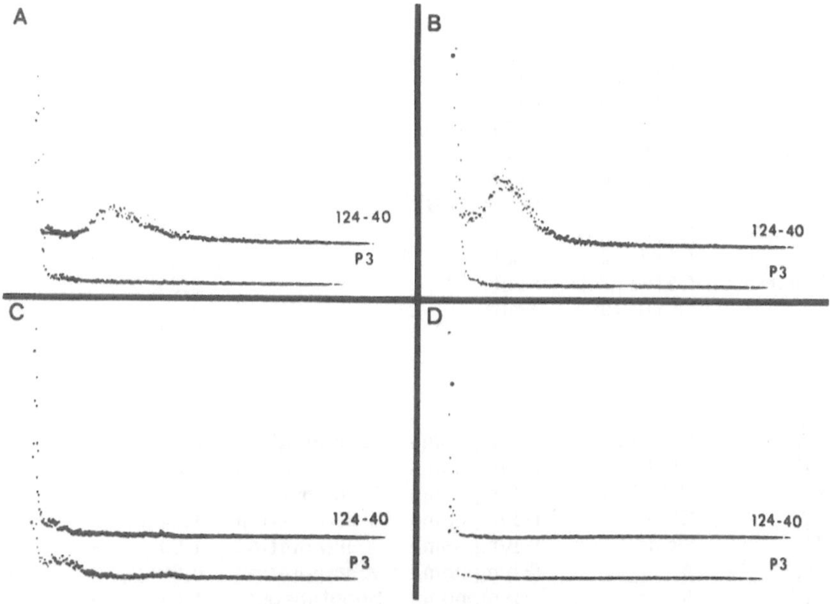

Figure 2. Flourescence histograms of lymphoid cells stained with 124-40. (A) Spleen cells from C6XL bearing mouse, (B) thymus cells from C6XL bearing mouse, (C) Normal C57B1/6 splenocytes, (D) normal C57B1/6 thymocytes were incubated with MAb 124-40 or P3 IgG1 followed by FITG-GAMIG and analyzed by flow microfluorometry.

tion reactive with MAb 124-40. Furthermore, no radioactive material was precipitated by MAb 124-40 from detergent-solubilized cells (Table I), indicating that failure to detect the antigen on normal lymphocytes by the other methods was not the result of masking. While it is difficult to exclude the possibility that Ag 124-40 is expressed at a very low level by a small subpopulation of normal lymphocytes, these data suggest that the antigen is, within the limits of these assays, tumor-associated.

TABLE I

Reactivity of MAb 124-40 with normal and malignant lymphoid cells

Cells	Strain	Cell Type	Origin	Assay Method		
				BA[a]	FMF[b]	RIP[c]
Normal						
Thymus	C57BL/KA	Adult		0.91	-	-
	C57BL/6	Adult		0.84	-	-
	C57BL/6	Fetus (18D)		0.92	-	-
Spleen	C57BL/KA	Adult		1.03	-	-
	C57BL/6	Adult		1.12	-	-
	C57BL/6	Fetus (18D)		1.01	-	-
Lymph node	C57BL/KA	Adult		0.94	-	-
	C57BL/6	Adult		0.96	-	-
Bone marrow	C57BL/KA	Adult		0.94	-	-
	C57BL/6	Adult		0.93	-	-
Lymphoma						
C6XL[e]	C57BL/Ka	T-lymphoma	X-irradiation	9.30	+	+
ERLD[f]	C57BL/6	T-lymphoma	X-irradiation	1.03	-	-
EL-4[g]	C57BL/6	T-lymphoma	Benzopyrene	1.03	-	N.D.[d]
XL-1[h]	C57L	T-lymphoma	X-irradiation	1.04	-	-
NIXT[i]	Swiss	T-lymphoma	X-irradiation	1.09	-	-
RADA1[j]	A	T-lymphoma	X-irradiation	0.98	-	-
ASL-1[k]	A	T-lymphoma	Spontaneous	1.04	-	-
920[l]	AKR	T-lymphoma	Spontaneous	1.03	-	-
1210[l]	AKR	T-lymphoma	Spontaneous	1.13	-	-
225[l]	AKR	B-lymphoma	Spontaneous	1:08	-	-

[a]Binding assay Binding ratio: $\frac{\text{cpm, 124-40 supernant}}{\text{cmp, P3 supernatant}}$

[b]Flow microfluorometry
[c]Radioimmunoprecipitation
[d]N.D., Not determined
[e]Kaplan and Brown, 1952
[f]Old et al., 1963
[g]Old et al., 1965
[h]Allison and McIntyre, submitted
[i]Fischinger et al., 1981
[j]Obata et al., 1978
[k]Old and Stockert, 1977
[l]Howell et al., submitted

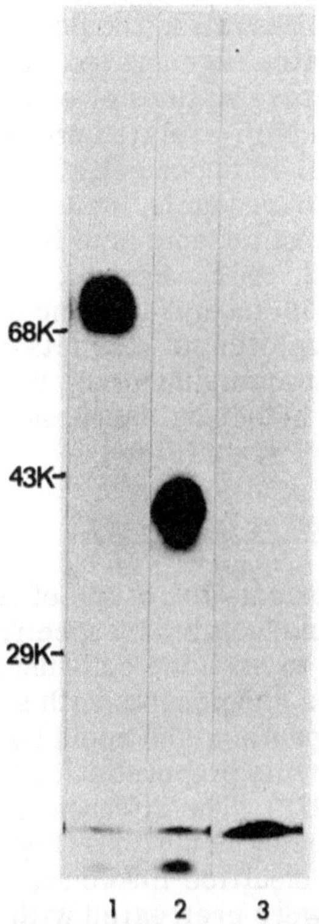

Figure 3. SDS-PAGE analysis of C6XL lymphoma antigen precipitated with MAb 124-40. NP-40 extracts of C6XL cells were precipitated with MAb 124-40 (lanes 1 and 2) or P3 IgG (lane 3) and analyzed on 10% polyacrylamide gels. Lane (1) nonreducing, lanes (2,3) reducing.

In order to determine whether the antigen is expressed by other lymphoid tumors, MAb 124-40 was tested for reactivity with a panel of spontaneous and X-ray or chemically induced lymphomas. These lymphomas were uniformly negative in quantitative binding assays, by indirect immunofluorescence (as analyzed by flow microfluorometry), and by radioimmunoprecipitation (Table I). This was an unexpected finding, since lymphoma cells are not generally considered to express individually specific tumor-associated antigens (Old and Stockert, 1977). Nonetheless, MAb 124-40 appears to be highly specific for an epitope expressed by C6XL, but not other lymphoma cells.

The observation that endogenous MuLV env genes undergo recombination (Elder et al, 1977) has raised the possibility that tumor specific antigens might be novel epitopes generated by recombinant MuLV env proteins (Lennox, 1980). However, several lines of evidence indicate that MAb 124-40 is not reactive with MuLV-related proteins. First, the antibody was unreactive with a panel of tumor cells (Table I) which collectively expressed a variety of MuLV products, including those related to gag genes as well as ecotropic, xenotropic, and recombinant env genes (O'Donnell et al, 1978, Fischinger et al, 1981). Second, coating of C6XL cells with excess antibody to MuLV p30 and gp70 had no effect on subsequent binding of MAb 124-40. Finally, depletion of extracts of radiolabeled C6XL cells of MuLV antigens by repeated immunoprecipitation with antibodies to MuLV p30 and gp70 had no effect on the subsequent precipitation of the antigen reactive with MAb 124-40 (Fig. 4).

3.4. Characterization of Ag 124-40

In order to determine the origin of Ag 124-40, we first sought to characterize the molecule isolated by specific immunoprecipitation from C6XL cells, and then to examine cells unreactive with MAb 124-40 for expression of surface components with similar structure. Ag 124-40 was found to be a glycoprotein, and could be absorbed by and specifically eluted from affinity absorbents prepared with Lens culinaris hemagglutinin (Hayman and Crumpton, 1972). The difference in mobility of the antigen under reducing and nonreducing conditions (Fig. 3) indicated that the molecule was composed of disulfide-linked subunits. Since identical results were obtained if the cells were pretreated with and extracted in the presence of 100 mM iodoacetamide, it is unlikely that the disulfide linkage was generated by oxidation of free sulfhydryl groups during extraction. Two-dimensional IEF:SDS-PAGE allowed the resolution of the molecule into two major components of 39Kd and 41 Kd, both of which exhibited microheterogeneity (Fig. 5). These data indicate that Ag 124-40 is displayed on the surface of C6XL lymphoma cells as a disulfide-bonded heteroduplex.

3.5. Expression of similar cell surface component by normal T cells and thymocytes

The presence of interchain disulfide bonds in the antigen was fortuitous, since proteins containing such bonds can be readily resolved from proteins not containing interchain disulfide bonds by a two-dimensional SDS-PAGE technique in which electrophoresis in the first dimension is carried out under non-reducing conditions, and in the second dimension under reducing conditions. Non-disulfide linked proteins will have the same mobility in both dimensions and will therefore lie on a diagonal. Disulfide-linked proteins migrate according to intact size in the first dimension, and according to subunit size in the second dimension, and therefore lie below the diagonal. As shown in Fig. 6A, the antigen isolated from C6XL cells by immunoprecipatation with MAb 124-40 lies at the expected position below the diagonal. As seen in Fig. 6B-6D, very few disulfide-linked molecules are expressed as surface components of lymphoid cells. On B lymphocytes, the major disulfide-bonded molecules recognizable in whole extracts of iodinated cells are the heavy and light chains of the surface immunoglobins (Fig. 6D). Thymocytes (Fig. 6B) displayed

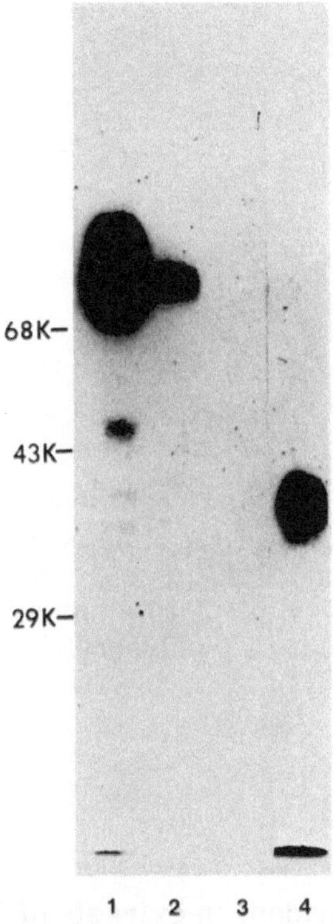

Figure 4. Immunodepletion analysis of C6XL antigens. NP-40 extract of ^{125}I-labeled C6XL cells was precipitated sequentially with anti-MuLB gp70 three times (lanes 1-3) and the supernatant precipitated with MAb 124-40 (lane 4). The precipitates were analyzed by SDS-PAGE under reducing conditions.

two major groups disulfide-bonded surface components, the Ly-2/3 complex and an additional diffuse component similar in mobility to AG 124-40. Interestingly, the major disulfide-bonded component of splenic T lymphocytes migrated to a position coincident with Ag 124-40 (Fig. 6C, at arrow

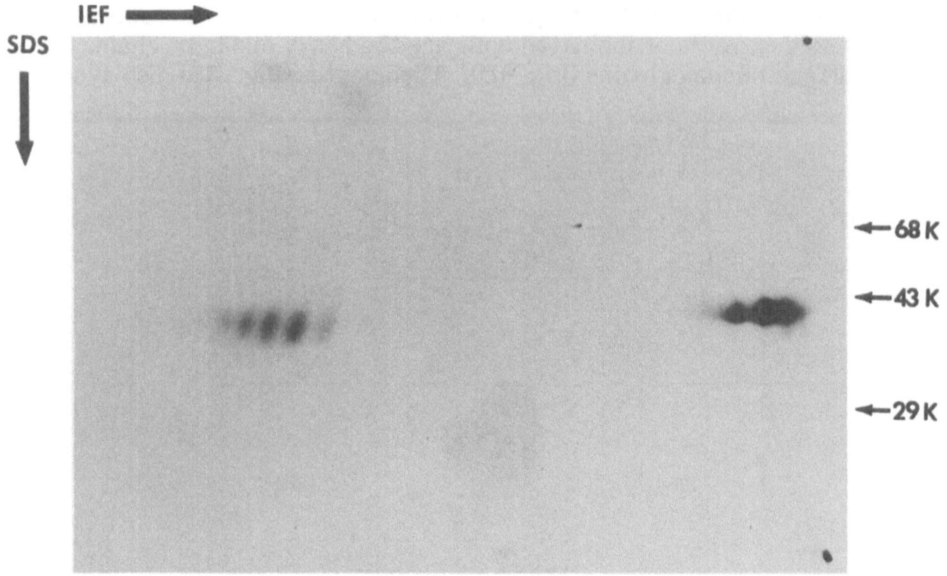

Figure 5. Two-Dimensional gel electrophoresis of C6XL antigen precipitated with MAb 124-40. The first dimension was isoelectric focused under reducing conditions in gels containing 1.6% pH 5.7 ampholytes and 0.4% pH 3-10 ampholytes. The second dimension was SDS-PAGE on a 10% polyacrylamide gel.

T). This component is also evident in extracts of T lymphomas, but not B lymphomas. The coincident mobilities of these components in the two-dimensional electrophoretic technique indicates that the molecules from the different T cells are very similar, if not identical, in gross structure. These observations suggest that MAb 124-40 defines a unique epitope carried by a T cell specific surface component.

At least two explanations may be offered for the origin of the epitope. One is that mutations directly or indirectly involved in the malignant transformation altered the genes encoding the subunits of the molecular, resulting in the production of a novel epitope. However, also consistent with the data is the possibility that the epitope represents a variable region determinant of a T cell specific surface component that contains constant and variable regions. Clonal neoplastic expansion of a T cell expressing a particular epitope would result in presentation of that epitope as a tumor specific antigen, just as the immunoglobulin idiotypes of B cells are presented as individually specific tumor antigens.

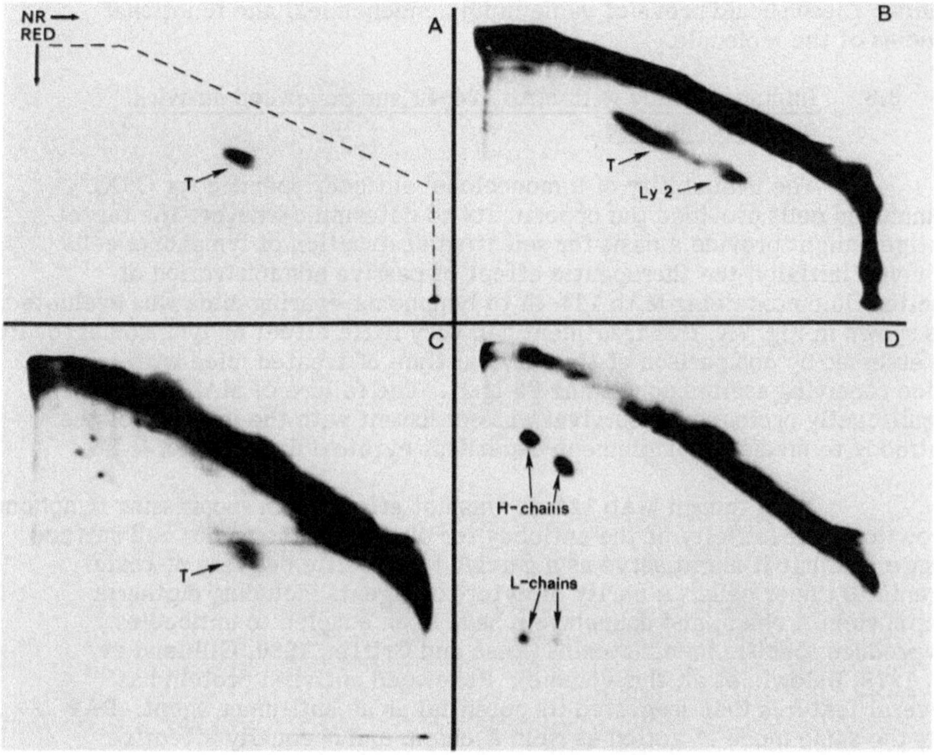

Figure 6. Diagonal electrophoresis of ^{125}I-labeled cell surface antigens. SDS-PAGE in the first dimension was on 7.5% polyacrylamide gels under nonreducing conditions. SDS-PAGE in the second dimensions was on 10% polyacrylamide gels under reducing conditions. (A) Ag 124-40 obtained from C6XL cells by precipitation with MAb 124-40. (B) Whole extract of C57BL/6 thymocytes. (C) Whole extract of splenic T-cells. (D) Whole extract of splenic B-cells.

While this possibility must be considered speculative at present, it is particularly intriguing, since the analogy with B lymphocyte immunoglobulin would imply that the molecule may function as the T lymphocyte antigen receptor. We are currently attempting to characterize the components obtained from different T lymphoma lines in order to determine if there is heterogeneity in primary structure, and we are attempting to raise monoclonal xenogeneic antibodies to possible framework determinants. These should prove of value in immunochemical and functional studies of the molecule.

3.6. Immunotherapy with MAb 124-40 and pokeweed antiviral protein conjugate

The availability of a monoclonal antibody specific for C6XL lymphoma cells provided the opportunity to determine whether the target antigen might provide a basis for selective destruction of lymphoma cells in vivo. Initially, the therapeutic effect of passive administration of ascites fluid containing MAb 124-40 to lymphoma-bearing mice was evaluated. As shown in Fig. 7A, this treatment had very little effect on lymphomagrowth, as assessed by comparison of the survival time of treated mice with control mice receiving ascites containing P3 IgG1. The failure of MAb 124-40 to significantly prolong host survival was consistent with the inability of the antibody to mediate complement-dependent cytotoxicity in vitro.

Even though MAb 124-40 was not effective in suppressing lymphoma growth, the specificity of the antibody for binding to the tumor cell surface suggested that it might serve as a carrier for specific delivery of toxic agents to tumor cells. A varity of cytotoxic agents including diptheria toxin, ricin A chain, and daunomycin have been coupled to antibodies to produce specific immunotoxins (Raso and Griffin, 1980, Gilliland et al, 1978, Baldwin et al, this volume). Pokeweed antiviral protein has several features that suggested its potential as an antitumor agent. PAP has the same mode of action as ricin A chain, and is equally effective on a molar basis in in vitro inhibition of protein synthesis (Gessner and Irvin, 1980). Unlike ricin, however, PAP exists as a single chain (Irvin, 1975) and does not require separation from a binding subunit to prevent nonspecific cytotoxicity caused by interaction with ubiquitous toxin receptors on cells.

PAP was conjugated to purified MAb 124-40 using the heterobifunctional reagent N-succinimidyl 3-(2-pyridyldithio) propionate (SPDP, Carlsson et al, 1978) as shown in Fig. 8. Briefly, SPDP was added at a molar ratio of 4:1 to MAb 124-40 and PAP in order to introduce 2-pyridyl disulphide protected thiol groups into the proteins. The 2-pyridyl disulphide groups of PAP were then removed by reduction with dithiothreitol, and after removal of excess reducing agents, the PAP-SH was added to the 2-pyridyl disulphide-containing antibody to allow thiol disulphide interchange. The covalently linked MAb-PAP conjugates were separated from reactants by chromatography on Sephacryl S-300. The conjugation procedure had no effect on the ability of the antibody to recognize tumor cells

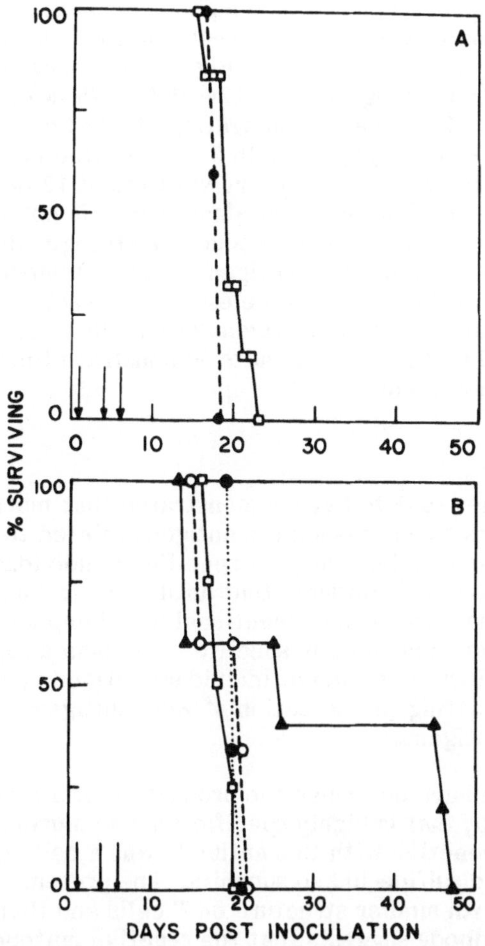

Figure 7. Effect of MAb 124-40 and pokeweed antiviral protein conjugate on survival of C6XL lymphoma bearing mice. On day 0, C57BL/6 mice (6 per group) received 2 x 10^3 C6XL lymphoma cells. (A) On days 1, 3 and 7, mice received 100 l of 124-40 ascites, titer 1:2 x 10^5 (o—o) or P3 ascites, ●—●. (B) On days 1,3 and 7, mice received phosphate buffered saline (□-□); 25 μg MAb 124-40 (●—●); 2.5 μg PAP (o—o), or 27 μg MAb-PAP conjugate containing 2.5 μg PAP (▲—▲).

in the binding assay, and there was no significant change in the ability of the PAP to inhibit cell-free protein synthesis.

In order to determine whether conjugation to PAP increased the ability of MAb 124-40 to inhibit C6XL lymphoma growth in vivo. MAb-PAP conjugate and unconjugated MAb 124-40 and PAP were administered to lymphoma-bearing mice. As shown in Fig. 7B, the survival times of mice receiving PAP (3 x 2.5 ug) or MAb 124-40 (3 x 25 ug) were not different from untreated mice. Significant prolongation of survival was seen in mice receiving MAb-PAP conjugate, with 2 of 6 treated mice surviving as long as 47 days (relative to a mean survival time of 18 days for untreated mice). No long-term survivors were obtained, and it is not known whether higher doses of conjugate would have produced a stronger therapeutic effect or whether small numbers of antigen-negative lymphoma cells might have escaped treatment. Nonetheless, these results clearly demonstrate that the C6XL lymphoma specific Antigen 124-40 can serve as a target for immune attack. Further, these results demonstrate the potential of PAP as an antitumor agent.

4. CONCLUSION

Serological studies to date have demonstrated that murine T lymphoma cells generally express tumor-associated antigens related to murine leukemia virus or to the TL complex, but antigens specific to individual lymphomas have not been reported. It is possible that antibodies to immunodominant MuLV and TL antigens present in conventional anti-lymphoma sera obscure minor reactivity toward individually specific lymphoma antigens. The hybridoma method allows selection of individual antibody-producing clones in isolation thus permitting the detection of weak antigens, including lymphoma specific antigens.

In this report we have described the production of a monoclonal antibody, MAb 124-40, that is highly specific for one murine X-ray lymphoma, C6XL. The antigen reactive with this antibody was a cell-surface glycoprotein composed of disulfide-linked subunits. The presence of a cell surface component with similar structure on T cells and thymocytes not reactive with the antibody suggests that the reactive epitope arose by mutation in a structural gene for a T cell surface protein. Alternatively, the epitope may represent a clonally expressed T cell antigen analogous to B-cell idiotypic determinants. Additional studies of the structure and function of the heteroduplex will be necessary to resolve the question of the origin of the epitope.

The exquisite specificity of MAb 124-40 for C6XL lymphoma cells suggested its potential for directing a cytotoxic agent, pokeweed antiviral protein to tumor cells in vivo. This potential was confirmed by the ability of MAb-PAP conjugates to prolong survival of lymphoma-bearing mice. These studies also demonstrated the potential of pokeweed antiviral protein as an antitumor agent.

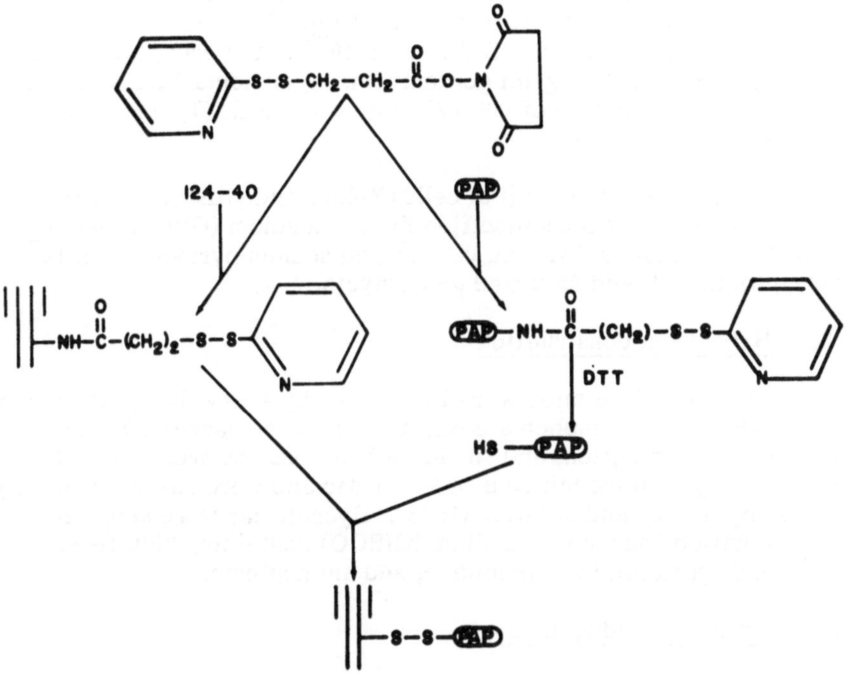

Figure 8. Conjugation of MAb 124-40 and pokeweed antiviral proteins. Convalent conjugates of MAb 124-40 and PAP were prepared using the heterobifunctional reagent N-succinimidyl 3-(2-pyridyldithio) propionate as described by Carlsson, et al, 1978.

5. EXPERIMENTAL PROCEDURES

5.1. Mice

Young adult AKR/J, A/HeJ, BALB/cJ, C57BL/6J, and C57L/J mice were purchased from the Jackson Laboratories (Bar Harbor, ME). C57BL/Ka mice were bred at the University of Texas Science Park from a stock obtained from Stanford University (Palo Alto, CA). Mice were used at 6-15 weeks of age.

5.2. Cell lines

The tumors used in this study are listed in Table I. C6XL, was obtained from Dr. B.G. Sanders (University of Texas, Austin, TX) and was maintained by transplantation of splenic tumors in C57BL/6J or C57BL/Ka mice. ASL1, ERLD, RADA1, 225, 216 and 316 were maintained by transplantation of splenic tumors in syngeneic hosts. EL-4, NIXT, 920, and 1210 were grown in suspension culture in RPMI-1640 (GIBCO) containing 10% fetal calf serum (GIBCO, 5×10^{-5} M 2-mercaptoethanol, 2 mM L-glutamine, and, 25 μg/ml gentamycin. Lymphoma XL1 was induced by fractionated X-irradiation of C57L/J mice 4 x 175 RAD), and was examined as a primary tumor.

P3X63Ag8 myeloma (P3) cells (Kohler and Milstein, 1975) were maintained in Dulbecco's modified Eagle's medium (GIBCO) supplemented with 10% fetal calf serum, 0.11 mg/ml sodium pyruvate, 5×10^{-5} M 2-mercaptoethanol, and 25 μg/ml gentamycin.

5.3. Hybridoma construction

Female Balb/c mice were immunized by 4 weekly i.p. injections of 1×10^{7} viable C6XL lymphoma cells. Spleens were harvested 4 days after the final immunization, and single-cell suspensions prepared. P3 cells and splenocytes were mixed at a 1:10 ratio and were fused essentially as described by Kohler and Milstein (1975). Hybridomas were selected in Iscove's Modified Dulbecco's medium (GIBCO) containing 20% fetal calf serum and hypoxanthine, thymidine, and aminopterin.

5.4. Antibody binding assay

Antibodies in culture supernatants or ascites fluid were detected in binding assays (Brown et al 1977) using rabbit anti-mouse IgG (Kirkegaard and Perry) and ^{125}I-labeled *Staphylococcus aureus* protein A (Pharmacia). Target cells (5×10^{5}) in 50 μl phosphate buffered saline were added to wells of vinyl microtiter plates (Dynatech) followed by 100 μl of hybridoma supernatant or monoclonal antibody dilution. After 1 hour, the cells were washed three times and incubated successively with rabbit anti-mouse IgG and ^{125}I-protein A, with intervening washes. After the final washes, wells were removed with a hot wire and counted. Assays were performed in triplicate and data presented as cpm bound. Binding ratios were calculated from the formula:

$$\text{Binding ratio} = \frac{\text{mean cpm Hybridoma supernatant}}{\text{mean cpm P3 supernatant}}$$

5.5. Flow microfluorometry

Cells were prepared for flow microfluorometic analysis by indirect immunofluorescent staining using MAb 124-40 or P3 supernatant and affinity-purified fluorescein-conjugated goat anti-mouse IgG as described (Lanier and Warner, 1981). Cells were analyzed with an Ortho series 50H cytofluorograf using the 488-nm argon laser line and the standard green fluorescence filter pack. The instrument was calibrated daily using fluorescent micropheres (Herzenberg and Herzenberg, 1978). Appropriate gates were set by light scatter to exclude debris from the analysis. Fifty thousand cells from each sample were accumulated for each analysis.

5.6. Radioiodination

Protein A was labeled by Chloramine-T catalyzed radioiodination as described by McConahey and Dixon (1966). Cells were labeled by lactoperoxidase-catalyzed radioiodination as described by Keski-Oja, et al (1977). Viability of the labeled cells, as determined by trypan blue dye exclusion, always exceeded 90%.

Washed cells were lysed by resuspension in 1.0 ml of 0.1M Tris-HC1, pH, 8.0, 20 KIU/ml aprotinin, and 0.02% sodium azide. After 30 min at 4°C, nuclei and insoluble material were removed by centrifugation at 35,000 x g for 20 min. When extracts are to be analyzed under non-reducing conditions, the washed cells were incubated for 10 min in PBS containing 10 mM iodoacetamide before solubilization in lysis buffer containing 100 mM iodoacetamide.

5.7. Immunoprecipitation

Immunoprecipitations were performed with formalin-fixed Staphaureus (Kessler, 1975). Radiolabeled cell extracts (50 μl, 2.5 x 10^6 cell equivalents) were precleared by incubation with 2 μl rabbit anti-mouse immunoglobulin RAMIG and 50 μl of 10% (v/v) SACI in RIP buffer (0.01M Tris-HC1, pH 8.0, containing 0.5% NP-40, 0.15M NaC1, 1 mM EDTA, 1 mg/ml ovalbumin, 0.02% NaN_3) for 30 min at 4°. The bacteria were then pelleted by centrifugation at 7,000 x g for 1 min, and the cleared supernatant used for immunoprecipitation. Since the monoclonal antibody developed and used in these studies did not directly bind to SACI, a two-step procedure was used to obtain antibody-loaded SACI: 1) 3 μl RAMIG were added to 50 μl 10% SACI and incubated 30 min at 4°. The bacteria were pelleted by centrifugation at 7,000 x g for one min. and washed twice with RIP buffer. 2) The washed SACI were then resuspended in 1.0 ml hybridoma or P3 supernatant and incubated for 4 hrs at 4°. The bacteria were pelleted, washed three times with RIP buffer to remove unbound antibody, and added to the precleared cell extracts. After incubation at 4° for 6-16 hrs, the bacteria were pelleted and washed three times with RIP buffer, and bound antigen eluted with the appropriate sample buffer. After elution, SACI were pelleted and the supernatants analyzed by electrophoresis.

5.8. Polyacrylamide gel electrophoresis

Radiolabeled membrane proteins and specific immunoprecipitates were analyzed by 3 different electrophoresis techniques. One-dimensional sodium dodecyl-polyacrylamide gel electrophoresis (SDS-PAGE) was carried out on vertical 10% polyacylamide gels using the discontinuous buffer system described by Laemmli (1970). The molecular weight standards BSA (69,000), ovalbumin (43,000) and carbonic anhydrase (29,000) (Sigma) were visualized by staining with Coomassie blue.

Two-dimensional electrophoresis, employing separation by charge in the first dimension and size in the second dimension, was performed as described by O'Farrell (1975). The first-dimensional gels contained 1.6% pH 5-7 and 0.4% pH 3-10 ampholytes (Bio-Rad). The second dimension was carried out on 10% polyacrylamide gels.

Two-dimensional SDS-PAGE, or "diagonal" electrophoresis, employin SDS-PAGE under nonreducing conditions in the first dimension, followed by SDS-PAGE under reducing conditions in the second dimension (Takemoto et al, 1977), was used to resolve proteins containing interchain disulfide bonds from non-disulfide bonded proteins.

Gels were dried and autoradiographed for 3 to 48 hrs. at -70° C on Kodak XR film with intensifying screens (Cronex Lightning Plus, Dupont).

Acknowledgements

We would like to thank Cynthia Crawford and Jo Anne Lund for their excellent technical assistance. This work was supported by NCI grants CA 26321 and CA 26891.

REFERENCES

Allison, J.P., and Kendall, C., 1980, Immunochemical analysis of viral antigen expression by an X-ray induced lymphoma, Fed. Proc. 39:1154.

Beatty, P.G., Kim, B.S., Rowley, D.A., and Coppleson, L.W., 1976, Antibody against the antigen receptor of a plasmacytoma prolongs survival of mice bearing the tumor, J. Immunol. 116:1391-1395.

Brown, J.P., Kitzman, J.M., and Hellstrom, K.E., 1977, A microassay for antibody binding to tumor cell surface antigens using ^{125}I-labeled protein A from Staphylococcus aureus, J. Immunol. Methods 15:57-66.

Canaani, E., and Aaronson, S.A., 1979, Restriction enzyme analysis of mouse cellular type C viral DNA: Emergence of new viral sequences in spontaneous AKR/J lymphomas, Proc. Natl. Acad. Sci. USA 76:1677-1682.

Carlsson, J., Drevin, H., and Axer, R., 1978, Protein thiolation and reversible protein-protein conjugation, Biochem. J. 173:723-737.

Elder, J.H., Gautsch, J.W., Jensen, F.C., Lerner, R.A., Hartley, J.W., and Rowe, W.P., 1977, Biochemical evidence that MCF murine leukemia viruses are envelope (env) gene recombinants, Proc. Natl. Acad. Sci. USA 74:4676-4680.

Fischinger, P.J., Thiel, H.J., Ihle, J.N., Lee, J.C., and Elder, J.H., 1981, Detection of a recombinant murine leukemia virus - related glycoprotein on virus-negative thymoma cells, Proc. Natl. Acad. Sci. USA 78(3):1920-1924.

Gessner, S.L., and Irvin, J.P., 1980, Inhibition of elongation factor 2-dependent translocation by the pokeweed antiviral protein and ricin, J. Biol. Chem. 225:3251-3253.

Gilliland, D.G., Collier, R.J., Moehring, J.M., and Moehring, T.J., 1978, Chimeric toxins: Toxic, disulfide-linked conjugate of concanavalin A with fragment A from diptheria toxin, Proc. Natl. Acad. Sci. USA 75(11):5319-5323.

Hayman, M.J., and Crumpton, M.J., 1972, Isolation of glycoproteins from pig lymphocyte plasma membrane using Lens culinaris phytohemagglutinin, Biochem. Biophys. Res. Comm. 47:923-930.

Haughton, G., Lanier, L.L., Babcock, G.F., and Lynes, M.A., 1978, Antigen-induced murine B cell lymphomas. II. Exploitation of the surface idiotype as tumor specific antigen, J. Immunol. 121:2358-2364.

Herzenberg, L.A., and Herzenberg, L.A., 1978, In: Handbook of Experimental Immunology, Weir, D.M. ed; Blackwell Scientific, London, 22.1-22.21.

Irvin, J.D., 1975, Purification and partial characterization of the antiviral protein from Phytolacca americana which inhibits eukaryotic protein synthesis, Arch. Biochem. Biophys. 169:533, 528.

Kaplan, H.S., and Brown, M.B., 1952, A quantitative dose-response study of lymphoid tumor development in irradiated C57 black mice, J. Natl. Cancer Inst. 13:185-194.

Keski-Oja, J., Mosher, D.F., and Vaheii, A., 1977, Dimeric character of fibronectin, a major cell-surface associated glycoprotein, Biochem. Biophys. Res. Commun. 74:699-706.

Kessler, S., 1975, Rapid isolation of antigens from cells with a Staphylococcal protein A-antibody absorbent: parameters of the interaction of antibody-antigen complexes with protein A, J. Immunol. 115:1617-1624.

Kohler, G., and Milstein, C., 1975, Continuous cultures of fused cells secreting antibody of predefined specificity, Nature 256:495-497.

Krolick, K.A., Isakson, P.C., Uhr, J.W., and Vitetta, E.S., 1979, BCL1, a murine model for chronic lymphocytic leukemia: use of the surface immunoglobulin idiotype for the detection and treatment of tumor. Immunol. Rev. 48:81-106.

Laemmli, U.K., 1970, Cleavage of structural proteins during the assembly of the lead of bacteriophage T4, Nature 227:680-685.

Lanier, L.L., and Warner, N.L., 1981, Paraformaldehyde fixation of hematopoietic cells for quantitative flow cytometry (FACS) analysis, J. Immunol. Met. 47:25-30.

Lennox, E.S., 1980, The antigens of chemically induced tumors, In: Fourth International Congress of Immunology - Immunology 80, Fourgereau, M., and Dausset, J., ed. Academic Press, N.Y., pp. 659-667.

McConahey, P.J., and Dixon, F.J., 1966, A method of trace iodination of proteins for immunologic studies, Int. Arch. Allergy 29:185-189.

Miller, R.A., Maloney, D.G., Warnke, R., and Levy, R., 1982, Treatment of B cell lymphoma with monoclonal anti-idiotype antibody, New Engl. J. Med. 306:517-522.

Nowell, P.C., 1976, The clonal evolution of tumor cell populations, Science 194:23-28.

Obata, Y., Stockert, E., O'Donnell, P.V., Okubu, S., Synder, H.W., Jr., and Old, L.Y., 1978, G(RADA1): a new cell surface antigen of mouse leukemia defined by naturally occurring antibody and its relationship to murine leukemia virus, J. Exp. Med. 147:108-1105.

O'Donnell, P.V., Stockert, E., Obata, Y., DeLeo, A.B., and Old, L.J., 1980, MuLV-related cell surface antigens as serological members of AKR ecotropic, xenotropic, and dualtropic MuLV, Cold Spring Harbor Symp. Quant. Biol. 44:1255-1264.

O'Farrell, P.H., 1975, High resolution two-dimensional electrophoresis of proteins, J. Biol. Chem. 250:4007-4021.

Old, L.J., Boyse, E.A., and Stockert, E., 1963. Antigenic properties of experimental leukemias. I. Serological studies in vitro with spontaneous and radiation-induced leukemias, J. Natl. Cancer Inst. 31:977-986.

Old, L.J., Boyse, E.A., Stockert, E., 1965, The G (Gross) leukemia antigen, Cancer Res. 25:813-820.

Old, L.J., and Stockert, E., 1977, Immunogenetics of cell surface antigens of mouse leukemia, Ann. Rev. Genet. 11:127-160.

Price, M.R., and Baldwin, R.W., 1975, Immunobiology of chemically induced tumors. In: Cancer, a Comprehensive Treatise, Becker, F.F., ed., Plenum Press, N.Y., p. 209-236.

Raso, V., and Griffin, T., 1980, Specific cytotoxicity of a human immunoglobulin-directed FAb-ricin A chain conjugate, J. Immunol. 125:2160-2616.

Takemoto, L.J., Miyakawa, T., Fox, C.F., 1977, Analysis of membrane protein topography of Newcastle disease virus and cultured mammalian fibroblasts. In: Cell Shape and Surface Architecture, David, J.P., Henning, V., and Fix, C.F., ed. Alan R. Liss, Inc., N.Y. p. 605-614.

Wysocki, L.J., and Sato, V.L., 1978, "Panning" for lymphocytes: a method for cell selection, Proc. Natl. Acad. Sci. USA 75:2844-2848.

PART V.

NEW AND RELEVANT METHODOLOGY

14

HUMAN HYBRIDOMAS

Carlo M. Croce

The Wistar Institute of Anatomy And Biology
36th and Spruce Street
Philadelphia, PA 19104

1. SUMMARY

We have hybridized a hypoxanthine phosphoribosyltransferase (HPRT)-deficient human B-cell line derived from a patient suffering from multiple myeloma with peripheral lymphocytes obtained from a patient with subacute sclerosing panencephalitis (SSPE).[1] These hybridomas were found to secrete human IgM specific for measles virus nucleocapsids.

2. INTRODUCTION

Since the discovery by Kohler and Milstein (1975) that somatic cell hybrids (hybridomas) between mouse myeloma cells and lymphocytes derived from immunized animals produce monoclonal antibodies of predefined specificities, rodent monoclonal antibodies produced by intra- or interspecific hybridomas have been used for a vast number of different purposes including studies of cell differentiation (Ferrero et al., 1982) and of tumor associated antigens (Koprowski et al., 1981) and for human diagnostics (Koprowski et al., 1978).

The use of monoclonal antibodies in human immunotherapy, however, has been hampered by the fact that the available monoclonal antibodies against pathogens or tumor cells are of rodent origin and elicit an immune response in the hosts where they are injected. As a result, not only the injected antibodies are neutralized by the antibodies raised by

[1]Abbreviations: SSPE, subacute sclerosing panecephalitis; EMS, ethylmethane sulfonate; NP, nucleocapsid polypeptide.

the host, but anaphylactic shock of serious consequences might develop. Therefore it is extremely important to attempt to produce hybridoma secreting human monoclonal antibodies in order to avoid these complications and to open new avenues in human immunotherapy.

3. RESULTS

One of the possible ways to obtain hybrids secreting human immunoglobulin is to hybridize mouse myeloma cells with human lymphocytes (Croce et al., 1979). These hybrids, however, segregate human chromosomes and might lose the human chromosomes carrying the human immunoglobulin genes. We have attempted, therefore, to determine the chromosomal location of the human immunoglobulin genes and to establish whether these chromosomes are retained preferentially in mouse x human hybridomas. Analysis of these hybrids for the expression of human immunoglobulin chains and the presence of human chromosomes indicates that the gene for heavy chains are located on chromosome 14 (Croce et al., 1979), the genes for lambda chains are located on chromosome 22 (Erikson et al., 1981) and the genes for κ chains are located on chromosome 2 (McBride et al., 1982; Erikson and Croce, 1982). Study of a large number of hybrid cells indicates that human chromosomal loss in mouse x human hybridomas is not random and that the heavy chain gene cluster containing chromosome 14 is preferentially retained in the hybrids (Table I) (Croce et al., 1980). Chromosome 2, however, is preferentially lost from the hybrid cells (Table I). It seems difficult, therefore, to obtain stable hybrids that produce human κ chains for long periods of time.

Because of the instability of the mouse x human hybridomas, we have decided to attempt to produce human x human hybridomas that should, like most of the intraspecific hybrids (Croce and Koprowski, 1974), be more stable.

We have selected for hypoxanthine phosphoribosyltransferase (HPRT) deficient mutants of GM 1500 human myeloma derived cells in the presence of 30 μg/ml of 6-thioguanine, following mutagenization with ethylmethane sulfonate (EMS; Croce et al., 1980).

The mutant cells were found to secrete human IgG ($\gamma 2, \kappa$), as did the parental GM 1500 cells (Fig. 1, lane 1).

Heparinized blood plasma was obtained from a patient suffering from SSPE. This patient was a 19 year-old female who had first developed SSPE at age 9. The disease lasted for 2 years, leading ultimately to recovery with the persistence of only some residual symptoms. At the age of 18 years, she married and became pregnant. Then in the fourth month of pregnancy SSPE recurred with great severity and she is now in a chronic vegetative state. Measles virus was isolated from a brain fragment during the first tissue culture. Serum from the patient, diluted $1{:}10^6$, bound in radioimmunoassay (RIA) with measles-infected target

TABLE I.

Expression of human isozyme markers in mouse x human hybridomas

Human chromo-some	Isozymes tested	Total No. of clones analyzed	No. of Clones containing human chromosome	% of clones containing human chromosome
1	ENO-1	56	5	8.9
2	IDH_S	42	0	0
3	β-GAL	22	5	23.0
4	PGM_2	40	16	40.0
5	HEX_B	41	29	70.7
6	GLO-1	32	11	34.3
7	β-GUS	23	4	17.4
8	GSR	44	11	25.0
9	$ACON_S$	45	5	11.1
10	GOT_S	44	8	18.1
11	LDH_A	54	26	48.0
12	LDH_B	54	12	22.2
13	Est-D	43	5	11.6
14	NP	46	46	100.0
15	MPI	47	20	42.5
16	APRT	33	11	33.3
17	GK	46	19	41.3
18	PEP A	53	14	26.4
19	GPI	57	13	22.8
20	ADA	48	9	18.7
21	SOD-1	44	14	31.8
22	ARS	21	13	61.9
X	G6PD	30	16	53.3

cells. These data, as well as typical histological lesions of the brain, confirmed the clinical diagnosis of SSPE.

Using established procedures, we fused 10^7 GM 1500 6TG cells with Ficoll-purified lymphocytes derived from peripheral blood (10 ml) taken from the patient in the presence of polyethylene glycol 1000 (ref. 7). The fused cells were distributed into 24 wells of a Linbro FB-16-24 TC plate in the presence of HAT selective medium. Growth of hybrids was

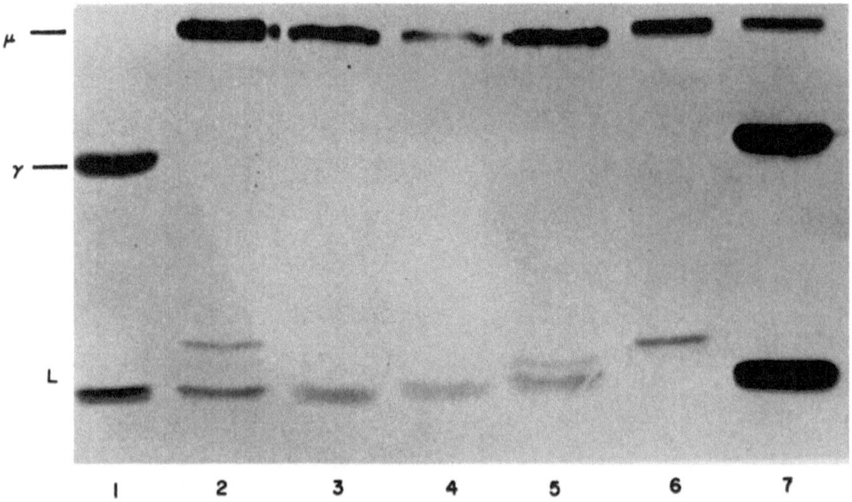

Figure 1. Immunoprecipitation and 10% SDS-polyacrylamide gel electrophoresis of secreted human immunoglobulin chains produced by human hybridomas. Lane 1, immunoprecipitates of the immunoglobulin produced by GM 1500 6TG-A12 cells after reaction with an anti-human γ antiserum; lanes 2-6, immunoprecipitates of immunoglobulin chains secreted by human x human hybridomas, after reaction with anti-human μ and γ antiserum.

detected in 20 of the 24 wells. These hybrids were propagated in HAT-RPMI 1640 medium and cloned by limiting dilution. Each independent clone (derived from an independent well) was tested for the expression of human immunoglobulin chains and for the ability to immunoprecipitate measles virus proteins.

As shown in Fig. 1, six hybrid clones, including the two that reacted with measles virus (discussed later in Fig. 3), expressed human μ chains, as determined by immmunoprecipitation of the immunoglobulins secreted by the hybrids with rabbit anti-human μ antiserum. Some hybridomas also expressed a light chain (L) that migrated differently from the light chain expressed by the GM 1500 line. The clone of the hybridoma illustrated in lane 6 (Fig. 1) lost the ability to produce the light chain expressed by GM 1500. The hybridoma culture fluid in lane 7 was immunoprecipitated with both anti-human μ and anti-human γ chain antisera - the hybrid expressed two heavy chains, the γ chains of GM 1500 parent and the μ chain of human SSPE B-cell parent.

The specificity of the antibodies for measles virus was determined by immunoprecipitation using methods similar to those described elsewhere. Culture fluids from cultures D3 and C5 only precipitated the virus nucleocapsid polypeptide (NP) and varying amount of its cleavage fragments (Fig. 2). The antibodies in the culture fluids seemed to be specific for this one polypeptide which is the major polypeptide of the virus nucleocapsid. NP is extremely sensitive to proteolytic cleavage, thus the long incubation periods used in the immunoprecipitation procedure could be partly responsible for the high level of cleavage of the polypeptide. The level of cleavage is particularly evident in Fig. 2. The specificity of the antibodies for NP was confirmed by the demonstration that subclones of hybrid clone D3 could also precipitate this polypeptide (Fig. 3).

In collaboration with Dr. G. Eisenbarth we have also hybridized peripheral lymphocytes derived from a patient with juvenile diabetes Type I with the GM 1500-6TG cells and obtained hybridoma cells that secrete auto antibodies specific for the islet cells of the pancreas (Eisenbarth et al., 1982).

4. DISCUSSION

These results indicate that it is possible to obtain human B-cell hybrids which continuously secrete human antibodies, particularly antibodies specific for a human pathogenic virus. The availability of a human continuous B-cell line with appropriate drug resistance markers represents a breakthrough in work towards the possible application of this technology to human immunotherapy. In addition, lymphocytes from patients with human autoimmune disease such as myasthenia gravis and Graves' disease could be fused with human B-cell lines to produce hybridomas that secrete the autoantibodies responsible for the disease. Once such monoclonal autoantibodies are available, the production of anti-autoantibodies could provide a cure for patients with these autoimmune diseases.

In addition we have also developed an ouabain resistant mutant of the GM 1500-6TG cells (GM 1500-6TG-OUB) that is at least ten-fold more efficient than the parental line to produce human hybridomas. The use of this double mutant should improve considerably our ability to obtain human hybridomas secreting antibodies of the desired specificity.

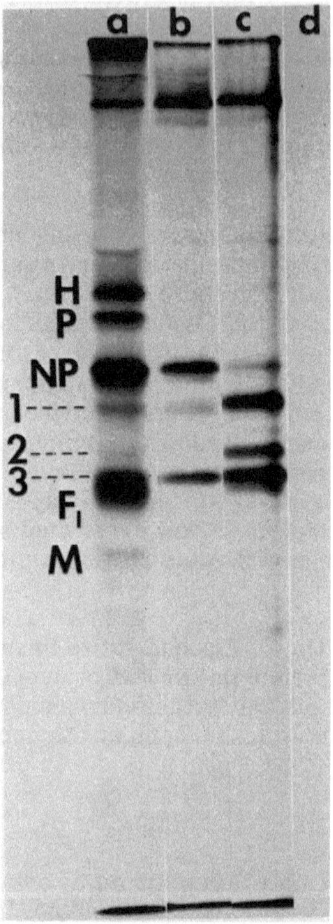

Figure 2. Polyacrylamide gel electrophoretic analysis of the measles virus polypeptides precipitated by the various monoclonal antibodies. a, Virus polypeptides precipitated by convalescent serum of a patient with atypical measles were used as markers. The polypeptides are as follows: H, the virus haemagglutination; P, a polypeptide associated with the internal nucleocapsid structure; NP, the major structural polypeptide of the nuclecapsid; F, the polypeptide responsible for cell fusion and haemolytic activities; M, the non-glycosylated membrane polypeptide. Bands 1, 2 and 3 are not unique virus polypeptides, but represent proteolytic cleavage fragments of the NP polypeptide. b, Virus polypeptides precipitated by antibody D3. d, Culture fluid from the human GM 1500 6TG cell line. Culture fluids of b, c and d were concentrated 20-fold by freeze-drying

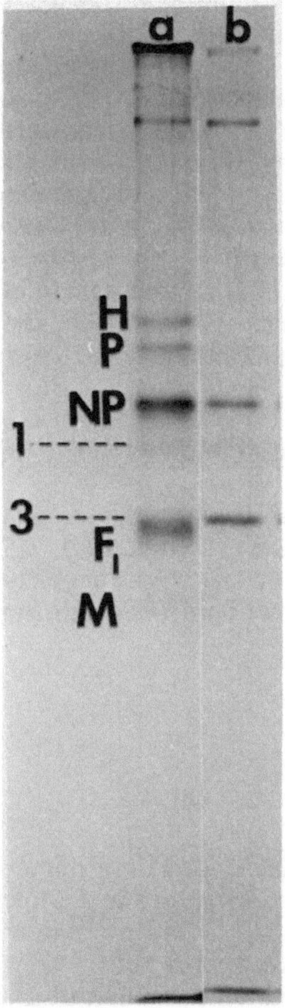

Figure 3. Lane a represents the virus polypeptides precipitated by the same atypical measles serum as in Fig. 2a. b, Polyacrylamide gel electrophoretic analysis of the measles virus polypeptides precipitated by culture fluid from subclone D3M2. The conditions of the experiments were as described in the legend to Fig. 2, except that culture fluid was unconcentrated.

5. EXPERIMENTAL METHODS

5.1 Selection of somatic cell hybrids

Either P3 x 63 Ag 8 (Kohler and Milstein, 1975) or P326 Bu 4 (Margulies et al., 1976) mouse myeloma cells deficient in hypoxanthine phosphoribosyltransferase or thymidine kinase, respectively, were fused with human peripheral lymphocytes (HPL) or GM 1500 human myeloma cells, or GM 607 or GM 1056 human lymphoblastoid cells in the presence of polyethyleneglycol 1000 (Pontecorvo, 1975), according to established procedures (Croce et al., 1979). Fused cells were seeded in 24-well Linbro plates (Linbro Chemicals, New Haven, CT) with selective medium. Mouse myeloma x HPL hybrids were selected in hypoxanthine/aminopterin/thymidine (HAT) medium (Littlefield, 1964), and mouse myeloma x human myeloma or lymphoblastoid cell hybrids were selected in HAT medium containing 10^{-4} M ouabain (Croce, 1977). Each hybrid clone derived from a different Linbro well. Following their isolation, the hybrids were kept in nonselective medium. Human x human hybrids were obtained as described (Croce et al., 1980).

5.2 Production of human immunoglobulin (Ig) chain

Parent and hybrid cells were grown in media containing 100 μCi/ml (=3.7 MBq) (^{3}H) leucine (70 Ci/mmol) for 24 hours. Cells were pelleted, and supernatants (500 μl) were incubated with 50 μl of either rabbit anti-human μ or γ heavy chain-specific sera and/or rabbit anti-mouse γ-globulin

Figure 2 (Continued)
before use. The procedures used in immunoprecipitation, similar to those described elsewhere, were as follows: lysates of virus-infected CV1 cells labelled with ^{35}S-methionine were used as antigen. Aliquots (25 μl) were mixed with 100 μl concentrated culture fluid and incubated at 37°C for 90 min then at 4°C for 4 h. Rabbit total anti-human antibody (25 μl) was then added and the incubation period repeated. Precipitated polypeptides were collected by centrifugation in an Eppendorf centrifuge for 20 min. at 10,000 r.p.m. The visible pellet was resuspended and washed three times. After the final washing the pellet was suspended in lysis buffer, boiled for 3 min. and electrophoresed on a 10% SDS-polyacrylamide gel in conditions described elsewhere. After fluorography, dried gels were exposed to Cronex X-ray film.

for 1 h in ice. One hundred µl of a 10% suspension of fixed Staphylococcus aureus (Croce et al., 1979; 1980) was added to each reaction for 30 min, or 50 µl of goat anti-rabbit γ-globulin serum was added for 16-18 h at 4°C. Double antibody precipitates were collected by centrifugation through a 0.2 ml pad of 1 M sucrose. All immunoprecipitates were washed with 5% sucrose, 15 mM Tris HCl, pH 7.4, 0.5 M NaCl, 5 mM EDTA, 1% Nonidet-P 40 and 2 mM phenyl methyl sulfonyl fluoride. The immunoprecipitates were resuspended in Laemmli buffer and analyzed on discontinuous 10 and 11% reducing sodium dodecyl sulfate polyacrylamide gels, embedded in PPO and subjected to fluorography for 5-14 days (Croce et al., 1979; 1980a, 1980b). Human hybridoma cultures were labelled with 100 µCi ^{3}H-leucine (70 Ci per mmol) per ml for 12 h. The human immunoglobulin chains were immunoprecipitated with rabbit anti-human heavy chain antigen using established procedures, then separated by 10% SDS-polyacrylamide gel electrophoresis as described elsewhere.

5.3 Isozyme analysis

Hybrid cells were tested for the expression of enzyme markers whose genes are located on each of the different human chromosomes by starch or cellulose acetate gel electrophoresis. ENO-1 = enolase 1 (E.C.4.2.1.11); IDH_S = isocitrate dehydrogenase (E.C.1.1.1.42); β-gal = β-galactosidase (E.C.3.2.1.23); GPX = glutathionine peroxidase (E.C.1.11.-1.9); PGM_2 = phosphoglucomutase (E.C.2.7.5.1); HEX_B = hexosaminidase (E.C.3.2.1.30); GLO-1 = glyoxylase (E.C.4.4.1.5); β-GUS = β-glucoronidase (E.C.3.2.1.31); GSR = glutathione-reductase (E.C.1.6.4.2); $ACON_S$ = soluble aconitase (E.C.4.2.1.3); GOTS = glutamate-oxaloacetic transaminase (E.C.2.6.1.1); LDH_A = lactate dehydrogenase A (E.C.1.1.1.27); LDH_B = lactate dehydrogenase B (E.C.1.1.1.27); Est-D = esterase-D (E.C.3.1.-1.1); NP = nucleoside phosphorylase (E.C.2.4.2.1); MPI = mannose phosphate isomerase (E.C.5.3.1.8); APRT = adenine phosphoribosyltransferase (E.C.2.4.2.7); GK = galactokinase (E.C.2.7.1.6); PEP A = peptidase A (E.C.3.4.11.13); GPI = glucose phosphate isomerase (E.C.5.3.1.9); ADA = adenosine deaminase (E.C.3.5.4.4); SOD-1 = superoxide dismutase (E.C.1.-15.1.1); ARS = arylsulfatase (E.C.3.1.6.1); ACO_M = mitochondrial aconitase (E.C.4.2.1.3); G6PD = glucose 6 phosphate dehydrogenase (E.C.1.1.-1.49).

5.4 Chromosome analysis

Karyologic analyses of the hybrid cells were done after banding by trypsin-Giemsa staining according to a modification of the method described by Seabright (1971) and Croce et al., (1973). Chromosomes of some clones containing the human chromosome 14, and one or two other human chromosomes were also banded by the G 11 method (Bobrow and Cross, 1974).

REFERENCES

Bobrow, M., and Cross, J., 1974. Differential staining of human and mouse chromosomes in interspecific cell hybrids. Nature 251:74-79.

Croce, C.M., Kieba, I., and Koprowski, H. 1973. Unidirectional loss of human chromosomes in rat-human hybrids. Exp. Cell Res. 79:315-318.

Croce, C.M., and Koprowski, H., 1974. Positive control of the transformed phenotype in hybrids between SV40-transformed and normal human cells. Science 184:1288-1289.

Croce, C.M., 1977. Assignment of the integration site for Simian virus 40 to chromosome 17 to GM54VA, a human cell line transformed by Simian virus 40. Proc. Natl. Acad. Sci. USA 74:315-318.

Croce, C.M., Shander, M., Martinis, J., Circurel, L., D'Ancona, G., Dolby, T.W., and Koprowski, H., 1979. Chromosomal location of the genes for human immunoglobulin heavy chains. Proc. Natl. Acad. Sci. USA 76:3416-3419.

Croce, C.M., Shander, M., Martinis, J., Circurel, L., D'Ancona, G., and Koprowski, H. 1980a. Proferential retention of human chromosome 14 in mouse x human B cell hybrids. Eur. J. Immunol. 10:486-488.

Croce, C.M., Linnenbach, A., Hall, W., Steplewski, Z., and Koprowski, H. 1980b. Production of human hybridomas secreting anti-measles virus antibodies. Nature 288:488-489.

Eisenbarth, G.S., Linnenbach, A., and Croce, C.M. 1982. Human hybridomas secreting anti-islet antibodies. Nature (In Press).

Erikson, J., Martinis, J., and Croce, C.M. 1981. Assignment of the genes for human immunoglobulin chains to chromosome 22, Nature 294:173-175.

Erikson, J., and Croce, C.M. 1982. Secretion of human immunoglobulins by mouse myeloma x Daudi somatic cell hybrids. Eur. J. Immunol. 12:697-701.

Ferrero, D., Pessano, S., Pagliardi, G.L., and Rovera, G. 1982. Induction of differentiation of human myeloid leukemias. Surface changes probed with monoclonal antibodies. Blood. (In Press).

Kohler, G., and Milstein, G. 1975. Continuous cultures of fused cell secreting antibody of predefined specificity. Nature 256:495-597.

Koprowski, H., Sears, H.F., Herlyn, M., and Steplewski, Z. 1981. Specific antigen in serum of patients with colon carcinoma. Science 212:53-55.

Koprowski, H., Steplewski, Z., Herlyn, D., and Herlyn, M. 1978. Study of antibodies against human melanoma produced by somatic cell hybrids. Proc. Natl. Acad. Sci. USA, 75:3405-3409.

Littlefield, J. 1964. Selection of hybrids from mating of fibroblasts in vitro and their presumed recombinants. Science 145:709-710.

Margulies, D.H., Kuehl, W.M., and Scharff, M.D. 1976. Somatic cell hybridization of mouse myeloma cells. Cell 8:405-415.

McBride, O.W., Hieter, P.A., Hollis, G.F., Swan, D., Otey, M.C., and Leder, P. 1982. Chromosomal location of human kappa and lambda immunoglobulin light chain constant region genes. J. Exp. Med. 155:4180-1490.

Pontecorvo, G. 1975. Production of mammalian somatic cell hybrids by means of polyethylene glycol treatment. Som. Cell Gen. 1:397-400.

Seabright, M. 1971. A rapid binding technique for human chromosomes. Lancet 2:971-972.

15

PROCEDURES FOR IN VITRO IMMUNIZATION AND MONOCLONAL ANTIBODY PRODUCTION

Christopher L. Reading

Department of Tumor Biology
The University of Texas System Cancer Center
M.D. Anderson Hospital and Tumor Institute
Houston, Texas 77030

1. SUMMARY

In this paper, procedures are described for immunization of mouse spleen cells in culture and their subsequent fusion with myeloma cells to produce antibody-secreting hybridomas, and for screening antibody reactivity with cellular and soluble antigens using a modified enzyme-linked immunosorbent assay (ELISA)[1]. Using these procedures, we have produced monoclonal antibodies reactive with with small haptens, with highly conserved proteins such as actin and calmodulin (Pardue, et al., 1981)., carbohydrates (Reading, 1981), metastatic murine tumor cells (Miner, et al. 1981), and human leukemic cells (Dicke, et al, 1981). In addition, methods for characterizing monoclonal antibody reactivity with cells and with tissue sections and cytocentrifuge preparations using an immunogold technique, for cell separation based on monoclonal antibody binding using immunogold reagents, and for identification of cell surface antigens using a high-pressure liquid chromatography ELISA assay are described. In addition, a new process for biologically producing bifunctional antibody molecules is discussed. The new process involves production of two separate hybridomas secreting monoclonal antibodies to distinct antigens and their hybridization to form a "quadroma" which secretes monoclonal antibodies which are formed by intercellular recombination of the parental heavy and light chains. The construction of "recombinant monoclonal antibodies" may allow new immunological reagents for diagnosis and therapy.

---

[1]Abbreviations: ELISA, enzyme-linked immunosorbent assay; RMA, recombinant monoclonal antibodies; GED-ELISA, gel electrophoresis-derived ELISA; FMF, flow microfluorometry; IIF indirect immunofluorescence.

2. INTRODUCTION

Monoclonal antibody production requires the successful marriage of immunology and cell biology. The mechanisms involved in immunization in vivo are not yet perfectly understood, and in this respect, remain an art rather than a science. Immunization in vivo can often lead to restricted responses to immunodominant groups while tolerized determinants remain unreactive. Tolerance exists to highly conserved proteins, polysaccharides and glycoconjugates, and highly restricted responses are often elicited by in vivo immunizations with normal and tumor cells.

Most of the recent advances in the understanding of cellular interactions during the elaboration of an immune response have been elucidated by using in vitro culture systems. Culture systems developed in the late 1960's by Mishell and Dutton (1967) and by Marbrook (1967) demonstrated the participation of several cell types in the elaboration of a humoral response. Precursor bone-marrow lymphocytes (B cells), helper and suppressor thymus-derived lymphocytes (T cells), and macrophages (monocytes) all participate in the initiation of antibody production. These and similar culture systems led to the discovery that many cellular interactions were mediated through soluble products elaborated by the lymphocytes (lymphokines) or the monocytes (monokines). Recently, Luben and Mohler (1980) demonstrated that medium conditioned by thymocytes from 10 day-old mice could replace the immune T cell requirement for initiation of a primary humoral response in culture. Murine spleen cells immunized in this sytem were used for fusion with myeloma cells to produce monoclonal antibodies to the lymphokine osteoclast activation factor. We have modified this system of immunization in culture, and have produced monoclonal antibodies to haptens, highly conserved cellular proteins (Pardue, et al., 1981), soluble glycoproteins and polysaccharides (Reading, 1981), and to urine (Miner et. al. 1981), and human (Dicke, et. al., 1981) tumor cells. This procedure has several advantages over immunization in vivo: 1) the immunization procedure requires only five days, 2) it is possible to maintain defined concentrations of antigen during the immunization, 3) it is possible to test the effects of various monokines and lymphokines at defined concentrations, 4) periodic sampling of the same immunization is possible during the course of an experiment, and 5) it is possible to produce antibodies to "self" antigens which are not produced in vivo due to suppression or tolerance.

Somatic cell hybridization of immune B cells with myeloma cells was first reported in 1975 by Kohler and Milstein. In addition to the immunization procedures, investigations of the other steps involved have led to improved protocols and understanding. These steps include the selection of the myeloma cell line used for hybridization, the use of polyethylene glycol for the fusion, biochemical selection for the hybridomas, improved culture conditions and supportive feeder cells to enhance the growth and survival of the hybridomas, improved screening procedures for the detection of monoclonal antibodies, an appreciation for the necessity of early cloning of monoclonal antibody-producing cells, improved methods for large scale monoclonal antibody production, and an awareness of the problem of contamination by mycoplasma.

3. RESULTS

The immunization of murine spleen cells in culture results in the formation of large blast cells which secrete antibody specific for the immunogen. Fusion of these immunized cells with the non-secreting myelomas such as the Sp2/0 (Schulman, et al., 1978) or P_3X63Ag8.653 (Kearney, et al., 1979) cell lines results in hybridomas which continue to secrete these antibodies. The results of several representative fusions are presented in Table I. Between 12 and 40 percent of the cultures from each fusion produce antibodies which react with the immunogen in an enzyme-linked immunosorbent assay (ELISA). This method has been successful with cellular antigens, haptens, glycoproteins, and highly conserved proteins.

We have used several methods to characterize the reactivity of monoclonal antibodies with soluble proteins and with cellular components. Radioimmunoprecipitation is an excellent analysis for soluble protein antigens, and we have developed a competition radioimmunoassay for calmodulin using monoclonal antibodies (Pardue, et al., 1981). We are developing a new technique for the analysis of monoclonal antibody reactivity with soluble and cellular antigens. This assay, the gel permeation-high pressure liquid chromatography enzyme-linked immunosorbent assay (HPLC-ELISA), allows resolution of cell surface membrane antigens according to their molecular weight after denaturation in sodium dodecyl sulfate (SDS) and 2-mercaptoethanol (2-ME). The entire separation requires only 20 min, and fractions of the separated proteins are collected and used for an ELISA. Figure 1 demonstrates the use of this method with monoclonal antibody to the hapten rhodamine. Samples of 10 ug of bovine serum albumin (BSA) or rhodamine isothiocyanate-BSA (RITC-BSA) were denatured in SDS and 2-ME and applied to a gel permeation column on a HPLC equipped with a UV detector. The top panel represents the absorbance at 280 nm. Fractions were collected, and prepared for the ELISA assay. The reactivity of the separated fractions of RITC-BSA and of BSA with the monoclonal antibody Z64.1 specific for rhodamine was measured in the ELISA. The lower panel represents the specific absorbance of 414 nm from the ELISA.

Analysis of the reactivity of monoclonal antibodies with cell surface antigens requires methods for the detection of antibody binding to individual cells. We have used indirect immunofluorescence (Purdue, et al., 1981), complement-mediated lysis (Dicke, et al., 1981; Miner, et al., 1981), indirect immunoperioxidase, rosette formation, and immunogold staining (DeMay, et al., 1981) to analyze the reactivity of cellular specific antibodies.

Table I

Frequency of antibody producing hybridomas reacting

with the immunogen after in vitro inmunization

Immunogen	Frequency	Reference
X2180-1A Yeast Mannan	18/96	Reading, 1981
RAW117 lymphosarcoma cells	23/96	Miner, et al., 1981
Human AML cells	33/96	Dicke, et al., 1981
Calmodulin	30/96	Pardue, et al., 1981
Fluorescein	12/96	Reading and Zuckerbrod unpublished
Actin	20/96	Pardue and Reading unpublished
BN myeloid leukemia cells	18/45	Vellekoop and Reading unpublished

We have developed a procedure for the elimination of antibody reactive cells using immunogold. Immunogold reagents are produced by absorbing purified immunoglobulins onto colloidal gold particles. In the separation procedures, the density of antibody reactive cells is increased to greater than 1.080 g/ml by indirect immunogold binding, and the reactive cells are separated from the nonreactive cells on a cushion of Ficoll-hypaque (d = 1.077). Table 2 describes the basis of the separation. It is calculated that $9.3x10^5$ 20 nm particles are required to increase the density of acute leukemic cells (d = 1.050) to 1.077 g/ml. If 50 nm particles are used, this number is reduced to $6x10^4$ particles. We are investigating the use of 140 nm particles, but we have effected separation of monoclonal antibody-reactive cells with 20 nm particles. The density of the reactive cells can also be increased by the application of a second immunogold reagent which contains bound mouse immunoglobulin (DeMays, et al., 1981).

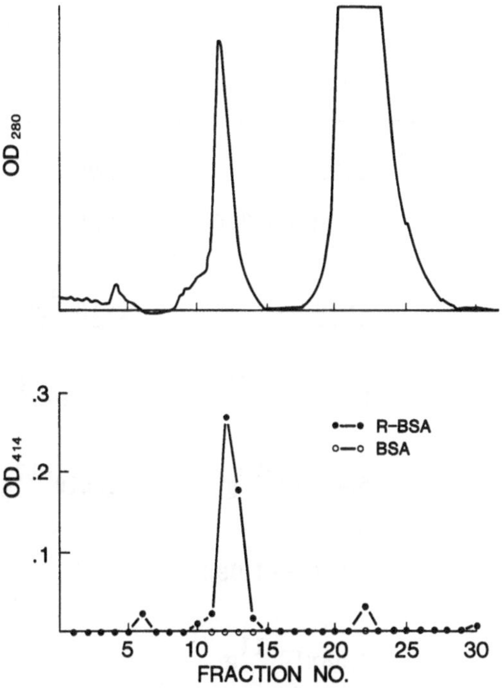

Figure 1. HPLC-ELISA. Samples of either rhodamine-isothicyanat labeled bovine serum albumin (RITC-BSA), or BSA were separated by high resolution gel-permeation high-pressure liquid chromatography and the absorbance at 280 nm was recorded using a flow thru spectophotometer in line with a HPLC and fractions were collected. The absorbance of the RITC-BSA sample was aligned with the fractions and is reproduced in the top panel. Samples from each fraction were analyzed for reactivity with a monoclonal antibody specific for the hapten rhodamine, using a modified ELISA procedure as described in the text. The bottom panel demonstrates the absorbance at 414 nm of the colored enzyme product from the HPLC-ELISA when RITC-BSA (closed circles), or BSA (open circles) was separated by gel permeation chromatography.

We are also developing a new type of bifunctional monoclonal antibody "recombinant monoclonal antibodies" (RMA). These new reagents are produced from "quadromas" (see Figure 2) which arise from the fusion of two hybridomas. Conventional hybridomas are hybridized to yield quadromas. If nonproducer myeloma lines are used for the construction of the hydridomas, they will only make the spleen cell-specific immunoglobulin chains, and the quadromas will produce the immunoglobulin chains of each parent hybridoma. Antibody chains recombine in hybrids to form

Table II

Immunogold separation

	Cells	Gold	
Density (g/ml)	1.050 - 1.070	17.0	
Diameter	$1x10^5$m	$2x10^{-8}$m	$5x10^{-8}$m
Volume	5.23×10^{-10}ml	$10^{-18}cm^3$	$1.56x10^{-17}cm^3$
Weight	$5.49x10^{-10}$g	$1.7x10^{-17}$g	$2.65x10^{-16}$g
Increase in Density	1.050 — 1.080		
Increase in Weight	$1.58x10^{-11}$g		
# of particles needed/cell		$9.3x10^5$	$6.0x10^4$

new species of mixed chains (Cotton and Milstein, 1973), and in the case of quadromas, some of the RMA's formed will have a combining site from each of the original spleen cell antibodies. In Table III several examples of RMA's and their projected functions are listed. The expected frequency based on the assumption of random recombination of immunoglobulin chains intercellularly is presented in Table IV. Quadroma formation from hybridomas producing immunoglobulins with two combining sites per molecule (IgG, IgA, IgD, and IgE) are expected to produce antibody with one of each of the parental combining sites with a frequency of one in eight molecules. For several of the applications suggested in Table III, affinity chromatography of the antibodies such as IgG might be necessary to enrich for the desired molecules. If IgM producing hybridomas are used for the quadroma formation, 89% of the secreted RMA molecules should contain at least one of each parental combining site, based on the assumption of random association of antibody chains. If there is preferential association of the paired heavy and light chains, these yields should be increased.

4. DISCUSSION

The production of monoclonal antibodies is one of the major advances

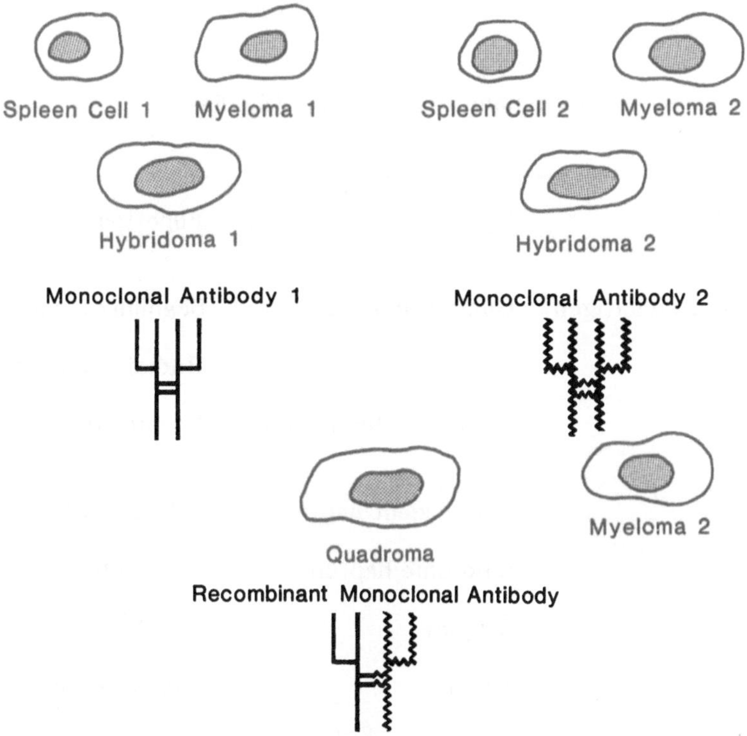

Figure 2. Quadroma Formation. Spleen cell 1 making antibody 1 and myeloma cell 1 producing no antibody are fused to produce hybridoma 1 which produces monoclonal antibody 1. Spleen cell 2 making antibody 2 (with a specificity different from antibody 1) is fused with myeloma cell 2 which produces no antibody to form hybridoma 2 which produces antibody 2. Hybridoma 1 and 2 are fused together to produce a new biological entity, a quadroma, which produces heavy and light chains of both parental types. The immunoglobulin heavy and light chains recombine to form "recombinant monoclonal antibodies", some of which will have one combining site specificity from each parental spleen cell.

in modern biology and medicine. The only limitations to this advance appear to be technical problems related to the cell biology aspects of these procedures and the lack of humoral response or selective response when certain antigens are used for immunization in vivo. Antibody production in vivo to T-dependent antigens requires T-helper cells which

Table III

Recombinant Monoclonal antibodies

Hybridoma 1	Hybridoma 2	Function
Tumor associated antigen	Gamma-labeled hapten	Scanning localization
"	"	Radioimmunoassay
"	Alpha-labeled hapten	Localized radiotherapy
"	Fluorescent hapten	FMF detection, IIF
"	Liposome hapten	Drug delivery
"	Enzyme	ELISA
"	Toxin	Tumor cell killing
Molecules of biological and pharmacological interest	Gamma-labeled hapten	Radioimmunoassay
"	Enzyme	ELISA
"	Fluorescent hapten	Fluorescent assay

interact with B cells over short distances to enhance humoral immunity. T-suppressor cells may act to limit humoral responses by inhibiting T-helper cells. The in vitro immunization procedure utilizes a soluble lymphokine preparation to stimulated antibody production. The mixed-thymocyte culture apparently generates T cell replacing factors which may be identical with, or related to soluble factors produced by T-helper cells during in vivo immunization. The factors present in the thymocyte-conditioned medium are antigen independent, and the effect on B lymphocyte stimulation in vitro may be analogous to the in vivo nonspecific activation of "bystander" B cells during stimulation of specific humoral responses.

TABLE IV

Recombinant Monoclonal Antibody (RMA)

IgG, IgA, IgD, IgE	IgM
Parental species[a] lhhl, LHHL	
Recombinant pairs:	
lh, lH, lh, LH, 2 chains	lH, lH, Lh, LH, 10 chains
Frequency of useful RMA's:	
2/16 = 1/8	p = 89%
Recombinant antibodies:	
lhhl, lhHl, lhhL, lhHL	
lHhl, lHHl, lHhL, lHHL	
Lhhl, LhHl, LhhL, LhHL	
LHhl, LHHl, LHhL, LHHL	
Useful RMA's after one affinity step:	
2/7	94%
Useful RMA's after two affinity steps:	
100%	100%

[a]Consider one antibody molecule with light chain l and heavy chain h, and another antibody with light chain L and heavy chain H.

In vitro responses have been produced to antigens which elicit either limited or no humoral response in vivo such as ornithine decarboxylase (R. Luben, personal communication), and bovine and human terminal deoxynucleotidyl transferase (Steve Zimmer, personal communication). In addition to the lack of restricted response in vitro, the immunization procedure is also simpler than most in vivo immunization protocols and more rapid (5 days total).

We have used the immunogold staining technique of De Mey and coworkers (1981) for light microscopic analysis of monoclonal antibody reactivity with cells in complex mixtures such as bone marrow, and have adapted the technique for preparative separation of monoclonal antibody-binding cells. We intend to develop this technique for the elimination of monoclonal antibody-reactive residual leukemic cells from remission marrow samples prior to autologous bone marrow transplantation of acute leukemic patients.

The HPLC-ELISA is a modification of the gel electrophoresis-derived enzyme-linked immunosorbent assay (GED-ELISA) described by Lutz and coworkers (1979). This modification eliminates the necessity of elution of the proteins from the polyacrylamide gel after separation, since they are collected in solution from the effluent of the gel permeation chromatography. Direct coating of the separated protein samples to microtiter plates was not possible due to the SDS in the solution, but fixation of the samples after air drying onto the plates allowed removal of the SDS and retention of the antigen. We are adapting this technique for the analysis of monoclonal antibodies to cell surface antigens on human acute leukemic cells. This procedure will probably require purified plasma membranes as does the GED-ELISA procedure.

The utility of recombinant monoclonal antibodies may depend ultimately upon the ease with which they can be produced and the stability of the quadromas. Bifunctional monoclonal antibodies have been produced previously by chemical reduction and reoxidation of disulfides in mixtures of two antibody specificities (Raso and Griffen, 1931). The quadroma process, for which various patents have been filed and assigned to The University of Texas System Cancer Center, M.D. Anderson Hospital and Tumor Institute, may allow the biological production of a wide variety of recombinant monoclonal antibodies for the analysis of numerous drugs and macromolecules, and for cancer diagnosis and therapy.

5. METHODS

5.1. Culture conditions

Cells were cultured in a supplemented Dulbecco's modified Eagles' medium (SDMEM) which contained the high glucose (1 g/l) formulation of DMEM (Gibco) supplemented with 1% MEM non-essential amino acids (Gibco), 1 mM sodium pyruvate, 50 μM 2-mercaptoethanol, 30 μM hypoxanthine, 30 M thymidine, 26 mM sodium bicarbonate, and 18 mM HEPES (Flow Labs). The pH was adjusted to 7.3 and the osmolarity to 290 mOsmoles/liter prior to sterile filtration. All cultures were maintained in 2% type 100 rabbit serum (Quadroma, Escondido, Calif.) in a humidified incubator in an atmosphere of 5% CO_2 in air at 37 °C. Myeloma cells were maintained in 75 cm^2 tissue culture flasks containing 30 ml of SDMEM and were subcultured when the density approached 10^6 cells/ml. The myeloma cells were maintained in 100 μM 8-azaguanine except for the two subcultures prior to a hybridization.

5.2. Immunization in culture

The procedure for in vitro immunization utilizes a mixed-thymocyte culture-conditioned medium as a source of lymphokines to stimulate immunization. Thymocytes from Balb/c mice and from a strain differing at the major histocompatability locus (e.g., C57Bl/6) are cocultured at 2-4x10^6 cells/ml in SDMEM with 2% rabbit serum for 48 hours. The cells and debris are removed by centrifugation and the medium is stored at -70 °C in 10 ml aliquots in tightly capped vials.

To perform the immunization in culture, a spleen is removed from a non-immunized female Balb/c mouse using sterile technique. A single cell suspension is prepared by pressing the cells through a sterile 50 mesh stainless steel screen with the plunger from a 12 ml plastic syringe. The capsule remains in the screen and most of the splenocytes are collected in a 60 mm petri dish containing 5 ml of SDMEM with 10% rabbit serum. The clumps are pipetted with a 5 ml pipet until a single cell suspension is achieved (5-6 times). The cells are diluted to 20 ml with serum-free SDMEM containing 30-1000 μg/ml of soluble antigen or 10^7 irradiated (2500 rad) cells. A vial of thymocyte-conditioned medium is thawed rapidly at 37 °C and added to the culture. The low rabbit serum concentration appears to reduce background antigenic stimulation. The mixture is placed in a 75 cm^2 flask and placed in the incubator for 5 days. Immunization with particulate antigens (such as cells or membranes) is facilitated by placing the flask on a rocking platform after equilibration of the gas phase. With a successful immunization, numerous large blast cells can be observed by phase-contrast microscopy.

5.3. Hybridization

The immune spleen cells are dislodged from the flask by sharply striking the side of the flask, transferred to a 50 ml conical polypropylene centrifuge tube, and centrifuged. The medium is aspirated and 30 ml of myeloma cells in log growth phase (5-8x10^5 cells/ml) are added and the tube is centrifuged again. The medium is aspirated, and the cells are resuspended in 30 ml of serum-free SDMEM. The cells are washed by centrifugation, the medium is aspirated, and the pellet is placed in a 37 °C water bath for 2 min. The immune spleen cells and myeloma cells are fused (Galfre, et al., 1977; Oi and Herzenberg, 1980) in a solution containing polyethyleneglycol 1540 M.W. (Polysciences) at 47% (v/v) (Portecorvo, 1975), and dimethylsulfoxide at 7.5% (v/v) (Norwood, et al., 1976) in tricine-buffered Hanks balanced salt solution (Gibco) pH 8.0. The fusion medium is prepared fresh daily. Containers with fusion medium, 10 ml of serum free SDMEM, 10 ml of SDMEM containing 10% rabbit serum, and 90 ml of SDMEM containing 2% rabbit serum and $4x10^{-7}$ M aminopterin are each warmed to 37 °C, and a single cell suspension of spleen cells from an unimmunized animal are prepared as described above and $3\text{-}4x10^{7}$ cells are left at room temperature in DSMEM. The mixed pellet of immune spleen cells and myeloma cells are removed from the water bath and resuspended in 1 ml of fusion medium just removed from the water bath. The solution is added over one min, and the cells are continuously stirred with the pipet tip. Stirring is continued for another min and then 2 ml of serum-free SDMEM at 37 °C is added over the next three min with stirring. Seven ml of SDMEM containing 10% rabbit serum are added over the next three min with stirring. The cells are diluted to 100 ml with the prewarmed SDMEM containing 2% rabbit serum and aminopterin, and the spleen cells are added as feeder cells. The suspension is diluted to 96 wells (16 mm diamter) of 24 well plates (Linbro). Cells in the process of fusion can be observed at this point by phase contrast microscopy. Due to the feeder cells only a single medium change at 8-10 days after the fusion is necessary, reducing the chances of contamination during the two to three week growth period in the one ml cultures.

5.4. Enzyme-linked immunosorbent assay (ELISA)

The ELISA assay was modified from the procedure of Saunders (1979). To screen cultures for antibodies reactive with cell surface components, the cells are immobilized on the surface of microtiter plates. The wells are treated with 50 ul of 5% glutaraldehyde in 0.1 M $NaHCO_3$ for 30 min at 25 °C. The plates are washed three times by filling the wells with 0.01 M Na_2HPO_4 in 0.1 M NaCl, pH 9.0 (PBS-9), and flicking out the solution. A washed target cell suspension (10^7 cells/ml) in HEPESbuffered Hanks balanced salt solution (HHBSS) is added (50 µl/well), the plates are spun at 500 x g for three min, and 200 µl of 1% formaldehyde is added to each well. After 15 min at 25 °C, the plates are spun again, and the liquid is flicked out. The plates are washed three times with PBS-9 and 50 µl of 1% bovine serum albumin (BSA), in PBS-9 is added to each well. After 15 min at 25 °C, 50 µl of each hybridoma medium sample is added to duplicate wells, and 2% RS in SDMEM is added to eight wells as controls. The plates are incubated for 90 min at 25 °C or overnight at 4 °C and the plates are washed 10 times with 0.05% Triton-X-100 in distilled water. Horseradish peroxidase-conjugated IgG fraction of goat anti-mouse immunoglobulins (Cappel Labs) diluted 1:300 in 0.5 M NaCl containing 0.5% Triton-X-100, 0.01 M Na_2HPO_4 and 1 mg/ml BSA is added (50 μl/well), and the plates are incubated at 25 °C for 10 min. The plates are washed 10 times with 0.05% Triton-X-100 and 100 μl/well of substrate solution ((0.4 mM 2,2'-azino-di-(3-ethylbenzathiazoline sulfonic acid) diammonium salt (ABTS) and 0.3% hydrogen peroxide in 0.1 M sodium citrate buffer, pH 4.0)) is added. After 30 min at 25 °C, the absorbance of the enzyme product is measured at 414nm in a plate reader (Flow labs). The readings of the medium controls are averaged for each plate, and samples are considered positive if they are greater than the mean of the controls + 2 S.D. For soluble antigens, the same protocol is used, except that the plates are prepared by air drying (at 37 °C) 50 μl/well of a solution of 1-100 μg/ml of antigen in water. Before use, the plates are washed six times with PBS-9 and unbound sites on the plastic are blocked with BSA as above. A parallel ELISA test on plates prepared with an irrelevant soluble antigen or cell type is essential in the early screening of hybridoma cultures to obtain the desired specificities.

5.5. HPLC-ELISA

Protein samples (0.5 mg/ml) were heated for 10 min at 10° C in sample buffer (Maizel, 1971), and after cooling to 23 °C, 20 μl was injected onto a TSK 3000 gel permeation high pressure chromatography column (300x7 mm) (Beckman Instruments) fitted with a precolumn (80x7 mm) of the same material. The samples were chromatographed in (0.9% sodium chloride) and fractions (0.7 ml) were collected. The eluate was analyzed using a uv detector with a 280 nm filter. The eluted fractions were tested for reactivity with monoclonal antibodies by air drying 50 μl of each fraction in duplicate wells of 96 well polystyrene microtitration plates at 37 °C. When the samples were dry, the protein was fixed onto the plates, and the SDS removed with 50 μl of 10% acetic acid in 25% isopropanol for 30 min at 23 °C. The plates were washed six times with PBS-9 and nonspecific binding sites were blocked with BSA as dascribed above. The plates were then tested by the standard ELISA procedure, except that the antibody sample used was the same for all of the wells. The column buffer was used as a control for the HPLC-ELISA.

5.6. Immunogold preparation

Colloidal gold reagents coated with antibodies were prepared by a modification of the procedures described by Geohegan and Ackerman (1977). Hydrogen tetrachloroaurate (Alpha-Ventron) (0.01 was dissolved in 100 ml of boiling distilled water, and various amounts of 1% trisodium citrate was rapidly pipetted into the stirred solution. The volume of citrate added determines the final diameter of the colloidal gold particles. For the preparation of 18-20 nm diameter particles, 2.5 ml of citrate is added, but particles as large as 147 nm can be produced by using less citrate (Frens, 1973). The solution was cooled to 23 °C and the pH was adjusted to 8.0 using 0.2M K_2CO_3 and pH paper, since the colloidal gold may clog the electrode of pH meters. The IgG fraction of goat anti-mouse immunoglobulin (GAMI) (Quadroma, Inc.) was dissolved in water and dialyzed against 2 mM potassium borate buffer, pH 8.0. Four mg of GAMI was added to the colloidal gold and this amount was sufficie to prevent floculation when a drop of the solution was mixed with a drop of 1% NaC1 solution. After one min, one mg of PEG 20,000 MW (Sigma) was added to stabilize the immunogold preparation (Horisberger, et al., 1975), and the suspension was centrifuged at 5000g min to remove large aggregates. The immunogold was purified by three cycles of centrifugation at 840,000g min and resuspension in 25 ml of HEPES-buffered (18 mM) Hanks balanced salt solution containing 1% BSA, $2x10^{-2}$ M NaN_3 and 100 ug/ml polyethyleneglycol (20,000 M.W.), pH 7.4.

5.7. Immunogold staining

Cells ($2-3x10^6$) were washed in HEPES-buffered Hanks balanced salt solution containing 1% BSA, $2x10^{-2}$M NaN_3, pH 7.3. The cells were pelleted at 200g for 5 min and incubated with 0.3 ml of monoclonal antibody containing medium from hybridoma cultures, or purified monoclonal antibody diluted in the same buffer for 1 hr at 4^oC. The cells were washed once with the above buffer, pelleted, resuspended in 0.25 ml of the goat anti-mouse immunogold reagent, and incubated for 45 min at 23^oC. The cells were washed once with the above buffer and cytocentrifuge preparations were prepared from samples and controls in duplicate. Staining could be enhanced by a second incubation with mouse immunoglobulin immunogold (De Mey, et al., 1981). The reactive cells appeared, colored when viewed by light microscopy, the color being dependent on the size of the immunogold particles. After air drying, the cells often appeared gold in color rather than orange to violet.

5.8. Immunogold separation

For cell separation, the light density cells were collected from human peripheral blood on a cushion of Ficoll-hypaque as previously described (Dicke, et al., 1931) and treatad as described above for immunogold staining. After the final wash, the immunogold-treated cell mixture was again subjected to centrifugation on Ficoll-hypaque. The cells with immunogold bound to their surface were collected in the pellet, and the unreactive cells were collected from the Ficoll-hypaque interface.

ACKNOWLEDGEMENTS

We wish to thank Dr. Raymond Ivatt for help with the gel-permeation high pressure chromatography, and Le Foster for help with the manuscript. The quadroma concept was initiated from suggestions by Tim Updyke.

REFERENCES

Cotton, R.G.H., and Milstein, C., 1973. Fusion of two immunoglobulin producing myeloma cells. Nature, 244:42.

De Mey, J., Moeremans, M., De Waele, M., Geuens, G., and De Brabander, M., 1981. The IGS (immuno gold staining) method used with monoclonal antibodies. Peptides Biological Fluids, (in press).

Dicke, K.A., Tindle, S.E., Davis, F.M., and Reading, C.L., 1981. Elimination of myeloid leukemic cells in human bone marrow after treatment with monoclonal antibodies to cell surface determinants. J. Supramol. Struct., (in press).

Frens, G., 1973. Controlled nucleation for the regulation of the particle size in monodisperse gold suspensions. Nat. Phys Sci. 241:20.

Galfre, G., Howe, S.C., Milstein, C., Butcher, G.W., and Howard, J.C., 1977. Antibodies to major histocompatibility antigens produced by hybrid cell lines. Nature, Lond., 266:550.

Geoghegan, W.D., and Ackerman, G.A., 1977. Adsorption of horseradish peroxidase, ovomucoid and anti-immunoglobulin to colloidal gold for the indirect detection of concanavalin A, wheat germ agglutinin and goat antihuman immunoglobulin G on the cell suraces at the electron microscopic level: A new method, theory and application. J. Histochem. Cytochem. 25:1187.

Horisberger, M., Rosset, J., and Bauer, H., 1975. Colloidal gold granules as markers for cell surface receptors in the scanni electron microscope. Specialia, 15:1147.

Kearney, J.F., Radbrunch, A., Liesegana, B., and Rawjewsky, K., 1979. A new mouse myeloma cell line which has lost immunoglobulin expression but permits the construction of antibody-secreting hybrid cell lines. J. Immunol., 123:1548.

Kohler, G., and Milstein, C., 1975. Continous cultures of fused cells secreting antibody of predetermined specificity. Nature, Lond., 256:495.

Luben, R.A., and Mohler, M.A., 1980. In vitro immunization as an adjunct to the production of hybridomas prodcuing antibodies against the lymphokine osteoclast activation factor. Molec. Immunol., 17:635.

Maizel, J., 1971. Polyacrylamide gel electrophoresis. Methods Virol., 5:179.

Marbrook, J., 1967. Primary immune response in cultures of spleen cells. Lancet, ii:1279.

Mishell, R.I., and Dutton, R.W.., 1967. Immunization of dissociate spleen cell cultures from normal mice. J. Exp. Med., 126:423.

Miner, K.M., Reading, C.L., and Nicolson, G.L., 1981. In vivo and in vitro production and detection of monoclonal antibodies to surface components on metastatic variants of murine tumor cells. Invasion Metastasis (in press).

Norwood, T.H., Ziegler, C.J., and Martin, G.M., 1976. Dimethyl sulfoxide enhances polyethylene glycol-mediated somatic cell fusion. Somatic Cell Genet., 2:263.

Oi, V.T. and Herzenberg, L.A., Immunoglob-lin-producing hybrid cell lines; in Mishell, Shiigi, Selected methods in cellular immunology, pp. 351-397 (Freeman, New York 1980).

Pardue, R.L., Brady, R.C., Dedman, J.R. and Reading, C.L., 1981. Monoclonal antibodies to calmodulin produced by in vitro immunization of mouse spleen cells. J. Cell Biol. 91:81a.

Portecorvo, G., 1975. Production of mammalian somatic cell hybrids by means of polyethyleneglycol treatment. Somatic Cell Genet., 1:397.

Raso, V., and Griffin, T., 1981. Hybrid antibodies with dual specificity for the delivery of ricin to immunoglobulin-bearing target cells. Cancer Res., 41:2073.

Reading, C.L., 1981. Immunization in culture for monoclonal antibodies to mannan. Proc. VIth Intl. Symp. on Glycoconjugate Japan Scient. Soc. Press, Tokyo pp. 332-232.

Schulman, M., Wilde, C.D., and Kohler, G., 1978. A better cell line for making hybridoma secreting antibodies. Nature Lond., 276:269.

16

LIPOSOME FACILITATED IN VITRO INDUCTION OF PRIMARY CELL-MEDIATED IMMUNITY TO HUMAN CANCER ANTIGENS: POTENTIAL ADJUNCT TO HYBRIDOMA TECHNOLOGY

Baldwin H. Tom, Leonard Raphael, Jie-shi Liu* and Jayati Sengupta

Departments of Surgery and Biochemistry and Molecular Biology, The University of Texas Medical School, MSMB 6240, 6431 Fannin and the University of Texas Graduate School of Biomedical Sciences in Houston, Houston, Texas 77030; *Lanzhou Medical School, Gansu, People's Republic of China

1. SUMMARY

Liposomes, prepared with human colon or breast cancer membrane vesicles, have been shown to successfully induce primary cell-mediated xenoimmune tumor-specific cytotoxic responses following in vitro culture. These results support the use of murine xenogeneic responses for studying human tumor antigens. However, comparable experiments utilizing human peripheral blood lymphocytes in coculture with colon cancer antigens only resulted in generating blastogenic and not cytotoxic cells in vitro. Nevertheless, the ability to induce primary immune responses in vitro provides a means to characterize tumor antigen-lymphocyte interactions and the possibility to selectively isolate effector T or B lymphocytes in culture with hybridoma techniques.

2. INTRODUCTION

The availability of monoclonal antibodies and their subsequent use in isolating unique antigenic determinants have led to new directions for cancer research. Consequently, the potential for deliberate immunization of patients with "pure" antigens exists. An ultimate goal of such a study would be to immunize patient lymphocytes in vitro against their tumors for reinfusion in an active, specific, immunotherapy protocol. A logical component of the latter studies would be in analyzing the cellular mechanisms of immunity by using hybridoma techniques in immortalizing the lymphoid cells during the course of in vitro immunization. However, a most important deterrent to early success with the former suggestion is the lack of availability of an in vitro immunization system. Thus, research emphasis in our laboratory has been directed toward developing an in vitro immunization technique capable of utilizing isolated, "pure" antigens and of producing a primary cell-mediated immune response.

2.1. In vitro immunization

The possibility of sensitizing lymphocytes in vitro to nonlymphoid target cells was first examined, systematically, in the xenogeneic system (Ginsburg and Sachs, 1965). Rat lymphocytes sensitized to cultured mouse embryo cells in vitro produced lymphocytes with a stronger immune response than those sensitized in vivo (Berke, et al, 1971). Wagner and Rollinghoff (1973) reported that mouse lymphoid cells sensitized in vitro to syngeneic and allogeneic plasma cell tumors inhibited the growth of the same tumors in syngeneic mice. In an immunotherapy model using in vitro sensitized syngeneic mouse lymphoid cells (Treves, et al, 1975), more than 75% of the mice who received the sensitized lymphocytes resisted lethal lung metastases from a Lewis lung tumor while only 30% of the control mice survived. Comparable human studies are difficult to design.

When in vitro human lymphocyte-tumor cell interaction studies were first performed using lymphocytes from cancer patients and normal individuals, there was a greater target cell destruction with lymphocytes from cancer patients than normal donors (Bloom and Collnaghi, 1974; Hellstrom, et al., 1971). However, lymphocytes from normal individuals were later shown to exhibit natural cytotoxicity on cultured tumor cells (Takasugi, et al., 1973). When the cytotoxic activity of lymphocytes from melanoma patients to melanoma-associated antigen was boosted in vitro (Golub and Morton, 1974), a cytotoxic activity detectable against autochthonous and allogeneic melanoma cells, but not to other tumor cells or autochthonous fibroblasts was shown. While histiotypic specificity is expressed in some systems, nonspecificity appears in others. Sharma and Terasaki (1974) and Sharma (1976), immunized human lymphocytes using a five day incubation period. They showed that normal lymphocytes could be induced to exhibit cytotoxic responses to the sensitizing cell types. However, as should be expected when employing a heterogeneous, cellular antigen, specificity was unconvincing, and germane to this study, no colon cancer lines were used. Sharma has noted that some colon cancer cell lines tested in his system were unable to induce cytotoxic cells (personal communication). Schecter and colleagues (1976) also reported in vitro sensitization of human lymphocytes against tumor cells. Whitson et al., (1981) examined the generation of cytotoxicity after a six day in vitro sensitization of human lymphocytes to two human multiple myeloma lines. Cytotoxic lymphocytes were generated at moderate levels with considerable cross-reactivity between immune lymphocytes induced from different myeloma lines. Thus while whole cell approaches can provide a cytotoxic effector cell, it is doubtful that any deliberate control or molecular characterization of the specific immunogens is possible. Cell-free (molecular) approaches are needed.

2.2. Cell-free immunization

Attempts to induce in vitro immunity with cell-free antigens have heretofore been marginally effective (Table I.). For example, Lemonnier, et al. (1978) demonstrated that while murine P815 membranes generated a strong secondary cytotoxic response, only a weak primary response was elicited. Indeed, membrane immunogens were shown to be immunosuppressive by Maier, et al. (1980). Corley et al. (1975) showed that subcellular fractions of human lymphoblasts were capable of stimulating only weak proliferative, and weak cytotoxic responses. Utilizing human tumor cell extracts (but not including colon), Sharma (1977) reported the generation of nonspecific, cytotoxic human peripheral lymphoid cells.

2.3. Liposome-mediated immunity

Liposomes, containing purified HLA-A and HLA-B (Engelhard, et al. 1978), and HLA-DR molecules (Burakoff, et al. 1980), have been successfully used to elicit xenogeneic mouse cytotoxic T lymphocytes (CTL)[1] by culturing them with previously in vivo primed spleen cells. Secondary anti-Sendai CTLs using liposomes composed of hemagglutinin-neuraminidase glycoproteins of Sendai virus and the $H\text{-}2K^k$ glycoprotein have been induced (Hale, et al., 1980). In studies by Herrmann and Mescher (1981), the inclusion of a detergent-insoluble, actin-rich membrane matrix into liposomes containing purified $H\text{-}2K^k$ enhanced the stimulation of secondary CTLs over liposomes without the matrix. Liposomes bearing viral and H-2 antigens have been reported to generate primary cytotoxic T lymphocytes, but only in the presence of Concanavalin A supernates (Hale and McGee, 1981). Not only do liposomes serve as antigen carriers, but also they serve as adjuvants. For example, the immunogenicity of human serum albumin and bovine gamma globulin are enhanced when they are attached to the surface of liposomes, but not when internalized (Van Rooijen and Van Nieuwmegan, 1980). Supporting evidence suggesting that cell bound or surface presented antigens are more effective immunogens than unbound antigens has been described (Nakao et al. 1981), in which antigens absorbed to autologous or xenogeneic cell surfaces provided enhanced antibody responses.

Liposomes are thus endowed with key properties which proffer them as appropriate vehicles for generating in vitro immune responses, especially their amenability for specific tailoring, e.g., selecting for cellular versus humoral immunity, or for characterizing roles for macromolecules (e.g., HLA,DR) necessary for inducing immune responses (see Tom and Six, 1980). Indeed, with the availability of monoclonal antibodies to purify antigens, molecular analysis of T-cell activation can be undertaken. For example, Herrmann and colleagues (1982) studied the two-signal requirement for T-cell activation using liposomes and $H\text{-}2K^k$ antigens.

[1] Abbreviations: CTL, cytotoxic T lymphocyte; CMI, cell-mediated immunity; MLV; multilamellar liposomes; Mn, tumor cell membrane; MLV-Mn/Tryp, liposomes made with tumor membranes and trypsinized.

TABLE I.

In vitro cell-free immune responses

Type of cytotoxic response	Murine	Human
Primary response	YES, WEAK: (Lemonnier, et al, 1978; Hale and McGee, 1981; Treves, et al, 1975)	NONE
Secondary response	YES, STRONG: (Lemonnier, et al, 1978; Hale, et al, 1980; Herrmann and Mescher, 1981)	YES, STRONG: (Englehard et al, 1978; Burakoff, et al, 1980)
Alloantigen	YES: (Herrmann and Mescher, 1981)	YES, SECONDARY ONLY: (Englehard, et al, 1978; Burakoff, et al, 1980)
Tumor or viral antigen	YES: (Lemonnier, et al, 1978; Hale and McGee, 1981; Treves, et al, 1975)	NONE[a]

[a] Sharma (1977) has reported nonspecific cytotoxic responses to human tumor cell fractions.

Additional details and applications of liposomes in immunobiology have been reviewed (Tom and Six, 1980). Thus, the hypothesis to be tested in our studies was that in vitro primary induction of cell mediated immunity (CMI) by cell-free human tumor antigens could be accomplished with liposome carriers (Fig 1.). This hypothesis was proven true and is reported herein.

3. RESULTS

3.1. Summary of studies

We previously demonstrated that specific cell-mediated xenoimmune responses to human colon cancer could be induced in mice (Hayashi, et al., 1982). Liposomes were subsequently used to characterize their interaction with colon tumor cells (Tom, et al., 1983), and to provide effective carriers for colon cancer antigens (Raphael and Tom, 1982). The efficacy of the liposome-antigens was tested in the immunization of rabbits and was most effective in producing antibodies to human colon cancer (Tom, et al, 1982). Utilizing the xenoimmune approach, we established a cell-free in vitro immunization protocol using liposome carriers for colon cancer (Raphael and Tom, 1982) and recently for breast cancer cells.[2] The resultant primary and secondary cell-mediated xenoimmunity[3] was shown to resemble allogeneic and syngeneic cell-mediated reponses (Hayashi, et al., 1982). Furthermore, it was demonstrated that the physicochemistry of the lipid was important in inducing maximal anti-colon cancer cytotoxic responses in vitro.[4] Finally, the in vitro liposome-mediated xenoimmune system was adapted to allogeneic human peripheral blood mononuclear cells. The latter studies, discussed below, revealed that specific primary blastogenic responses were produced against the immunizing antigens. However, cytotoxic T-cells generated in the human allogeneic combinations were unimpressive.

3.2. In vitro xenoimmunization to human tumor antigens

The first studies were designed to demonstrate that in vitro CMI reactions in the xenogeneic tumor immune system would exhibit tumor, and not species specificities. BALB/c spleen cells immunized in vitro for 5 days with different whole, cultured human tumor cells exhibited

---

[2] Liu, J.S., Raphael, L. and Tom, B.H.: In vitro induction of primary xenoimmunity to human breast cancer cells. (submitted).

[3] Raphael, L. and Tom, B.H.: Liposome-facilitated xenogeneic approach for studying human colon. Carrier and adjuvant effect of liposomes (submitted).

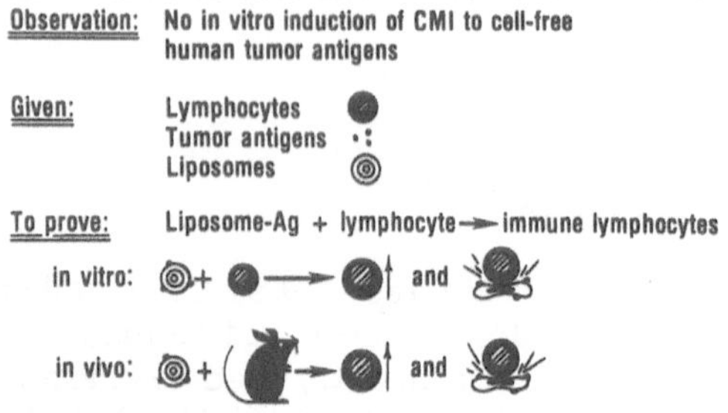

Figure 1. The problem. In vitro primary immunization to human tumor antigens was tested using liposomes and naive lymphocytes.

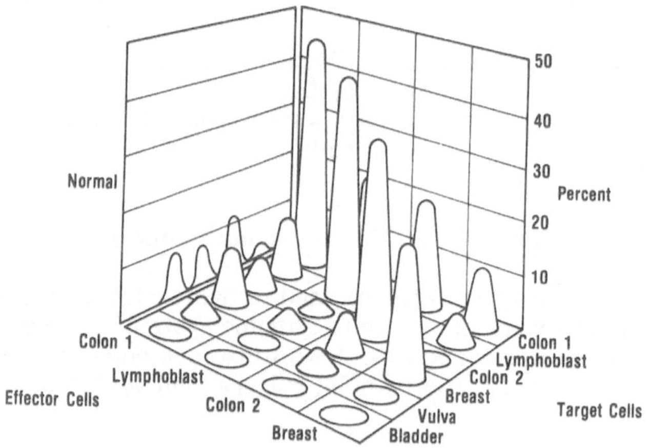

Figure 2. In vitro induction of cell-mediated cytotoxicity with whole tumor cells. Specificity of effector cell cytotoxicity corresponds with each homologous target cell. Normal spleen cell percent cytotoxicities on each target cell are depicted on the left wall panel.

specific CMI in a chromium release assay (Fig 2). Although cross-reactivities were exhibited in each of the effector spleen cell sets, the results confirmed the usefulness of the xenogeneic system to study human tumor immunity.

Histocompatibility antigens may play a role in the reactions, since cytotoxicity by effector cells immune to "colon 1" also were reactive on the lymphoblast target derived from the patient providing "colon 1". Thus"allospecificity" rather than nonspecific xenospecificity in CMI was demonstrated. Further studies utilizing liposome-antigens were thus justified.

When liposome-tumor membrane antigens were used to generate primary CMI responses (Raphael and Tom, 1982), the peak for cytotoxic activity was found on day 5 (Fig. 3). In this example, intact colon tumor cells, LS174T, were more active than the other antigen preparations in inducing cytotoxic cells. However, since a tumor specific molecule has not been identified, comparisons of cell-equivalents is not meaningful. The relevant comparison is between the "membrane" and "liposome-membrane" antigens. When this is done, liposome-antigens invariably induce the strongest cytotoxicity (Fig. 3).

Specificity of the different effector cells (all immunized to the same LS174T colon cancer cell antigens) is clearly depicted in

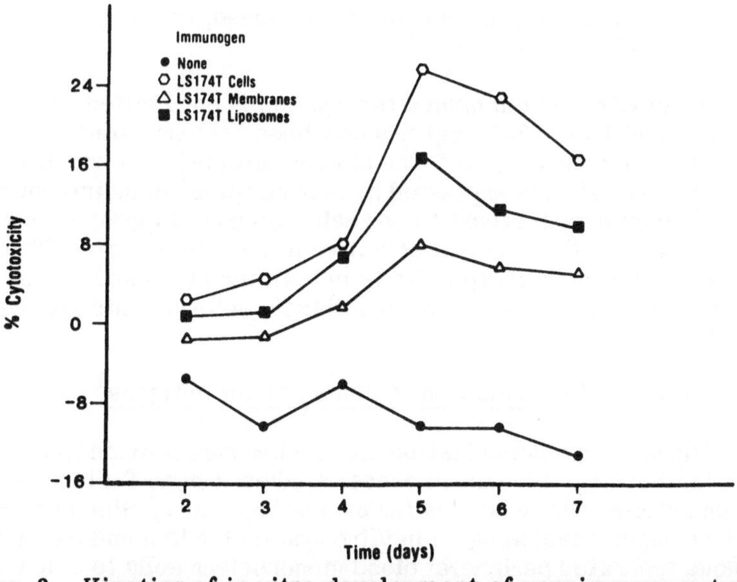

Figure 3. Kinetics of in vitro development of xenoimmune cytotoxic spleen cells. Maximum cytotoxicity against LS174T colon tumor cells is on day 5. (from Raphael and Tom, Cell. Immunol. 71: 224, 1982).

Fig. 4. When compared to the whole tumor cell, "colon 1", immune cells (presented also in Fig. 2), the non-cell antigen preparations also produced very similar cross-reactive cytototoxicities. The important point is that cytotoxicity is greatest against the colon 1 (LS174T) target cells. Also, the liposome-mediated in vitro immunization system was found to require that the tumor antigens be presented on the liposome surface.[3] Liposomes containing the same amount of entrapped proteins (with surface exposed proteins trypsinized) were inefficient antigens. The results in Fig. 4 illustrate these points. Whether liposome-antigens are prepared with tumor membranes (Mn) attached to preformed multilamellar (MLV) liposomes, i.e., MLV+Mn, or with Mn partially incorporated in the liposome preparation (i.e., MLV-Mn), cytotoxic cells were produced (40-45% cytotoxicity in this example). The amount of protein exposed on MLV+Mn preparations was 78-80% and on MLV-Mn preparations, 60% (Tom, et al., 1982). Tumor membranes alone were also effective immunogens. However, if the liposome surface antigens were removed by trypsin before use in immunization, then the production of cytotoxic effector cells is decreased 50% or more (see MLV-Mn/Tryp). In the latter studies, the amount of liposomes and thus liposomal lipid was increased in order to accomodate the increased protein concentrations. There is thus the possibility that the extra lipids acted to suppress the immunization. This was tested by preparing different sets of liposomes with increasing amounts of protein.[3] The results confirmed the observation that surface presentation was required for generating a cytotoxic immune cell (data not shown). Treatment of the liposome-antigen (MLV-Mn) immunized cytotoxic effector cells with anti-thymus cell antibody plus complement also reduced cytotoxic activity. This role of the T cell has recently been confirmed; indeed, the effector cell is Thy 1.2+, I-A-, Lyt 1-, and Lyt-2+.[3]

This in vitro xenoimmunization system is not limited to colon cancer (Raphael and Tom, 1982), but has now been demonstrated with human breast cancer cells (Fig. 5).[2] The studies revealed that specificity to the breast cancer cell was expressed by murine spleen cells immunized in vivo or in vitro by human breast tumor cell antigens. Liposome-carrier antigens were consistently the most effective in vitro immunogen. The question of tumor and/or MHC specificity in both the colon and breast cancer studies will remain until isolated antigenic determinants are evaluated in reconstitution studies.

3.3. In vitro alloimmunization to human tumor antigens[4]

Although xenoimmunization approaches may provide new insights into methodology for human cancer studies, successful human-human immunizations will be needed for evaluating future, clinical applications. We have begun developing an in vitro system for immunizing human Ficoll-Hypaque separated peripheral blood mononuclear cells to colon tumor antigens.

---

[4] Sengupta, J., Klesius, P.H. and Tom, B.H.: Liposome mediated, transfer factor enhanced in vitro human colon tumor immunity. (submitted).

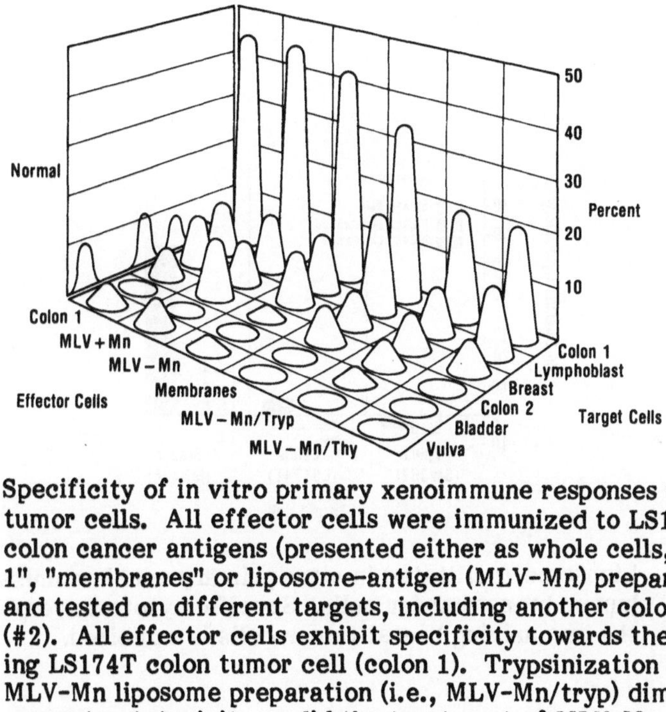

Figure 4. Specificity of in vitro primary xenoimmune responses to colon tumor cells. All effector cells were immunized to LS174T colon cancer antigens (presented either as whole cells, "colon 1", "membranes" or liposome-antigen (MLV-Mn) preparations) and tested on different targets, including another colon tumor (#2). All effector cells exhibit specificity towards the immunizing LS174T colon tumor cell (colon 1). Trypsinization of the MLV-Mn liposome preparation (i.e., MLV-Mn/tryp) diminished the percent cytotoxicity as did the treatment of MLV-Mn immunized effector cells with antithymocyte serum plus complement (MLV-Mn/Thy).

The first question asked in studying allogeneic responses was whether liposomes could be used as carriers to induce blastogenic responses and/or CMI in human lymphocytes in vitro. The representative results of this series of studies are presented in Fig. 6 and 7. Primary blastogenic responses in human peripheral blood lymphocytes (PBL) were readily produced by in vitro incubation with colon cancer membrane (Mn) vesicles, as well as with liposome-antigens (Mn+MLV and Mn-MLV), ranging from two to almost seven-fold over control on days 5,6, to 7. Empty liposomes (PBL+MLV) were not effective stimulants. As a means to determine the presence of specifically immune cells, secondary blastogenic restimulation was demonstrated with colon cancer, but not breast cancer cells,

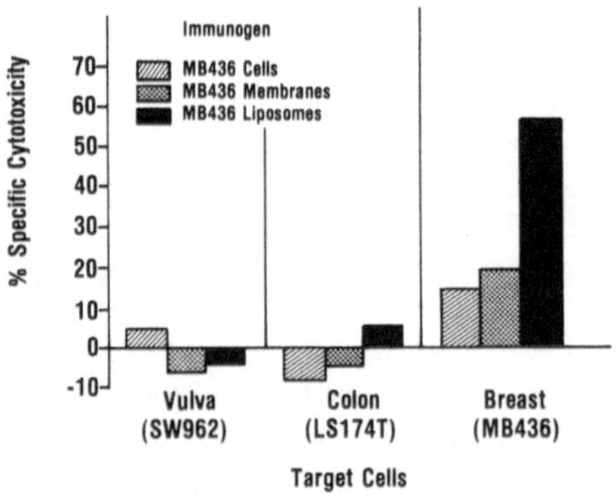

Figure 5. Liposome-facilitated in vitro primary xenoimmunization to human breast cancer cells. Specific cytotoxic responses to the immunizing breast tumor cells is exhibited.

MB436 (Fig. 7). A two to three-fold restimulation with the LS174T colon tumor occurred in the liposome-antigen immune (Mn-MLV) group significantly greater than the restimulation of the cells previously immunized to membrane (Mn) vesicles alone. Another colon tumor cell (Colon 2 or SW1083) was also effective in restimulating the immune cells. The colon tumor immune cells did not respond to breast tumor cells; stimulation indices were 1.4-2.86 in two experiments. The most critical question in this series of studies was whether cytotoxic lymphocytes were produced by in vitro immunization. Although blastogenic responses were readily generated, the results from four experiments revealed no significant cytotoxic responses (data not shown).

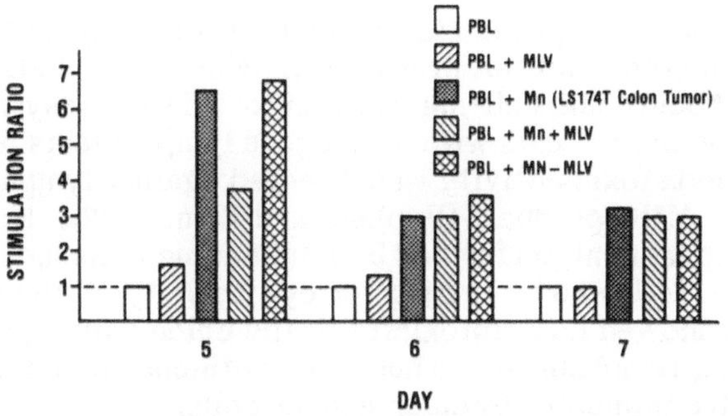

Figure 6. Liposome-facilitated primary immunization to colon cancer in a human allogeneic system. Maximal blastogenic responses to the immunizing LS174T colon tumor cells are seen on day 5. Tumor membranes (Mn) alone, as well as, liposome-membranes are effective inducers of blastogenesis.

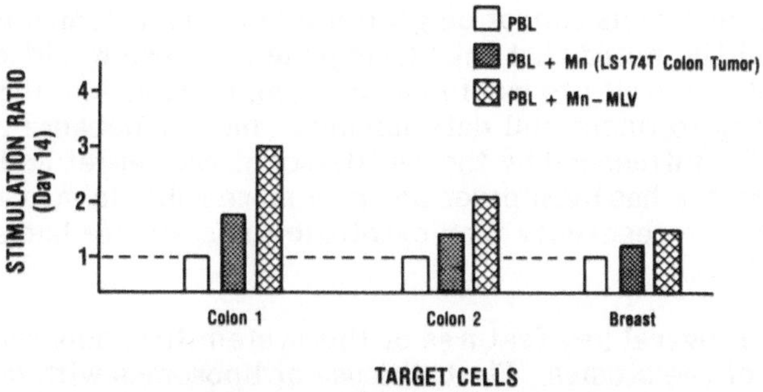

Figure 7. Secondary blastogenesis of in vitro primed human peripheral blood lymphocytes. Blastogenesis is greatest with colon 1 (homologous LS174T tumor) target cells used as secondary stimulators.

4. DISCUSSION AND CONCLUSIONS

An in vitro system to induce primary CMI reactions to human cancers in mouse spleen cells was established and partially characterized. This xenoimmune approach employing liposomes has the potential for use in dissecting the antigenic determinants important in human cancers. Specificity of the in vitro primed spleen cells was clearly demonstrated (p < 01) on human colon, but not breast or bladder tumor cells using both the primed lymphocyte and cell-mediated cytotoxicity assays. The results of competitive inhibition tests with autologous lymphoblasts demonstrated that 30% of the cytotoxic activity was directed against lymphocyte antigens, including possibly MHC products (Raphael and Tom, 1982). However, the reciprocal experiment performed by immunizing cells against the lymphoblast cells, followed by testing for cytotoxicity on the autologous colon tumor cell showed no cytotoxicity to the colon cancer, but 28% on the immunizing lymphoblasts. Thus MHC antigens played a minor role in this in vitro immunity to colon cancer cells.

On the other hand, the authors' efforts in establishing a similar in vitro system with human lymphocytes have not been equally successful. Although liposome-antigen systems induced blastogenesis in human lymphocytes, cytotoxic effector T cells have not been demonstrated. Note that the allogeneic immunizations of human lymphocytes with colon tumor should yield some anti-HLA activities. Only when isolated colon tumor antigens are used in the context of the responders own HLA will a syngeneic, non anti-HLA system be available, e.g., by reconstituting both HLA and tumor antigens into the same liposome bilayer. A question raised by the latter studies will be whether the tumor antigen-HLA reconstituted liposomes generate MHC-restricted cytotoxic T cells in vitro. If tumor specific cytotoxic T cells cannot be generated in human lymphocytes in vitro, it should be noted that the blastogenic response would still serve as a measure of reactivity to purify antigens, and a means to identify patient reactivity to tumor cell determinants. Indeed, because lytic assays are readily influenced by the sensitivity of the target cells, the blastogenic response has been described as a more reliable measure of specific lymphocyte reactivity than cytotoxicity assays for human studies (Vanky, et al., 1981).

There were several key features of this system that may shed light on the success of the studies. First, the use of liposomes with membrane vesicles as the antigen source has heretofore not been attempted by others. Tumor membrane vesicles provide a complex array of surface determinants for fusion to liposome bilayers. One might speculate that the association of several determinants may be needed for immunogenicity, a concept recently described by Brodsky and Parham (1982). Furthermore, the presence of actin filaments on inside-out membrane vesicles associating with the tumor cell antigens may have enhanced immune reactivity (Fig. 8). This latter need for actin-positive cell proteins for in vitro immunity was reported by Herrmann and Mescher (1981). Second, the use of negatively-charged (phosphatidic acid) liposomes and 8 mM $CaCl_2$ may have increased the interaction and fusion of membrane vesicles with the liposome bilayer. Although specificity to the colon tumor cell, LS174T, was demonstrated, one could still argue that the target of the CMI was MHC determinants on the cancer cell which were chemically or physically altered and were no longer identical to those expressed on the lymphoblasts. Furthermore, the above in vitro system lacked consistency and requires improvement; Approximately 3 of 10 comparable experiments showed no specificity, and specific cytotoxicities varied from 10-50%. The use of isolated, identified determinants for reconstitution into liposomes may resolve these problems.

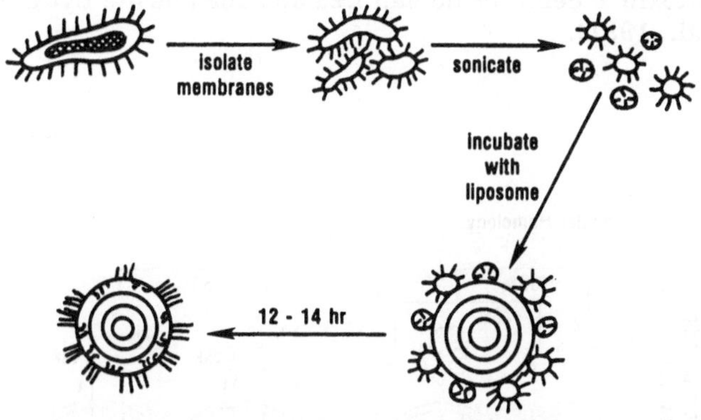

Figure 8. Tumor cell membrane vesicles in liposomes. It is assumed that both right-side out and inside-out vesicles are produced in the sonicate which fuse with the liposomes.

Although a systematic evaluation of cell-mediated xenoimmunity to human tumor cells has not yet been presented, cellular immunity to xenogeneic lymphoid cells (rats, mouse, human) has been described indicating that host lymphoid cells could recognize MHC antigens of xenogeneic lymphoid cells in vitro and in vivo (Burakoff, et al., 1977; Carnaud, et al., 1977; Ginsburg, 1968; Lindahl and Bach, 1975; Peck et al., 1967; Sondel and Bach 1975). Indeed Swain and colleagues (1982) recently reported that human anti-mouse cytotoxic responses were restricted to murine target cells of the homologous H-2 Class I alloantigens. Furthermore, IL-2 production from the immune human cells was similarly MHC restricted, in this case by murine Class II antigens. The human-mouse xenogeneic system has proven to be an ideal tool for the dissection of the molecular events in immunity to human Class I and II alloantigens (Engelhard, et al., 1980).

The foundation for the observed functional "allogenicity" resides in the structural and molecular homologies being revealed between murine and human antigens (Fig. 9). Thus, there exists 68% homology between beta-2 microglobulins (Gates, et al., 1981), as well as 70% homology of the amino terminal 275 amino acid residues in the heavy chain of Class I (H-2K, H-2D and HLA-A, B, C) histocompatibility molecules (Ploegh, et al., 1981). Furthermore, recent evidence suggests that the heavy chains of the Class II antigens (Ia, Dr) exhibit at least 50% homology (Dorf (ed)., 1981). Even the efficiency of induction of CTL in mice to Ia or Dr is equivalent (Billings, et al., 1977). Finally, there is size, charge, and subunit composition homology between the surface markers of helper and of suppressor/cytotoxic T cells for human Leu and for murine Lyt antigens (Ledbetter, et al., 1981).

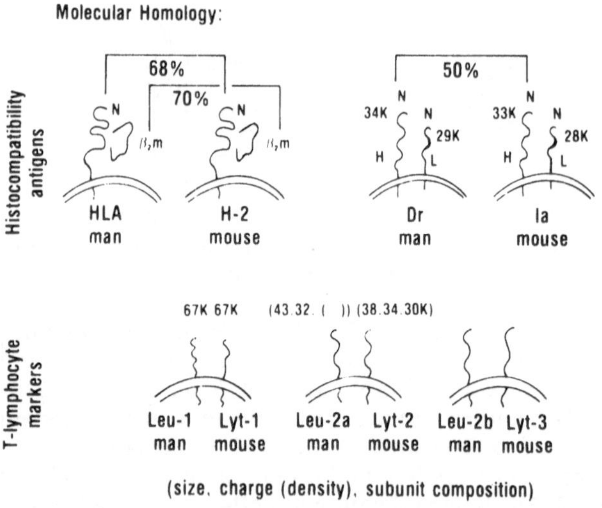

Figure 9. Molecular basis for "pseudoalloimmune" reactivities of mouse lymphocytes to human antigens. Structural homologies exist on comparable antigens between mouse and man which are known to be involved in cell-mediated immune responses.

In contrast to anti-species reactivities exhibited in polyclonal xenoantibodies, CMI responses express remarkable specificities toward the xenogeneic immunizing antigens. Indeed, the xenoimmune responses between human and murine systems appear to function mechanistically, as well as antigenically, as allogeneic and even syngeneic reactions (Lindahl and Bach, 1975; Peck, et al., 1967; Swain et al., 1982; Carnaud et al., 1980; Hayashi et al., 1982). The xenogeneic approach offers an immune system to develop methodology and to begin understanding cellular and molecular mechanisms underlying human tumor immunity. Until the in vitro human-human system is perfected, the xenogeneic approach will serve to develop new techniques, raise questions, foster new concepts in the understanding of tumor cell-lymphoid cell interactions. The utilization of an in vitro system allows for total flexibility in manipulating and characterizing the afferent, central, and efferent arms of cell-mediated immunity to human cancers. The use of liposomes provides the means to control presentation and density of antigenic determinants, to evaluate the requisite molecular entities in the immune response, and to serve as a vehicle for biological response modifiers. Finally, the potential for exploiting this in vitro immunization system for immortalizing effector cells is present.

5. EXPERIMENTAL PROCEDURES

5.1. Preparation of tissue culture cells

The LS174T colon tumor cell line, derived from a human colonic adenocarcinoma and previously described (Tom, et al., 1976; Tom, et al., 1977), was maintained in MEM-10 (Minimum Essential Medium supplemented with 10% fetal bovine serum, 100 units/ml penicillin, 100 microgram/ml streptomycin, and 10 mM Hepes, pH 7.2) at 37 C and humidified air. Other human tumor cell monolayers were maintained in a similar fashion:

SW1083 colonic adenocarcinoma (from Dr. O.D. Holton, Temple, TX)
MDA-MB436 breast carcinoma (from Dr. R. Cailleau, Houston)
SW780 bladder carcinoma (from Dr. O. D. Holton)
SW962 vulval carcinoma (from Dr. O.D. Holton)

Tumor cell suspensions for immunizations were prepared from stock culture flasks with 0.1% trypsin and 0.025% EDTA-buffer, washed with Hank's Balanced Salt Solution, pH 7.2 (HBSS) and resuspended in MEM-10 for suspension culture. After overnight (14-16 hr) growth in suspension on a slow magnetic stirrer, the tumor cells were washed twice with HBSS before use. Cell viability as determined by trypan blue dye exclusion usually ranged from 80-95% between preparations.

Normal peripheral blood lymphocytes from the patient providing the LS174T colon tumor cell line were converted into a lymphoblastoid cell line with EB virus by Dr. G. Moore's laboratory (Denver). The resultant lymphoblastoid line, COLO462V, was maintained in McCoy's 5a medium supplemented with serum and antibiotics as in MEM-10.

5.2. Preparation of tumor cell membranes

As previously described (Raphael and Tom, 1982), membranes were isolated by a modification of the method of Brunette and Till (1971). Twenty-four hours before beginning the procedure, 8×10^8 LS174T tumor cells were trypsinized, washed, and placed in a 25 ml aliquot of water supplemented with 1mM each of calcium, of magnesium and of protease inhibitor, PMSF, for 45 minutes. The cells were Dounce homogenized, the membranes collected, and resuspended in a two-phase polymer (4.6% Dextran T-500 and 3.9% PEG 6000 with sodium phosphate, pH 6.6). Following centrifugation, the membranes were removed from the interface formed by the two polymers. The harvested membranes were pooled, washed and resuspended in PBS, and stored at -70°C. The washed membranes were sonicated in PBS for 1 min at a setting of 2 in a cuphorn sonicator (Heat Systems, Inc., Plainview, NY) to produce small vesicles for use as antigens.

5.3. Preparation of liposome-membrane antigens

As schematized in Fig. 10, multilamellar vesicles (MLV) were prepared by the method described by Bangham, et al. (1965) with modifications (Raphael and Tom, 1982). The liposome (MLV) antigen complexes were prepared with the sonicated LS174T colon tumor cell membrane (MN) vesicles and either incorporated in the preparation of MLV liposomes (i. e., MN-MLV) or attached to preformed MLV liposomes (i.e., MN+MLV). Three mg of sonicated membrane vesicles in 1 ml were added to a dried lipid film of 7:2:1 molar ratios of phosphatidylcholine, cholesterol and phosphatidic acid followed by the addition of 0.5 ml 8 mM $CaCl_2$. This mixture was placed on a magnetic stirrer and incubated for one hour at 37°C under N_2 gas to produce MN-MLV antigens. The liposomes were then washed 2X with phosphate buffered saline to eliminate unincorporated membrane fragments. The concentration of a protein and phosphorus were determined on each preparation. The MN+MLV antigen preparation was produced by adding 3 mg MN to the same amount of preformed MLV liposomes in the presence of 8 m M $CaCl_2$. The rest of the preparation was processed as above.

5.4. Quantitation of protein antigens

Modified Lowry assay (Kruski and Narayan, 1972). A designated number of Whatman No. 1 filter paper discs of 2.0 cm diameter were placed on a sturdy sheet of aluminum foil. Aliquots of .05 ml of liposome samples were pipetted on each paper disc and allowed to dry in an oven at 60°C for 15 minutes. The pipetting of .05 mg of sample per disc was repeated until 25 to 100 microgram of protein was spotted on each disc. After the sample had dried on the paper disc, they were transferred to individual test tubes. Between 2.5 and 5 ml of chloroform was added for lipid extraction. The sample was the kept at room temperature

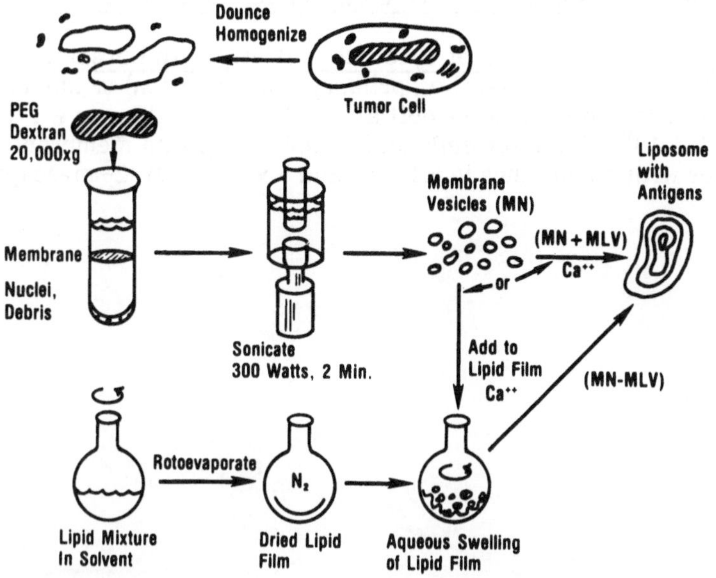

Figure 10. Preparation of liposome-antigens. Tumor cell membranes were collected by centrifugation in a two-phase polymer system following Dounce homogenization. Sonicated membrane (MN) vesicles were prepared with or added to multilamellar liposomes (MLV) to produce MN-MLV or MN + MLV liposome-antigens, respectively.

for 15 minutes. The solvent was carefully decanted off and the paper discs were air dried. After drying, the procedure described for protein determination of insoluble proteins was followed. To each test tube was added 0.2 ml of 0.5 N NaOH and left for two hours. The Lowry procedure for protein determinations was followed.

5.5. Quantitation of liposomes

Phosphorus determinations were used as a measure of the phospholipids (Bartlett, 1959). Evaporate the sample of liposomes or membranes by boiling in a water bath to dryness; add 0.5 ml of 70% perchloric acid and heat on heating block until solution is clear (about 20 minutes) while keeping temperature below 200 ^{o}C. After sample has cooled, add water to reach a 3.0 ml sample volume. Next 1 ml of 4% ammonium molybdate and 0.5 ml elon ($NaHSO_3$, p dimethyl amino phenol sulfate) were added and mixed very well. The tubes were placed in a boiling water bath for ten minutes and cooled to room temperature. The optical density was then read at 660 nm and compared with a sodium phosphate standard curve.

5.6. In vitro immunization (see Fig. 11 for flow chart)

5.6.1. Generation of cytotoxic activity with liposomes. Spleen cells (100×10^6) from normal (i.e., primary immunization) or immunized mice (i.e., in vitro booster) were cultured at 37°C, 5% CO_2 in a T75 flask with 2×10^6 stimulator cells (50:1), 100 microgram membranes, or liposomes with membranes in MEM10 and 5×10^{-5} M 2- mercaptoethanol. On day 3, 1 ml of a nutritional mix containing MEM, non-essential amino acids, essential amino acids, vitamins, and $NaHCO_3$ was added to each well. In addition, 10% CO_2, 7% O_2 gas was bubbled into the flasks for 20 seconds every 2 days. On day 5, splenocytes were removed, washed, and counted. The effector cells were then used in the cytotoxicity assay.

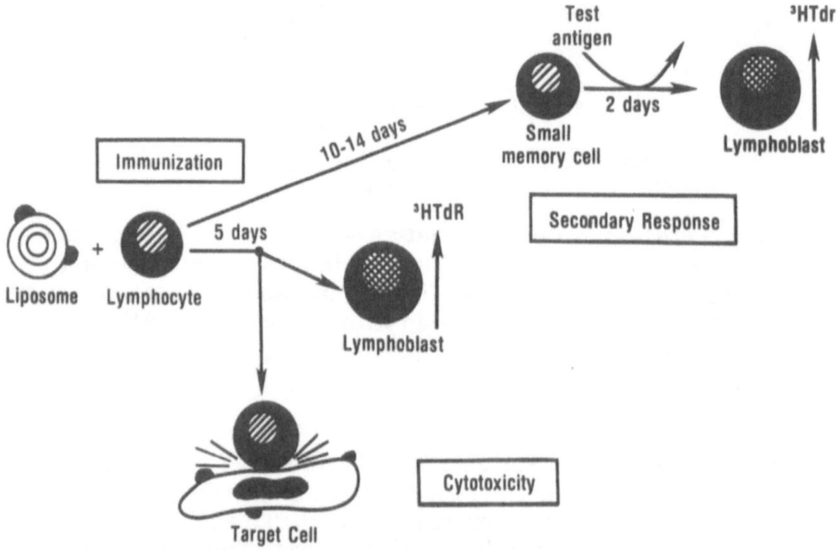

Figure 11. Molecular basis for pseudoalloimmune reactivities of mouse lymphocytes to human antigens. Structural homologies exist on comparable antigens between mouse and man which are known to be involved in cell-mediated immune responses.

5.6.2. Inhibition of cytotoxicity. Colon cancer target cells were labelled with 51Cr and cultured in flat bottom wells (Microtest II at 10^4 target cells per well) overnight at 37 °C, 5% CO_2. Cells (cold targets) to be tested for inhibition of cytotoxicity were trypsinized and placed in spinner culture overnight. Inhibitor cells were added to the monolayers of radiolabeled LS174T cells at effector to inhibitor cell ratios of 5:1, 10:1, 20:1, 50:1, 100:1. The effectors, inhibitors, and targets, in a total volume of 200 microliter, were incubated for 18 hours at 37 °C in 5% CO_2. After incubation the plates were centrifuged at 200 xg for 4 minutes and 100 microliter of media were removed from each well for counting in a Beckman Gamma 4000 counter. Percent specific release was calculated as described for the cell mediated cytotoxic assay. Immune spleen cells from in vivo or in vitro sensitization were added at an effector to target cell ratio of 200:1. This inhibition assay (i.e., cold target inhibition when using unlabelled homologous target cells) is useful in evaluating the specificity of effector cells; for example, the presence of anti-MHC activity in immune cells against colon tumor targets may be tested by inhibition (competition) with the patient's lymphocytes.

5.6.3. Cell-mediated cytotoxicity (CMC) assay. Cultured monolayer target cells were dispersed with 0.1% trypsin in 0.025% EDTA-buffer, washed twice, resuspended in MEM-10 and counted. One milliliter of cell suspension containing 5 x 10^6 tumor cells was incubated for 1 hr at 37 °C in humidified air with 0.1 mCi of $Na^{51}CrO_4$ (2 mCi/2.7 ug/ml; Amersham Corp.). After incubation the cells were washed three times and resuspended in MEM-10. One hundred microliters of cell suspension containing 1 x 10^4 labelled tumor targets were added to wells of Falcon 3040 flat bottom microtest plates and incubated for 24 hr at 37 °C in 5% CO_2 and humidified air. After incubation the labelled targets adherent to the bottom of plate were washed twice with HBSS to remove floating cells and recultured in 100 microliter of MEM-10 per well. Cell counts were obtained by trypsinizing and pooling the contents of 12 wells. Effector cells were added to the target cells based on these counts. Nonadherent lymphoid cell targets were washed once, resuspended in MEM-10 and counted. These suspended targets were labelled and used in the same fashion as the monolayer targets, but on the same day of preparation. One hundred microliters each of effector cells (6-10 x 10^5) and target cells at an effector to target ratio of 100:1 were added to each microtiter well containing target cells and incubated overnight (16-18 hr). The plates were then spun for five min at 100 xg and 100 microliter of the culture fluid were removed by autopipeter and counted to determine isotope release.

The percent cytotoxicity was calculated as follows:

$$\frac{E-N}{M-N} \text{ X100},$$

where E = isotopic release (cpm) from the well containing experimental combinations; M = maximum isotopic release from targets incubated with 4% citrimide; N = isotopic release (cpm) from the well containing normal lymphoid cells and tumor targets.

5.6.4. Elicitation of blastogenic response. Spleen cells (100 x 10^6) were added to MLV-membrane (25 microgram protein), membranes (25 microgram protein) or 2 x 10^6 mitomycin C treated stimulator cells (50:1, responder: stimulator) in a total volume of 200 ul in a Microtest II flat bottom well and incubated for 1-5 days at 37°C in 5% CO_2 and humidified air. Cultures were pulsed for 24 hours with 1 Microcurie per well of ^{3}H-thymidine (New England Nuclear, 1 Ci/ml) the day before harvest, followed by harvesting on fiberglass filter paper in a multiple automatic sample harvester. The filter discs were placed in 5 ml of spectrafluor scintillation fluid (Amersham Corp, Arlington Heights, Ill.) and counted in a LS8100 beta counter (Beckman Instrument, Irvine, CA). Secondary blastogenic responses or the primed lymphocyte assay provides a measure of specific responsiveness to antigens, and of the presence of memory cells. This assay was used in conjunction with in vitro primary immunization. Responder spleen cells from normal mice (100 X 10^6) in MEM-10 are coincubated with a) 2 x 10^6 mitomycin C treated stimulator cells at a 50:1 responder to stimulator ratio, b) membrane fragments with 100 microgram protein, or c) liposomes with 100 ug membrane protein in a T75 tissue culture flask (Falcon Plastics, Oxnard, CA) at 37°C, 5% CO_2. On days 3,5, and 7, 5ml of MEM-10 were added to each flask. On day 10, the spleen cells were removed and washed in HBSS. Then 5 x 10^5 viable, primed spleen cells in 100 microliter were incubated in microtiter wells with 100 microliter of 2 x 10^4 mitomycin C treated stimulator cells. On day 11, the cultures were pulsed with 1 microcurie per well of ^{3}H-thymidine and harvested after 24 hours.

Acknowledgements

The studies presented in this paper were supported by NCI/DHHS Grant CA 24024 and Research Career Development Award 5K04 CA 00579 (BHT).

REFERENCES

Bangham, A.D., Standish, M.M., Watkins, J.C. 1965. Diffusion of univalent ions across the lamellae of swollen phospholipids. J. Mol. Biol. 13:238-252.

Bartlett, G.R. 1959. Phosphorus assay in column chromatography. J. Biol. Chem. 2343:466-468.

Berke, G., Clark, W.R., and Feldman, M. 1971. In vitro induction of the sensitization of rat lymphocytes against mouse cells in vitro. Transplantation 12:237-248.

Billings, P., Burakoff, M., Dorf, M.E. and Benacerraf, B. 1977. Cytotoxic T lymphocytes specific for I region determinants do not require interaction with H-2K or D gene products. J. Exp. Med. 145:1387.

Bloom, G., and Collnaghi, M.G. 1974. Cellular and humoral immunity against human malignant melanoma. Int. J. Cancer 8:344-350.

Brodsky, F.M. and Parham, P., 1982. Monomorphic anti-HLA-A,B,C, monoclonal antibodies detecting molecular subunits and combinatorial determinants. J. Immunol. 128:129.

Brunette, D.M., and Till, J.E. 1971. a. rapid method for the isolation of L-cell surface membranes using an aqueous two phase polymer system. J. Membrane Biol. 5:215-224.

Burakoff, S.J., Englehard, V.H., Kaufman, J., and Strominger, J.L. 1980. Induction of secondary cytotoxic T lymphocytes by liposomes containing HLA-DR antigens. Nature 283:495-497.

Burakoff, S.J., Germain, R.N., Dorf, M.D. and Benacerraf, B. 1977, Specificity of rat xenogeneic cell-mediated cytolysis for the products of K and D loci of the mouse H-2 complex. J. Immunol. 118:1795-1798.

Carnaud, C., Attman, J., Errasti, P., Van der Gaag, R. 1980. Mechanisms of cell-mediated cytotoxicity in mice rejecting xenogeneic human lymphoblastoid cells. Scand. J. Immunol. 11:503.

Carnaud, C., Fadei-Ghotbi, M., LeSavre, P., Bach, J.F. 1977. Education of human lymphocytes against mouse cells: Specific recognition of H-2 antigens. Eur. J. Immunol. 7:81-85.

Corley, R.B., Dawson, J.R., and Amos, D.B. 1975. Lymphocyte stimulation in vitro: Generation of cytotoxic effector lymphocytes using subcellular fractions of a lymphoid cell line. Cell. Immunol. 16:92-105.

Dorf, M.E. ed. 1981. The role of the major histocompatibility complex in immunobiology. Garland STPM press, New York, p. 115.

Engelhard, V.H., Strominger, J.L., Mescher, M. and Burakoff, S. 1978. Induction of secondary cytotoxic T lymphocytes by purified HLA-A and HLA-B antigens reconstituted into phospholipid vesicles. Proc. Natl. Acad. Sci. U.S.A., 75:5688-5691.

Engelhard, V., Kaufman, J.F., Strominger, J.L., and Burakoff, S.J. 1980. Specificity of mouse cytotoxic T lymphocytes stimulated with either HLA-A and -B or HLA-DR antigens reconstituted into phospholipid vesicles. J. Exp. Med. 152:54s-64s.

Gates, F.T., Coligan, J.E., Kindt, T.J. 1981. Complete amino acid sequence of murine B_2-microglobulin:structural evidence for strain related polymorphism: Proc. Natl. Acad. Sci. U.S.A. 78:554.

Ginsburg, H. 1968. Graft versus host reaction in tissue culture. I.Lysis of monolayers of embryo mouse cells from strains differing in the H-2 histocompatibility locus by rat lymphocytes sensitized in vitro. Immunology 14:621-637.

Ginsburg, H., and Sachs, L. 1965. Destruction of mouse and rat embryo in tissue culture by lymph node cells from unsensitized rats. J. Cell. Comp. Physiol. 66:199-207.

Golub, S.H., and Morton, D.L. 1974. Sensitization of lymphoctes in vitro against human melanoma-associated antigens. Nature 251:161-163.

Hale, A.H., Ruebush, M.J., and Harris, D.T. 1980. Elicitation of anti-viral cytotoxic T lymphocytes with purified viral and H-2 antigens. J. Immunol. 1251:428-431.

Hale, A.H., and McGee, M.P. 1981. A study of the inability of subcellular fractions to elicit primary anti H-2 cytotoxic T lymphocytes. Cell. Immunol. 58:227-285.

Hayashi, H., Miller, A.L., Tom, B.H. 1982. Characteristics of the xenogeneic cellular immune responses to human colon cancer. Afr. J. Clin. Exptl. Immunol. 33:293.

Hellstrom, I., Hellstrom, K.E., Sjogren, H.O. 1971. Demonstration of cell mediated immunity to human neoplasms of various histiologic types. Int. J. Cancer 7:1-16.

Herrmann, S.H., and Mescher, M.F. 1981. Secondary cytolytic T lymphocyte stimulation by purified $H\text{-}2K^k$ in liposomes. Proc. Natl. Acad. Sci. U.S.A. 78:2488-2492.

Herrmann, S.H., Weinberger, O., Burakoff, S.J. and Mescher, M.F. 1982. Analysis of the two-signal requirement for precursor cytolytic T lymphocyte activation using $H\text{-}2K^k$ in liposomes. J. Immunol. 128:1968.

Kruski, A.W., and Narayan, K.A. 1972. A simplified procedure for protein determination of turbid lipoprotein samples. Analyst Bioc. 47:299 302.

Ledbetter, J.A., Evans, R.L., Lipinski, M., Cunningham-Rundles, C., Good, R.A., Herzenberg, L.A. 1981. Evolutionary conservation of surface molecules that distinguish T lymphocyte helper/inducer and cytotoxic/suppr subpopulations in mouse and man. J. Exp. Med. 153:310.

Lemonnier, F., Mescher, M., Sherman, L., and Burakoff, S. 1978. Induction of cytolytic T lymphocytes with purified plasma membranes. J. Immunol. 120(4):1114-1120.

Lindahl, K.F., and Bach, F.H. 1975. Human lymphocytes recognize mouse alloantigens. Nature 254:607-609.

Maier, T., Levy, J.G., and Kilburn, D.G. 1980. The Lyt phenotype of cells involved in the cytotoxic response to syngeneic tumor and of tumor-specific suppressor cells. Cell. Immunol. 56:392-399.

Nakao, M., Adler, F.L., Adler, L.T., and Fishman, M. 1981. Induction of immunity and specific unresponsiveness in vitro by cell-bound antigen. Cell. Immunol. 58:448-457.

Peck, A.B., Alter, J., and Lindahl, K.F. 1976. Specificity in T cell mediated lympholysis:identical genetic control of the proliferative and effector phase of allogeneic and xenogeneic reactions. Transplant. Rev. 29:189-221.

Perlmann, P., and Cerottini, J.C. 1979. Cytotoxic lymphocytes. In Sela, M(ed). The Antigens, Vol. V. Academic Press, N.Y., pp. 173-281.

Ploegh, H.L., Orr, H.T., Strominger, J.L. 1981. Major histocompatibility antigens: The human (HLA-A,-B,-C) and murine (H-2K, H-2D) Class I molecules. Cell 24:287.

Raphael, L. and Tom, B.H. 1982. In vitro induction of primary and secondary xenoimmune responses by liposomes containing human colon tumor cell antigens. Cell. Immunol. 71:224.

Schechter, B., Treves, A.H., and Feldman, M., J. 1976. Specific cytotoxicity in vitro of lymphocytes sensitized in culture against tumor cells. J. Natl. Cancer Inst. 56(6):975-979.

Schlager, S.I. Cell. 1982. Relationship between cell-mediated and humoral immune attack on tumor cells. II. The role of cellular lipid metabolism and cell surface charge in the outcome of immune attack. Cell Immunol. 66:300.

Sharma, B. 1976. In vitro lymphocyte immunization to cultured human tumor cells: Parameters for generation of cytotoxic lymphocytes. J. Natl. Cancer Inst. 57:743-748.

Sharma, B., and Terasaki, P.I. 1974. In vitro sensitization to cultured human tumor cells. Cancer Res. 34:115-122.

Sharma, B. 1977. In vitro immuniztion against human tumor cells with tumor cell fractions. Cancer Research 37:4660-4668.

Sondel, P.M., and Bach, F.H. 1975. Recognitive specificity of human cytotoxic T lymphocytes. I. Antigen-specific inhibition of human cell-mediated lympholysis. J. Exp. Med. 142:1339-1348.

Swain, S.L., Yamamoto, J., Schwab, R., Dutton, R.W. 1982. Two subsets of human T cells defined by function and ability of anti leu reagents to block activity, respond strongly to the allele-specific determinants of the class 1 and class 2 murine MHC antigens. Fed. Proc. 41:723(2647).

Takasugi, M., Mickey, M.R., and Terasaki, P. 1973. Reactivity of lymphocytes from normal persons on cultured tumore cells. Cancer Res. 33:2898-2902.

Tom, B.H., Rutzky, L.P., Jakstys, M.M., Oyasu, R., Kaye, C., and Kahan, B.D. 1976. Human colonic adenocarcinoma cells. I. Establishment and description of a new line. In vitro 12:180-191.

Tom, B.H., Rutzky, L.P., Oyasu, R., and Kahan, B.D. 1977. Human colon adenocarcinoma cells. II. Tumorigenic and organoid expression in vivo and in vitro. J. Natl. Cancer Inst. 58:1507-1512.

Tom, B.H. and Six, H.R. eds. 1980. Liposomes and immunobiology. Elsevier North-Holland, New York, 333 pp.

Tom, B.H., Goodwin, T., Sengupta, J., Kahan, B.D., and Rutzky, L.P. 1982. A new approach for the immunogenic presentation of membrane bound human colon tumor antigens. Immunol. Commun. 114:315-323.

Tom, B.H., Macek, C.M., Raphael, L., Sengupta, J., Cerny, E., Jonah, M.M. and Rahman, Y.E. 1983. Proc. Soc. Exp. Biol. Med. 172:16-28.

Treves, A.J., Cohen, I.R., and Feldman, M. 1975. Immunotherapy of lethal metastases by lymphocytes sensitized against tumor cells in vitro. J. Natl. Cancer Inst. 54:777-780.

Van Rooijen, N., and Van Nieuwmegan, R. 1980. Liposomes in immunology: evidence that their adjuvant effect results from surface exposition of the antigens. Cell. Immunol. 49:402-407.

Vanky, F., Argov, S. and Klein, E. 1981. Tumor biopsy cells participating in systems in which cytotoxicity of lymphocytes is generated. Autologous and allogeneic studies. Int. J. Cancer 27:273.

Wagner, H., and Rollinghoff, M. 1973. In vitro induction of tumor specific immunity. I. Parameters of activation and cytotoxic reactivity of mouse lymphoid cells sensitized in vitro against syngeneic and allogeneic plasma cell tumors. J. Exp. Med. 138:1-15.

Whitson, M.E., Griffin, G.D., Novelli, G.D., and Solomon, A. 1981. In vitro sensitization of human lymphocytes to a myeloma cell-related antigen. Cell. Immunol. 60:489-497.

17

COMPARISON OF THREE IMMUNOASSAYS FOR SCREENING ANTI-HEPATITIS B HYBRIDOMAS

Irina Ionescu-Matiu, Cynthia Kendall and Gordon R. Dreesman

Department of Virology and Epidemiology Baylor College of Medicine Houston, Texas 77030

1. SUMMARY

Three micro solid phase immunoassays were developed for screening secreting anti-hepatitis B hybridoma clones: a micro-solid phase radioimmunoassay (micro-SPRIA)[1] and two enzyme-linked immunosorbent assays (ELISA). In all three procedures, aliquots of hybridoma culture fluid were first added to antigen-coated wells. Goat anti-mouse IgG (GtαM) was used as second antibody. For micro-SPRIA, Gt α M was iodinated by the chloramine T method. For the ELISAs, we compared two methods of coupling the antibody to alkaline phosphatase (AP): with glutaraldehyde (ELISA-glutaraldehyde) and with N-succinimidyl 3-(2-pyridyldithio) propinate (SPDP) (ELISA-SPDP). All three tests were more sensitive than the commercially available AUSAB (Abbott), which could not detect low enough levels of antibody to be used for hybridoma screening. ELISA-glutaraldehyde was the least sensitive of the three and produced a number of false positive results. Micro-SPRIA was 5 times more sensitive than ELISA-glutaraldehyde, and ELISA-SPDP was 2.5 times more sensitive than micro-SPRIA. This higher sensitivity of ELISA-SPDP allowed us to detect a greater number of secreting clones at an earlier time after fusion.

---

[1]Abbreviations: micro-SPRIA, micro-solid phase radioimmunoassay; ELISA, enzyme-linked immunosorbent assay; SPDP, N-succinimidyl 3-(2-pyridyldithio) propinate.

2. INTRODUCTION

Screening for secreting hybridoma clones at an early time after fusion requires high sensitivity and specificity at an economical price. Commercial tests, if available, usually do not meet all of these requirements. A large variety of methods has been developed for detecting monoclonal antibodies, either in the culture medium or directly on the antibody-secreting cells (see review by Galfre and Milstein, 1981). The more commonly used method is the indirect binding assay, using a second antibody capable of reacting with the mouse monoclonal immunoglobulins. This second antibody may be directed against a whole immunoglobulin molecule (IgG, IgM) (Goldsby et al., 1980) or against Ig fragments Fab, $F(ab')_2$ (Kennett et al., 1980; Solomon and Jones, 1980) and can be labeled with isotopes or enzymes. Iodinated protein A has been used as an alternative to the anti-globulin reagents (Nowinski et al., 1980).

Three micro solid phase immunoassays have been developed in our laboratory for the detection of antibodies against hepatitis B surface antigen (anti-HBs): a micro solid phase radioimmunometric assay (micro-SPRIA) and two types of enzyme-linked immunoadsorbent assay (ELISA). A number of studies have shown that ELISA using antibody coupled to enzyme with glutaraldehyde is generally less sensitive than radioimmunoassay methods using the same reagents (Wei et al., 1977; Wisdom, 1976). We compared two methods of coupling the antibody to alkaline phosphatase (AP): the classical glutaraldehyde technique (ELISA-glutaraldehyde) and a coupling procedure utilizing N-succinimidyl 3-(2-pyridlydithio) propionate (SPDP) (referred to as ELISA-SPDP). SPDP, a heterobifunctional crosslinker, was used to introduce a controlled number of thiol groups on both the antibody and the enzyme molecules. The thiol group substituted proteins were then allowed to interact, resulting in a conjugate with a defined 1:1 molar ratio of antibody; enzyme. The sensitivity of the three tests was compared to that observed using a commercially available solid phase radioimmunoassay (AUSAB kit; Abbott Laboratories, North Chicago, Ill.).

3. RESULTS AND DISCUSSION

Twenty-three mouse anti-HBs sera were titrated using the AUSAB kit and the three micro immunoassays develped in our laboratory. The highest positive dilution of individual antisera, as determined in each test, is illustrated in Figure 1. The individual serum antibody titers, as determined by ELISA-glutaraldehyde, micro-SPRIA, and ELISA-SPDP, were 1-625, 25-15,625, and 125-15,625 times higher, respectively, than the AUSAB results. For each test, the geometric mean titer (GMT) of the 23 antisera was calculated and a relative sensitivity was determined by comparing each GMT to that of the AUSAB test. The relative sensitivity, as compared to the AUSAB test results, was 50.3 for ELISA-glutaraldehyde, 251.6 for micro-SPIRIA, and 625 for ELISA-SPDP. It was of interest that ELISA-glutaraldehyde was 5 times less sensitive while ELISA-SPDP was 2.5 times more sensitive than micro-SPRIA. As all reagents,

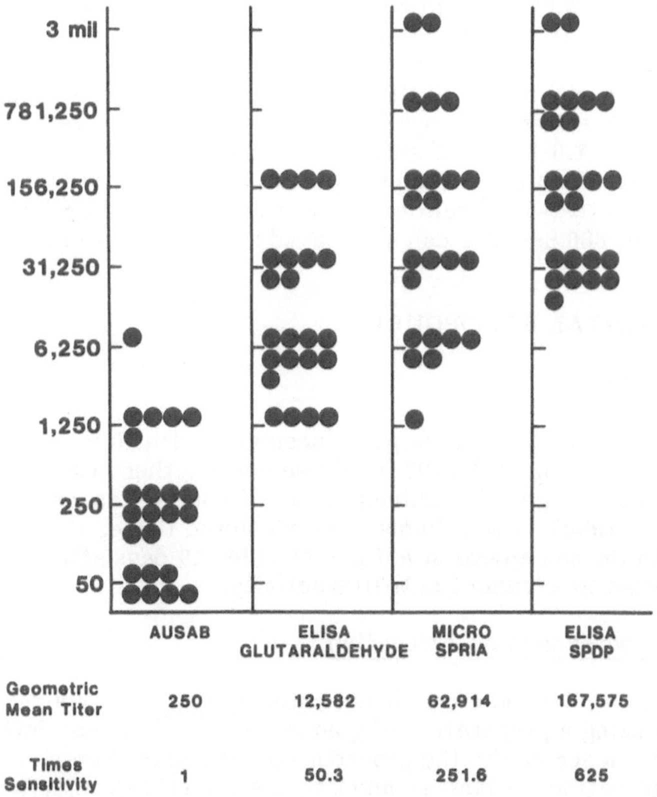

Figure 1. Highest positive dilution of 23 mouse anti-HBs antisera, as titrated in 3 micro solid phase immunoassays and in the commercial AUSAB test.

antibody dilutions, and protocols were identical, the 12.4-fold increase in sensitivity noted in comparing the two ELISA methods must have resulted from the method used to couple the second antibody to enzyme. Thus, a defined conjugate, with a controlled and homogenous molar ratio of antibody:enzyme, provides a reagent with a higher sensitivity than the randomly cross-linked aggregates obtained with glutaraldehyde.

To further substantiate the difference in sensitivity of the two types of ELISA, as well as their suitability for detection of secreting anti-HBs hybridoma clones, mock hybridoma detection tests were set up, using randomly spotted dilutions of monoclonal anti-HBs IgG. ELISA-glutaraldehyde consistently failed to identify low positive dilutions and occasionally gave a false positive reaction in negative wells, while ELISA-SPDP correctly identified the positive and negative wells.

Micro-SPRIA and ELISA-SPDP provide high sensitivity and specificity, and both are suitable for hybridoma screening. Micro-SPRIA can measure as low as 1.0-0.1 ng of antibody protein/ml, as determined by titrating an anti-HBs monoclonal preparation with known protein concentration. Both micro tests are practical for routine assay of large numbers of samples: 500-800 samples can be assayed by one person in 7-8 hours.

4. EXPERIMENTAL PROCEDURES

4.1 Antisera

Groups of BALB/c mice were inoculated with 10 μg of purified HBsAg/adw (Dreesman et al., 1972), administered either in saline suspension or adsorbed on a gel of aluminum potassium sulfate as described (Sanchez et al., 1980). The animals were boostered twice, at 2-week intervals, with the same dose of antigen and bled 10 days after each booster to obtain a broad spectrum of anti-HBs activity.

4.2 Antibody-Enzyme Conjugation

A glutaraldehyde cross-linked antibody-enzyme conjugate was prepared by linking a preparation of goat IgG with anti-mouse IgG activity (Gt α M) to AP according to the procedure described by Avrameas (1969) with the modifications of Madore and Baumgarten (1979). Briefly, 5 mg AP (type VIIS, Sigma Chemical Co., St. Louis, Mo.) was pelleted and resuspended in 0.2 ml of 0.1 M phosphate-buffered saline (PBS), pH 6.8, containing 1.25% glutaraldehyde. The reaction was allowed to proceed for 18 hr at room temperature. The mixture was then extensively dialyzed against 0.05 M carbonate-bicarbonate buffer, pH 9.5, and brought to 1 ml with the same buffer, Gt α M was diluted in 0.15 M NaCl, 0.1 M carbonate-bicarbonate buffer, pH 9.5, to a concentration of 2.5 mg/ml. The glutaraldehyde-activated AP and the Gt α M were then incubated at a 2:1 ratio (w/w) for 24 hr at 4°C. Excess lysine (10 μl of a 0.2 M solution) was added to the conjugate and incubation was continued for an additional 2 hr. The conjugate was then extensively dialyzed against 0.05 M PBS, pH 7.2, clarified by centrifugation at 20,000 g for 30 min at 4°C, and stored at 4°C in the dark until use.

For the antibody-enzyme conjugate crosslinked with SPDP, both Gt α M reagent and AP were thiolated with SPDP (Pharmacia Fine Chemicals, Piscataway, N.J.), following the procedure described by Carlson et al. (1978) (Figure 2). To conjugate 1 molecule of antibody to 1 molecule of enzyme, each protein was reacted with SPDP in the molar ratio previously shown to introduce 1 or 2 thiol groups per molecule. For IgG, this proportion was known to be 2.4 moles SPDP per 1 mole IgG (Fields et al., in press). We determined the substitution curve for AP by reacting

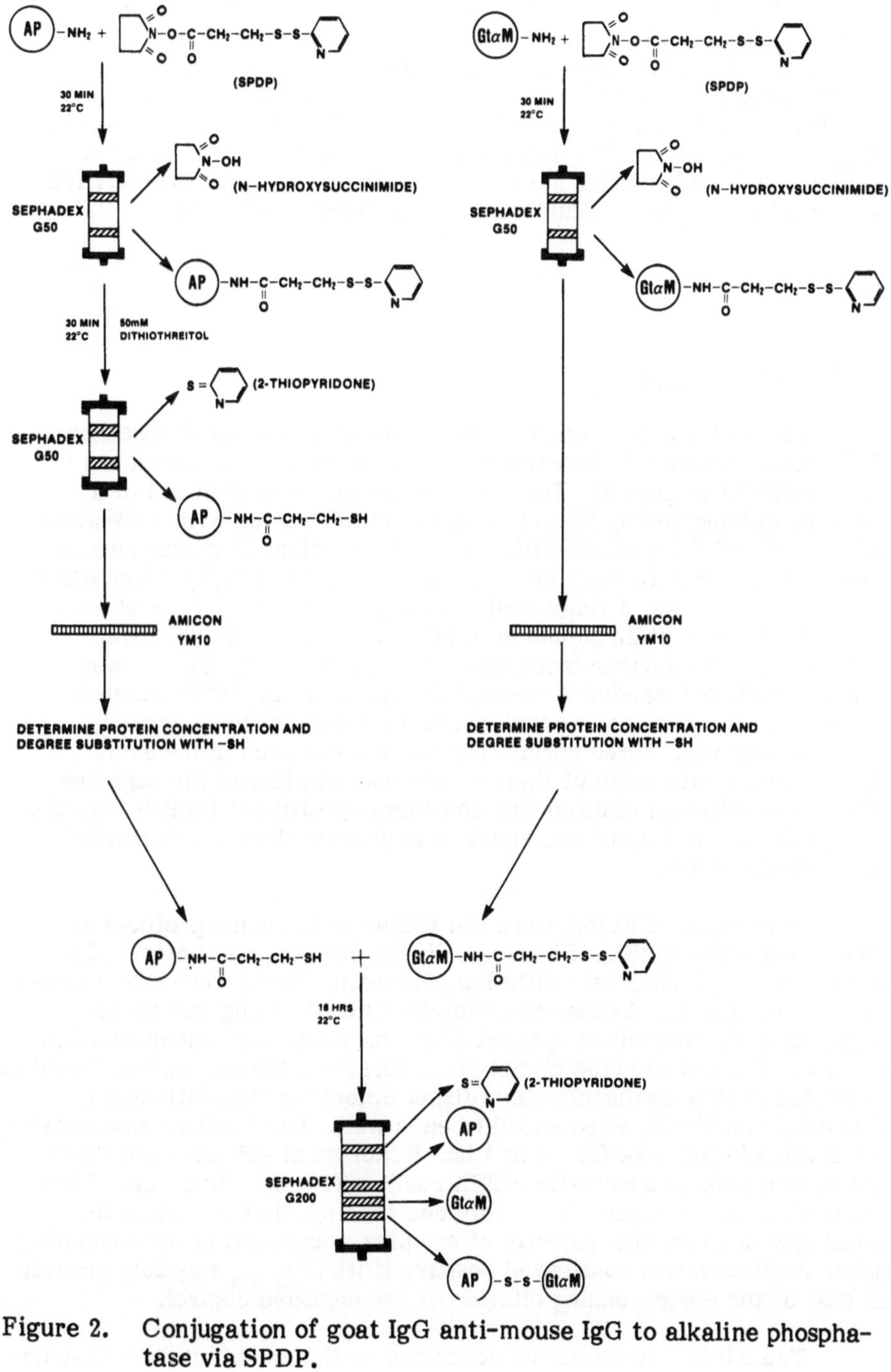

Figure 2. Conjugation of goat IgG anti-mouse IgG to alkaline phosphatase via SPDP.

SPDP with AP in different molar ratios reducing the 2-thiopyridyl-containing AP with50 mM dithiothreitol (DTT) and quantitating the reduced free thiol groups on AP by reaction with 2,2'-dithiopyridine, as described by Grassetti and Murray (1967). Under our experimental conditions, an SPDP:AP molar ratio of 100:1 provided a substitution rate between 1 and 2. The protected, 2-thiopyridil-containing Gt α M and the thiol-containing AP reduced with DTT were mixed together in a molar ratio of approximately 5 moles IgG/1 mole AP. The final conjugate was purified on a 2.6 x 90 cm Sephadex G200 column equilibrated with PBS.

4.3 Immunoassays

Flat-bottom, 96-well, flexible plates (polystyrene, Cooke Microtiter System, Dynatech Laboratories Inc., Alexandria, Va.) were used for micro-SPRIA (Figure 3). The plates were coated with 50 μl of a solution containing 200 ng HBsAg/well, then post-coated with 0.5% gelatin in 0.05M PBS, pH 7.2 (gelatin-PBS). Fivefold dilutions of mouse antisera were added in duplicate wells and incubated for 2 hr at 37°C, after which the plates were washed 3 times with 0.01% gelatin-PBS. ^{125}I-labeled Gt α M diluted to contain 50,000 cpm/50 μl was added to each well and the plates were further incubated for 2 hr. at 37° C. Gt αM was iodinated by the chloramine-T method (Hollinger et al., 1975; Sanchez et al., 1980). The plates were then washed 3 times as above, sealed, cut off and counted. Three normal mouse sera were run in the same test. The arithmetic mean of their counts was considered the negative control. An antiserum dilution was considered positive if its P/N (sample cpm/negative control cpm) was equal to or greater than an arbitrarily chosen cut-off of 2.1.

Both types of ELISA were run following the same protocol as described for micro-SPRIA (Figure 3). Rigid, 96-well, flat bottom plates (Linbro/Titertek, Linbro Scientific Inc., Hamden, Conn.) were used instead of the flexible plates. As second antibody, Gt α M conjugated to AP was used in a predetermined optimal dilution. After a 2-hour incubation 100 μl of AP substrate (104 phosphatase substrate tablets, Sigma Chemical Co.) diluted in 10% diethanolamine (Fisher Scientific Co., Pittsburgh, Pa.) and 0.5 mM $MgCl_2$ were added to each well. The reaction was allowed to develop in the dark for up to 4 hr. Readings at 405 nm were taken every 30 min with an automatic ELISA reader (Titertek, Multiscan, Flow Laboratories Inc., McLean, Va.). Only one reading, that providing the optimal P/N ratio for the majority of samples, was selected for computing results. A dilution was considered positive if its OD_{405} was 25% greater than that of the corresponding dilution of the negative control.

The AUSAB test was run according to the manufacturer's instructions.

FLAT-BOTTOM, 96-WELL MICRO PLATES

▼

COAT: 200 NG HBsAG/50 μL/WELL,
IN 0.3 M CARBONATE BUFFER, PH 9.5

18 HOURS ▼ 4 C

POST-COAT: 0.5% GELATIN-PBS/200 μL/WELL

30 MIN ▼ 22 C

WASH: 0.1% GELATIN-PBS/200 μL/WELL, 1X

▼

SAMPLE TO BE TESTED: 50 μL/WELL
(SERIAL DILUTIONS IN 0.1% GELATIN-PBS)

2 HOURS ▼ 37 C

WASH: 0.01% GELATIN-PBS/200 μL/WELL, 3X

MICRO-SPRIA	ELISA
2ND ANTIBODY: ^{125}I-GTαM DILUTED IN 0.1% GELATIN-PBS 50,000 CPM/50 μL/WELL	2ND ANTIBODY: GTαM-ALKALINE PHOSPHATASE, DILUTED IN 0.1% GELATIN-PBS, 50 μL/WELL
▼ 2 HOURS	37 C ▼

WASH: 0.01% GELATIN-PBS/200 μL/WELL, 3X

MICRO-SPRIA	ELISA
	SUBSTRATE: 0.1% P-NITROPHENYL PHOSPHATE IN 10% DIETHANOLAMINE, 0.5 MM $MgCl_2$, 100 μL/WELL
▼	▼
COUNT IN GAMMA COUNTER	MONITOR COLOR AT 405 NM

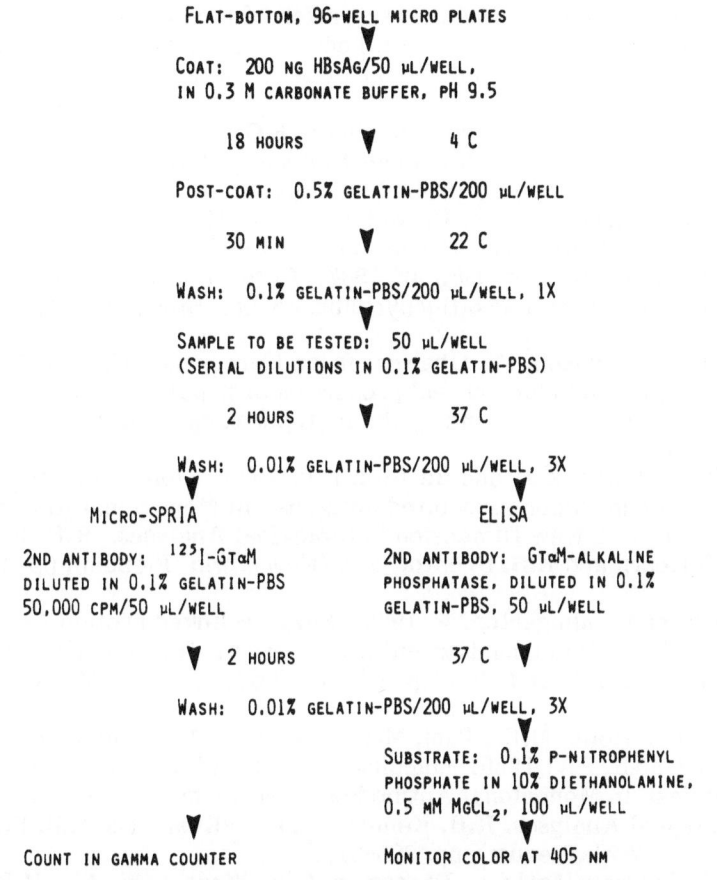

Figure 3. Micro solid phase immunoassays for antibody detection.

REFERENCES

Avrameas, S. 1969. Coupling of enzymes to proteins with glutaraldehyde. Use of the conjugates for the detection of antigens and antibodies. Immunochem. 6:43-52.

Carlson, J., Drevin, H. and Axen, R. 1978. Protein thiolation and reversible protein-protein conjugation. N-succinimidyl3-(2-pyridyldithio) propionate: a new heterobifunctional reagent. Biochem. J. 173:723-737.

Dreesman, G.R., Hollinger, F.B., McCombs, R.M. and Melnick, J.L. 1972. Production of potent anti-Australia antigen sera of high specificty and sensitivity in goats. Infect. Immun. 5:213-221.

Fields, H.A., Davis, C.L., Dreesman, G.R., Bradley, D.W. and Maynard, J.E. Enzyme-potentiated radioimmunoassay (EPRIA):a sensitive third-generation test for the detection of hepatitis B surface antigen. J. Immunol. Met. (in press).

Galfre, G. and Milstein, C. 1981. Preparation of monoclonal antibodies: strategies and procedures. In Methods in Enzymology, vol. 73, part B. J.L. Langone and H.van Vunakis, eds. (New York: Academic Press), pp. 3-46.

Goldsby, R.A., Osborne, B.A. and Engleman, E.G. 1980. Characterization of a human T-cell population identified and isolated by a monoclonal antibody. In Monoclonal Antibodies. Hybridomas: a New Dimension in Biological Analyses. R. H. Kennett, T. J. McKearn and K.B. Bechtol, eds. (New York: Plenum), pp. 121-135.

Grassetti, D.R. and Murray, J.F., Jr. 1967. Determination of sulfhydryl groups with 2,2'- or 4,4'-dithiopyridine. Arch. Biochem. Biophys. 119, 41-49.

Hollinger, F.B., Morrison, M., Chairez, R. and Dreesman, G.R. 1975. Immunological and biophysical properties of hepatitis B antigen labeled by the chloramine-T and by the lactoperoxidase methods. J. Immunol. Methods 8, 67-84.

Kennett, R.H., Jonak, Z.L. and Bechtol, K.B. 1980. Monoclonal antibodies against human tumor-associated antigens. In Monoclonal Antibodies. Hybridomas: A New Dimension in Biological Analyses. R.H. Kennett, T.J. McKearn and K.B. Bechtol, eds. (New York: Plenum), pp. 155-168.

Madore, H.P. and Baumgarten, A. 1979. Enzyme-linked protein A: an enzyme-linked immunoadsorbent assay reagent for detection of human immunoglobulin G and virus-specific antibody. J. Clin. Microbiol. 10:529-532.

Nowinski, R.C., Stone, M.R., Tam, M.R., Lostrom, M.E., Burnette, W.N. and O'Donnell, P.V. 1980. Mapping of viral proteins with monoclonal antibodies. In Monoclonal Antibodies. Hybridomas: A New Dimension in Biological Analyses. R.H. Kennett, T.J. McKearn and K.B. Bechtol, eds. (New York:Plenum), pp. 295-316.

Sanchez, Y., Ionescu-Matiu, I., Dreesman, G.R., Kramp, W., Six, H.R., Hollinger, F.B. and Melnick, J.L. 1980. Humoral and cellular immunity to hepatitis B virus-derived antigens: comparative activity of Freund complete adjuvant, alum and liposomes. Infect. Immun. 30:728-733.

Solomon, E. and Jones, E.A. 1980. Monoclonal antibodies as tools for human genetic analysis. In Monoclonal Antibodies. Hybridomas: a New Dimension in Biological Analyses. R.H. Kennett, T.J. McKearn and K.B. Bechtol, eds. (New York: Plenum), pp. 75-102.

Wei, R., Knight, G.J., Zimmerman, D.H. and Bond, H.E. 1977. Solid-phase enzyme immunoassay for hepatitis B surface antigen. Clin. Chem. 23:813-815.

Wisdom, G.B. 1976. Enzyme immunoassay. Clin. Chem. 22:1243-1255.

PART VI.

POSTER ABSTRACTS

1. T-LYMPHOCYTE SUBSETS OF RHESUS MACAQUE AS DETERMINED BY MONOCLONAL ANTIBODIES: Tγ, Tμ T-NULL, AND HISTAMINE RECEPTOR BEARING T-LYMPHOCYTES.

Larry R. Ellingsworth and Bennie I. Osburn
University of California
Davis, California 95616.

The distribution of helper (OKT4^{+}) and suppressor/cytotoxic (OKT8^{+}) cells was determined for rhesus monkey (Macaca mulatta) peripheral T-lymphocyte subpopulations using human T-lymphocyte specific monoclonal antibodies. Highly purified rhesus T-lymphocytes (> 90% E-rosetted) were spearated into subpopulations by Fc-receptors using IgG (Tγ) and IgM (Tμ) coated bovine red blood cells or by adsoption with insolubilzed histamine. Rhesus monkey T-lymphocytes were found to express cell-surface determinants recognized by OKT4 and OKT8 antibodies. Helper and suppressor/cytotoxic T-cells were found to be in approximately the same frequency as humans (50% OKT4 and 33% OKT8). The Tμ subpopulation was found to be comprised of 30% OKT4 type cells and to be depleted of suppressor cells (2% OKT8). The Tγ subpopulation, previously shown to have suppressor activity, consisted of 9% OKT4 and 2% OKT8 type cells. Our results suggest Tγ may represent an undifferentiated suppressor population. T-cells depleted of cells with Fc-IgG and Fc-IgM receptors, T-null, contained 47% OKT4 and 12% OKT8 type cells. T-cells adsorbed with insolubilized histamine had a reduced frequency of OKT8 type cells (70% reduction), while the frequency of OKT4 type cells was unaffected by histamine adsorption. These results suggested that the majority of suppressor cells (OKT8) express cell-surface receptors for histamine.

2. PRODUCTION OF HUMAN IMMUNOGLOBULIN SECRETING CLONES WITHOUT THE USE OF H.A.T. SENSITIVE FUSION PARTNERS.

A. J. Strelkauskas and G. B. Ferrara.
Medical University of South Carolina, Charleston, SC 29425, and Immunohematology Research Center, Ospedale Massa, and A.V.I.S., Bergamo, Italy.

We have utilized a long term B cell line derived from a patient with acute lymphocytic leukemia as a fusion partner in the production of human hybridoma clones. Our studies center around the production of human monoclonal antibodies to 1) T cell antigens, 2) defined HLA determinants, 3) Rh blood group antigens, and 4) the antigens associated with mammary carcinoma. Fusions are performed using peripheral blood from individuals shown to have an ongoing antibody response to these antigens. Each of the donors is defined with regard to specific HLA type. Twenty-four hours after fusion the cell mixtures are separated using a positive selection technique in which donor cells and clones are sensitized with highly specific human anti-HLA antibodies from planned immunizations (produced by Dr. Ferrara), followed by rosetting with red cells coated with purifed rabbit anti-human IgG antibodies. Screening for antibody specific HLA determinants was simplified by using cells from donor and recipient couples originally used for planned immunization. Positive results were obtained for anti-Rh antibodies as well as antibodies reactive with mammary carcinoma cell lines and T cell subsets.
(Supported by Grant AI-16651 from the NIH, and GNR #80.01542.96.)

3. SPECIES DISTRIBUTION OF ANTIGENIC SITES ON MOUSE LDH-C_4 DETERMINED BY MONOCLONAL ANTIBODIES.

Robin E. Goldman and Erwin Goldberg.
Northwestern University
Evanston, Illinois 60201.

LDH-C_4 is the form of lactate dehydrogenase unique to sperm and testes. Antibodies to LDH-C_4 do not bind the somatic forms of this enzyme. We are making monoclonal antibodies specific for antigenic determinants of mouse LDH-C_4 and are using these antibodies as probes to identify determinants is common among several species. Spleen cells from SJL/J mice immunized with mouse LDH-C_4 were fused with P3-X63-Ag8 myeloma cells and four hybrid cell lines generated using cell fusion techniques. Scatchard analyses established that the antibodies have high affinities ($K_a = 10^9 M^{-1}$) for LDH-C_4 and are monoclonal. These antibodies were tested for cross-reactivity with rat, hamster, rabbit and human LDH-C_4 using competitive binding assays. Monoclonal antibody (1)B-6:G1 detects a determinant common to all five species. Antibody (3)F-11:F4 is specific for a determinant unique to mouse LDH-C_4. Antibody (3)F-12:G8 binds to mouse, rat rabbit and human LDH-C_4 and antibody (3)A-8:H10 reacts with LDH-C_4 from the mouse, rat, hamster, and rabbit. These results verify that each antibody is directed against a different determinant of LDH-C_4 and, in addition, provide useful information for determining the amount of structural homology among different species of LDH-C_4. Supported by NIH Grants R01 HD05863 and T32 HD07068.

4. A NEW NON-IMMUNOGLOBULIN PRODUCING HUMAN MYELOMA CELL LINE FOR PRODUCTION OF HYBRIDOMAS.

Jerry W. Pickering and Frank B. Gelder,
Dept. of Microbiology/Immunology and Surgery
Louisiana State University Medical Center, Shreveport, La. 71130.

We selected an 8-azaguanine resistant mutant of a human myeloma cell line (RPMI 826) by cloning the parental cells on a feeder layer of mouse spleen cells in the presence of increasing concentrations of 8-azaguanine. Culture media and cell-free extracts of both the parental and mutant (8226 AR/NIP4-1) cell lines were assayed for immunoglobulin heavy and light chain by Ouchterlony double immunodiffusion and for lambda chain by radioimmunoassay. Secretion of free lambda chain by the parental cell line was confirmed. In contrast, no immunoglobulin heavy or light chains were detected in culture medium of the mutant cell line by either immunodiffusion or radioimmunoassay. No intracellular lambda chain could be detected in the mutant cells by radioimmunoassay ($< 100ng/4x10^7$ cells) of cell free extracts or by immunofluorescence of fixed cells. Hybridomas were produced by fusion of 8226AR/NIP 4-1 cells with lymphocytes from a mesenteric lymph node recovered at surgery from a hypertransfused (n=8) renal transplant recipient. Twenty hybrid culture media were assayed for immunoglobulin by double immunodiffusion and 15 contained either IgG (lamda) or IgG (kappa). None produced IgM or IgA. One hybridoma producing IgG (kappa) was shown by immunofluorescence not to reexpress lambda chain of the parental myeloma. Hybrid antibodies were screened for reactivity in a lymphocyte microcytotoxicity assay against a panel of 46 normal human donors. One hybrid antibody lysed 50% of peripheral lymphocytes from 4 of the donors. This antibody appeared to recognize an alloantigen present on a subpopulation of human lymphocytes.
Supported by NIH Grants CA27134 and AM21346.

5. MONOCLONAL ANTIBODIES TO HUMAN PROSTATE AND BLADDER TUMOR ASSOCIATED ANTIGENS.

James J. Starling and George L. Wright, Jr., Eastern Virginia Medical School, Norfolk, Va. 23501

Monoclonal antibodies to human prostate adenocarcinoma membrane antigens were produced by fusion of spleen cells from Balb/C mice immunized against the prostate cancer cells line DU145, and the P3X63/Ag8 mouse myeloma line. The hybrids were screened for antibody production using glutaraldehyde-fixed or live cells in an in vitro binding radioimmunoassay (RIA). Antibody binding specificity was also checked by quantitative adsorption studies. A hybridoma clone (83.21) was isolated that secreted antibodies which preferentially bound to prostate and bladder cancer cells but did not bind to a variety of other normal and malignant human cell lines. This antibody also reacted with a cytomegalovirus-transformed human embryonic lung cell line but not to normal human embryonic lung cells. Conventional and competitive binding studies indicate that the 83.21 monoclonal antibody does not bind to alpha-fetoprotein, carcinoembryonic antigen, prostatic acid phosphatase, HLA, β_2-microglobulin, fibronectin, prostate specific antigen or DR antigens. Preliminary immunoprecipitation data with ^{125}I-labeled, detergent soluble, DU145 membrane proteins and the 83.21 monoclonal antibody suggest that this antibody binds two proteins with molecular weights of 210 and 97 kilodaltons. (Supported by an N.I.H. Biomedical Research Support Grant #218 and a grant from the National Prostatic Cancer Project, NCI CA-26659.)

6. ANTIBODY PURIFICATION BY RECYCLING ISOELECTRIC FOCUSING.

N. B. Egen, M. Bier, L. S. Rodkey, and S. Binion. Biophysics Technology Laboratory, Unviersity of Arizona, Tucson, AZ 85721 and Division of Biology, Kansas State University, Manhattan, KS 66506.

Isoelectric focusing is widely used for the analysis of antibody preparations, even monoclonal antibodies exhibiting heterogeneity of isoelectric points. Preparative fractionation of clonotype antibodies (60-350 mg) has been accomplished in 2 to 4 hrs using the recycling isoelectric focusing apparatus (RIEF) (M. Bier & N. B. Egen, Electrofocus/78, p. 35, Elsevier 1978). The antibodies were elicited in a partially inbred strain of New Zealand rabbits, developed for its high antibody responses of restricted heterogeneity to several carbohydrate antigens. The rabbits were immunized intravenously with a vaccine of freeze-dried Micrococcus lysodeikticus cells. Serum samples from each rabbit, collected weekly, were pooled and the antibodies were purified by affinity chromatography on Sepharose-4B coupled to micrococcal carbohydrate. The antibodies were eluted with 0.2M acetic acid, stored in buffer and desalted by electrodialysis before fractionation. Several batches were fractionated in pH 7-9 range Ampholine and one sample of cryoimmunoglobulins in pH 3.5-10 range, the latter in presence of 4M urea. Narrower pH range Ampholines (7.3-8.2, & 7.5-8.5), were prepared using the RIEF apparatus. The resolution obtained was best with narrow pH range Ampholines, permitting the isolation of major single or doublet focused bands from the bulk of minor bands.

7. MONOCLONAL ANTIBODIES AGAINST HUMAN EJACULATED SPERM-APPLICATION IN ANTIBODY MEDIATED INFERTILITY.

Don P. Wolf and Kathleen B. Bechtol, UTHSC, Houston, Texas and The Wistar Institute, Philadelphia, Pa.

Antisperm antibodies occur commonly in the circulation of vasectomized males, in couples with idiopathic infertility and in animals hyperimmunized with human sperm. Many of these polyclonal antisperm antibodies are capable of inhibiting sperm function in vivo as well as in in vitro bioassays such as sperm fusion with the zona-free hamster egg. We are developing monoclonal antisperm antibodies (MASA) with the objective of defining a subset of surface antigens that are critical to sperm function and therefore could be involved in human infertility. Sperm membranes, employed as immunogens, were isolated from washed ejaculated sperm following nitrogen cavitation and repeated centrifugation. Spleen cells from immunized BALB/c mice were fused with HAT-sensitive myeloma cells (SP2/0-Ag14) and antisperm antibody activity was detected in the resultant hybridoma supernatants by a solid phase RIA using washed, intact sperm as the antigen source. Thirteen positive clones have been isolated, 6 producing immunoglobulins of the mu class and 7 of the gamma class. Characterization of several of these MASAs as to antibody and antigen concentration dependency in the RIA indicates that their behavior is highly individualistic. These MASAs are species but not tissue specific. MASA binding to human sperm has been localized to distinct surface domains by indirect immunofluoresence. The ability of these MASAs, individually or cooperatively, intact or as $F(ab)_2$ fragments, to inhibit sperm function is currently under examination. Supported by USPHS HD 6274, GM23892 and CA19815.

8. ANALYSIS OF PERIPHERAL BLOOD AND SALIVARY GLAND LYMPHOCYTES IN SJOGREN'S SYNDROME.

Robert Fox, Sherman Fong, Samuel Behar, Charles Robinson, Francis Howell, Thomas Adamson, Scott Carstens and John Vaughan. Scripps Clinic and Research Foundation, La Jolla, California 92037.

Sjorgren's syndrome (1°-SS) is characterized by dry mouth and dry eyes due to salivary and lacrimal gland infiltration by lymphocytes. Using monoclonal antibodies, we compared lymphocyte subsets in salivary glands (SG) and peripheral blood (PB). In SG biopsies stained with immunofluorescence or immunoperoxidase techniques, OKT4+ cells were the predominant subset; the OKT4/OKT8 ratio exceeded 3.0 in all 5 cases examined. In 1 case of pseudolymphoma, a lymph node (LN) biopsy contained 80% T-cells with an OKT4/OKT8 ratio of 3.2. By cytofluorographic analysis, PB lymphocytes in 15 patients with 1°-SS had a decreased number of OKT8+ cells as compared to age matched controls (353 ± 186 vs. $631 \pm 150/mm^3$) ($p < .001$). The number of OKT4+ cells was comparable in both groups. The ratio of OKT4/OKT8 reactive PBL was increased (> 2.4) in 67% of these patients (range from 1.0 to 6.4) (normal=1.8 ± 0.3). The decreased number of OKT8+ cells in 1°-SS was not due to circulating autoantibody, since patients' serum did not react with normal OKT8+ cells. T-helper cell activity for immunoglobulin synthesis was contained in the OKT4+ subset in both normals and 1°-SS patients' PBL. This study demonstrates predominance of a particular T-cell subset (OKT3+, OKT4+, OKT8-) at sites of inflammmation (SG, LN) as well as in PBL in patients with 1°-SS.

9. A MONOCLONAL ANTIBODY WHICH IDENTIFIES A SUBPOPULATION OF HUMAN B LYMPHOCYTES AND MONOCYTES.

M. Shepherd Munchus and Daniel Levitt. Univ. of Illinois Medical Center, Chicago, Illinois 60612, La Rabida Children's Hospital, Chicago, Illinois 60649.

Hybridoma antibodies were produced against the "pre-B" lymphoblatoid cell line, Nalm-6. A subclone of one of these hybridoma antibodies (I-96) identifies a determinant on the surface and in the cytoplasm of Nalm-6 and 6 of 11 human lymphoblastoid cell lines. The intensity of reaction of I-96 was greatest in cell lines with large numbers of cytoplasmic vesicles. I-96 identifies a different determinant than another monoclonal antibody specific for a common acute lymphoblastic leukemia antigen (J5). Furthermore, MABI-96 specifically reacts with a surface determinant on a subpopulation of sIg^+ and cIg^+ cells from peripheral blood, spleen, and tonsil, and with a subpopulation of G-10 adherent, OKMI cells. The I-96 antigen does not associate preferentially with any class of immunoglobulin and I-96 does not react with T cells. When PBMC are stimulated with the polyclonal B cell activators pokeweed mitogen and S. aureus (Cowans I) the number of MABI-96$^+$ sIg$^+$ cells decreases sharply. In unstimulated cultures the percentage of such cells markedly increases by day 4, followed by a gradual decrease toward the end of the culture period. It appears that this antibody defines a functional subpopulation of human B lymphocytes and monocytes. (Supported in part by the Chicago Leukemia Research Foundation, Inc. and Univ. of Illinois BRSG #80932, CRB-FY81).

10. IN VIVO LOCALIZATION OF TUMOR ASSOCIATED ANTIGENS WITH RADIOACTIVELY LABELED MONOCLONAL ANTIBODIES.

Richard D. Henkel and Gordon R. Dressman, Baylor College of Medicine, Houston, Tx. 77030.

Monoclonal antibodies were prepared against hepatitis B surface antigen (HB_sAg). These preparations were shown to react with HB_sAg secreted by human hepatoma cells from the cell line designated PLC/PRF/5. These cells and cells from another human tumor cell line (Hep-2) which is HB_sAg negative, were inoculated subcutaneously into BALB/c background nude mice. The anti-HB_sAg monoclonal antibodies and nonspecific monoclonal antibodies of the same isotype were purified by salt precipitation and gel filtration and labeled with ^{131}I-using the Hunter-Greenwood method. Three weeks after cell inoculation, the nude mice were injected with either the anti-HB_sAG antibodies or with the non-specific antibodies. Eight days later the mice were sacrificed and each major organ and tumor were weighed and counted to determine the amount of radioactivity per gram of tissue. The results indicated that: 1) more radioacitivity was found in the tumors bearing the HB_sAg than the Hep-2 tumor when anti-HB_sAg antibodies were used and; 2) no such difference could be seen when the nonspecific antibodies were used.

11. CHARACTERISTICS OF MONOCLONAL ANTIBODIES TO HERPES SIMPLEX TYPE 2.

Libi Zhang, Richard D. Henkel, Ronald C. Kennedy and Gordon R. Dreesman. Baylor College of Medicine, Houston, Tx. 77030.

Monoclonal antibodies were produced by fusing NS-1 cells and spleen cells from BALB/c mice which have been inoculated with either purified herpes simplex virus (HSV) type 2 (HSV-2) virion or the major envelope glycoprotein of HSV-2 with molecular weight 119,000 daltons, designated

VP 119. The specificity of each hybridoma antibody has been defined by: 1) western blot techniques employing viral infected cell lysates; 2) virus neutralization (NA); 3) radioimmunoassay (RIA) or; 4) immunoflorescence (FA) using both HSV type 1 (HSV-1) and HSV-2 antigens to determine viral type specificity. Two monoclonal antibodies designated M.186.1B.3 and M.186.2A.2 were produced against HSV-2 virions. One of these (M.186.1B.3) recognized only HSV-2 type specific determinants by RIA and was capable of reducing plaque forming units (PFU) by 70%. The second antibody (M.186.2A.2) recognized both HSV-1 and HSV-2 antigens as determined by either FA or RIA and reduced PFU by 20%. Two monoclonal antibodies designated HS2A and HS5 were generated against VP 119 and detected type specific glycoproteins by RIA. HS2A (anti-VP 119) was further characterized by NA and FA demonstrating partial viral neutralization (20%) and a nuclear staining reaction with HSV-2 infected cells.

12. TYPE SPECIFIC MONOCLONAL ANTIBODIES TO EQUINE INFECTIOUS ANEMIA VIRUS: DEVELOPMENT AND USE IN THE CHARACTERIZATION OF CANINE AND FELINE DERIVED VIRUSES.

C. V. Benton, J. S. Harshman and B. L. Brown, Bethesda Research Laboratories, Inc., Gaithersburg, Maryland 20760.

Monoclonal antibodies (MAb) were developed to the Wyoming strain of equine infectious anemia virus (EIAV) using the P3-X63-Ag8.653 BALB/c myeloma cell line hybridized to EIAV-immune BALB/c spleenic lymphocytes. The majority of monoclones produced antibodies to the envelope glycoproteins (gp48) of EIAV. All MAb were type specific for EIAV, showing no reactivity with other retroviruses, including M-MuLV, FeLV, BLV, and M7. These MAb were further used to demonstrate that the viruses produced in canine (Cf2Th) and feline (FEA) cell cultures, which we developed, were identical to the equine (EFK) derived EIAV, as confirmed by several other biochemical and immunological criteria.

13. BIOCHEMICAL AND BIOLOGICAL ANALYSIS OF VENEZUELAN EQUINE ENCEPHALOMYELITIS VIRUSES (VEE) USING MONOCLONAL ANTIBODIES.

John T. Roehrig, James H. Mathews, and Richard M. Kinney. Vector-Borne Viral Diseases Division Center for Infectious Diseases, CDC, PHS, Fort Collins, Colorado 80522.

Monoclonal antibodies directed against the surface glycoproteins of TC-83 virus were used to analyze the antigenic structure expressed on the virion surface. Three epitopes were identified on the gp56 and four epitopes were identified on the gp50 using a cross-reactivity assay with closely related VEE viruses and an ELISA competitive binding assay (CBA). Only antibodies directed against the gp56^c epitope effectively blocked viral hemagglutination and infectivity. Anti-gp50^b antibodies had very low levels of in vitro biologic activity. This activity was probably due to the close spatial arrangment this epitope has with the gp56^c epitope as demonstrated in the CBA. Antibodies directed against each epitope were used as a source for passive immunization of 3-week-old mice. Protection against subsequent VEE infection could be correlated to both epitope specificity and the ability to fix complement. The most protective epitopes were the gp56^c and gp50^b. As little as 5 μg of antibody administered i.v. could protect animals. Those animals that survived challenge demonstrated good anti-viral antibody when tested 25 days post-infection. Viremia studies indicated that the virus was replicating in the spleen, however little virus could be detected in the brain. Antibody fragment studies indicated that the Fc portion was important for in vivo protection.

14. MONOCLONAL ANTIBODIES SPECIFIC FOR HERPES SIMPLEX VIRUS TYPE 1 MEDIATE ANTIVIRAL EFFECTS IN VITRO AND IN VIVO.

James T. Rector, Robert N. Lausch, and John E. Oakes. University of South Alabama, Mobile, Alabama 36688.

Nine hybrid cell lines derived from four independent cell fusions produce monoclonal antibodies specific for herpes simplex virus type 1 (HSV-1). The reactivity of each monoclonal antibody was characterized by enzyme-linked immunoabsorbent assay (ELISA), virus neutralization, complement-dependent cytolysis (CDC) and antibody-dependent cellular cytotoxicity (ADCC). Results of these tests showed that ascites fluid from two of the nine lines (D8 and G8) neutralized virus infectivity. Hybridoma line G8 was also effective in mediating CDC and ADCC. Both lines were further tested for their ability to protect mice from lethal ocular infection with HSV-1. Four week old Balb/c mice infected ocularly with HSV-1 develop viral encephalitis and die in 10-12 days. Passive transfer of hyperimmune rabbit anti-HSV-1 serum does protect the host against virus challenge when given within 48 hours postinfection. Preliminary experiments have shown that passive transfer of monoclonal antibody from both G8 and D8 hybridomas independently confers protection against fatal herpetic encephalitis.

15. MONOCLONAL ANTIBODIES AGAINST ANTIGENS OF AMASTIGOTES AND EPIMASTIGOTES OF TRYPANOSMA CRUZI.

S. D. Sharma, F. G. Araujo, V.Tsai, P. Cox and J. S. Remington. Research Institute, Palo Alto, California 94301.

As part of our ongoing effort to isolate pure antigens of T. cruzi, we report here the development and characterization of monoclonal antibodies to antigens of amastigotes and epimastigotes of the parasite. Seventeen derived from mice immunized with the amastigote preparations, five produced antibody directed against amastigote antigens only and six produced antibody directed against antigens common to amastigotes and epimastigoes. Ten monoclonal antibodies were of the IgM class, and one that recognized a common antigen of amastigotes and epimastogotes was of the IgG_2 subclass. Six antibody-producing clones have been obtained from mice immunized with the epimastigote preparation; four of them are of the IgM class and recognize an antigen common to epimastigotes and amastigotes as determined by enzyme immunoassay and indirect immunofluorescent assay, whereas one of the IgG_1 and one of the IgG_2 subclass are directed against an antigen specific for epimastigotes.

Supported by grants from UNDP/World Bank/WHO Special Programme for Research and Training in Tropical Diseases, the Rockefeller Foundation, and NIH.

16. DIFFERENTIAL EFFECTS OF FOUR MONOCLONAL ANTIBODIES TO HUMAN β_2-MICROGLOBULIN ON IN VITRO LYMPHOCYTE FUNCTIONS.

S. A. Stern, R. A. Curry, and R. P. Messner. Department of Medicine, University of Minnesota, Minneapolis, MN. 55455.

Five clones secreting monoclonal antibodies to β_2-microglobulin (β_2m) were produced by cell fusion between the mouse myeloma P3-NS1-1-AG4-1 and spleen cells from BALB/c mice immunized with β_2m. Two of the monoclonal antibodies are IgG_1, two are IgG_2 and one is IgG_3. Three carry κ light chains and two carry λ chains. All five react with β_2m in ELISA assay. All except one react in indirect immunofluorescence assays with cell line 1788. None react with Daudi, a cell line which lacks cell-surface β_2m.

The antibodies are neither cytotoxic nor mitogenic when incubated with human peripheral blood monocytes (PBM's). They have been tested for their ability to block antigen and mitogen-induced proliferation and the one-way mixed lymphocyte reaction (MLR) and to inhibit E-rosette formation. Monoclonal antibodies 005 and 021 inhibit ConA and pokeweed mitogen-induced mitogenesis by 40-50% at a concentration of 3 g/ml. Antibody 003 inhibits this mitogenesis by 50% at 300 ng/ml. Mitogenesis induced by tetanus toxoid was inhibited by antibodies 003, 005, 006, and 021 by 60-70% at a concentration of 3 μg/ml. The one-way MLR was inhibited 32% by 005 at a concentration of 500 ng/ml, and 44% by 006 at the same concentration. Monoclonal 006 blocked the formation of E-rosettes by 35% at 70 ng/ml, and at 70 μg/ml 021 blocked rosetting by 56%.

Whether the differences among these antibodies are related to specificity for different epitopes, different affinities, or isotypes is under investigation.

17. SELECTIVE INHIBITION OF FUNCTIONAL SITES ON CELL BOUND HUMAN C3b BY MONOCLONAL ANTIBODIES.

J. D. Tamerius, M. K. Pangburn and H. J. Muller-Eberhard.
Research Institute of Scripps Clinic, La Jolla, California 92037.

Monoclonal antibodies (MoAb) to human C3b were generated from hybridomas produced by somatic cell fusion of splenocytes from C3b immune mice and murine myeloma cells. Several hundred positive clones from two separate fusions were detected by hemagglutination, RIA or ELISA. Four MoAb, which were shown to inhibit specific binding sites of C3b, have been examined in detail. Three MoAb, P-4, P-31 and X-163, blocked the binding of Factor P to EC3b cells by 96%, 84% and 92% and Factor B by 91%, 80% and 52%, respectively. Fab prepared from P-31 and X-163 similarly inhibited B and P binding. A fourth MoAb, X-14, blocked Factor H binding to EC3b by 76%, but B and P binding by less than 20%. MoAb X-14 (Factor H blocker) did not inhibit the binding of MoAb P-31 (Factor B and P blocker) to EC3b despite the fact that both antibodies bind to the C3c domain of C3, as shown in binding competition experiments using highly purified C3c and radiolabeled antibodies. Fab prepared from X-14 was not as efficient as the intact MoAb in blocking the binding of Factor H to EC3b (36% maximum inhibition). These results show that it is possible to interfere with one functional binding site on C3b without impairing other sites and that, using carefully defined MoAb, it may be possible to map topographically the C3b molecule with respect to its functional sites and immunochemical domains. (Supported by an Arthritis Fndn. Fellowship, Amer. Heart Assoc. E.I. No. 81-225 and USPHS Grants AI 17354, CA 27489 and HL 16411.)

18. GENERATION OF MONOCLONAL ANTIBODIES REACTING WITH HUMAN NORMAL AND NEOPLASTIC MAMMARY DUCT EPITHELIAL CELLS.

S. Menard, M. I. Colnaghi, E. Tagliabue, S. Canevari, G.Fossati, S. Sonnino. Istituto Nasionale Tumori, and Instituto Chimica Biologica Universita, Milano Italy.

A monoclonal antibody was obtained from a hybridoma line generated by the fusion of P3 myeloma cells with spleen cells of a mouse immunized with cell membrane extracts from the human breast cancer cell line MCF-7. The specificity was studied by isotopic immunoglobulin assay on several cell lines and cell membrane extracts from various surgical specimens and milk cells; by membrane immunofluorescence on cell lines and cell suspensions obtained from fresh tumors; by immunofluorescence and immunoperoxidase technique on histological slices from about 100 different normal or neoplastic tissues. The structure recognized by the monoclonal antibody was found on normal epithelial cells of the mammary duct and on 80% of the 50 breast carcinomas tested. The antigenic activity was unaffected by treatment with trypsin, V. C. neuraminidase and by heating the cells at 100°C, but was inactivated by treatment with methanol. When lipid extracts for MCF-7 membrane preparations were chromatographed on DEAE, Sephadex the antigenic activity was recovered only on the neutral ganglioside fraction.

19. RESOLUTION OF HUMAN LIVER MAO A AND B USING A MONOCLONAL ANTIBODY IMMUNOAFFINITY COLUMN.

Richard M. Denney, Richard R. Fritz, Nutan T. Patel and Creed W. Abell. Department of Human Biological Chemistry and Genetics, University of Texas Medical Branch, Tx.

Two forms of MAO which have long been distinguished by their substrate specificities can be separated using a newly isolated mouse monoclonal antibody. The antibody was elicited to human platelet MAO (B) which had been inactivated and radiolabeled by reaction with the MAO inhibitor, ^{3}H-pargyline. The antibody binds both ^{3}H-pargyline-labeled, inactivated MAO B and unlabeled, active MAO B, but does not precipitate them. Binding of the antibody to the enzyme does not inhibit catalytic activity. Attachment of mouse monoclonal anti-MAO B to Sepharose or Affigel supports results in a resin which binds 95% of MAO B, but no MAO A from crude mitochrondrial extracts of human liver. The only protein eluted from these columns with KSCN migrates in SDS-polyacrylamide gels as expected for MAO B, suggesting that binding to the columns is highly specific. The anti-MAO B immunoaffinity resin fails to bind MAO A from human placenta. In addition, the resin fails to bind MAO B from mouse liver, suggesting that the antibody may be useful in gene mapping studies of human MAO B in mouse-human somatic cell hybrids. (Supported by Grant #MH-34757 from the National Institute of Mental Health.)

20. ISOLATION OF ANTIBODIES THAT ARE SPECIFIC FOR HUMAN CARBOXYPEPTIDASE B BUT NOT FOR THE ZYMOGEN PROCARBOXYPEPTIDASE B BY AFFINITY CHROMATOGRAPHY.

Ann S. Delk and Corey Largman, UC Davis School of Medicine, Davis, CA 94616 and VA Medical Center, Martinez, CA 94553.

A major theory for the pathogenesis of acute pancreatitis is that the normally inactive precursors of pancreatic proteases become activated, with subsequent destruction of enzymes, polypeptide hormones and other proteins in tissue and blood. To investigate the role of proteases in the disease, we developed radioimmunoassays for a number of these enzymes. However, experiments with serum samples fractionated by gel filtration showed that the amount of immunoreactive protein is not directly related to the amount of active enzyme against which the antiserum was directed. For example, procarboxypeptidase B effectively crossreacts about 30% with carboxypeptidase B using antiserum to active enzyme. To obtain antibodies that are specific for active enzyme, we treated the antiserum with resin containing immobilized procarboxypeptidase B. Stepwise addition of the resin produced antiserum with progressively less reactivity towards the zymogen relative to the enzyme, with the final material having less than 1% crossreactivity. The titer of enzyme-specific antiserum was 1% that of the original; however, it could be estimated that 10% of the original antibodies were specific for the active enzyme. To increase the yield of enzyme-specific antibodies, we are currently trying to obtain a mouse hybridoma that produces the desired antibody. (Supported by the Medical Research Service of the Veterans Administration and grants from the NIH to C.L. and the Council of Tobacco Research to Dr. M. C.Geokas.)

21. DEXAMETHASONE EFFECTS ON HYBRIDOMA FORMATION.

C. Feit, A Bartal, M. Andreeff and Y. Hirshaut. Memorial Sloan-Kettering Cancer Center, New York, N.Y. 10021.

To assess the effects of sterioids on the formation of hybridomas, sheep red blood cell (SRBC) immunized spleen cells from Balb/c mice were fused with P_3U_1 myeloma cells. After fusion, cells were grown for 1 week in HAT medium containing 10^{-3}, 10^{-5}, 10^{-7}, 10^{-9} mM dexamethasone

(DM) or HAT alone. Subsequently spent media was removed and refeeding with HAT was commenced. Parameters measured were: clone number, size, antibody production and DNA content. At concentrations of DM given above, the number of clones produced was 16, 39, 24, 29 respectively as compared to 40 clones in control wells. At 10^{-3} and 10^{-5} mM DM parental cells were markedly suppressed. All clones were < 1 mm in size at 10^{-3} mM, whereas at 10^{-5} mM and in controls, 43% and 46% of the clones respectively were 1-4 mm. By micro hemagglutination it was found that only 5% of the clones at 10^{-3} mM and 10% at 10^{-5} mM produced anti-SRBC antibody. In contrast, at lower doses and in control plates 99% of clones were antibody producers. Computerized DNA-RNA flow cytometry using acridine orange indicated that DNA loss was more pronounced in clones exposed to DM than controls. This method also indicated a lower number of parental cells in cultures grown in the presence of DM. While clone number and size were not affected by 10^{-5} mM DM, the clones obtained were clearly different in that they did not produce antibodies. Further studies are needed to establish whether this reflects selected pressures on the type of cells fusing, suppressive effects on antibody production, or selective chromosome loss.

22. STUDIES ON THE MHC OF THE RAT: EVIDENCE FOR MORE THAN ONE CLASS I ANTIGENS USING MONOCLONAL ANTIBODIES.

Dhirendra N. Misra, Saad A. Noeman, Heinz W. Kunz and Thomas J. Gill III. Department of Pathology, University of Pittsburgh, School of Medicine, Pittsburgh, PA 15261.

Hybridomas secreting antibodies to rat class I antigens were produced using lymphocytes from WF rats immunized against the DA strain and P3-X63-Ag8.653 myeloma cells. A number of clones from 5 different hybridomas was generated and 5 monoclonal antibodies (mAbs), one from each clone, were used for the immunochemical study. All 5 mAbs precipitated molecules having 2 components: 45,000 and 12,000 daltons molecular weight. One mAb, 3-5-118(A-2), precipitated a unique species of molecules. The supernatant reacted with the mAb 3-1-3(B-1), 3-3-56(A-24) or 3-3-60(C-14) and precipitated a second set of molecules. The resulting supernatant when reacted with the mAb 2-1-155(G-4) or the WF anti-DA alloantiserum precipitated a third set of molecules. In reciprocal experiments, none of the mAbs could precipitate any molecule from the supernatant from the antigen preparation which was first reacted with the mAb 2-1-155(G-4). When the antigen preparation was first reacted with 3-1-3(B-1), 3-3-56(A-24) or 3-3-60(C-14), the supernatant did not contain any antigen reactive with 3-5-118(A-2), but it contained molecules reactive with 2-1-155(G-4). The results thus give evidence that there are 3 Class I molecules: all 3 possess a common determinant reactive with 2-1-155(G-4); 2 of them share determinants reactive with 3-1-3(B-1), 3-3-56(A-24) or 3-3-60(C-14). One of the last two classes of molecules also contains a determinant reactive with 3-5-118(A-2). Supported by CA 18659.

23. AN IMMUNOHISTOCHEMICAL METHOD FOR LARGE SCALE SPECIFICITY SCREENING OF MONOCLONAL ANTIBODIES.

D. Sedmak, A. Fishleder, R. Tubbs. Cleveland Clinic, Cleveland, Ohio 44106.

A study using mouse monoclonal antibodies was undertaken to correlate immunohistochemical tissue staining with serum antigen levels in patients with various pathologic conditions. This necessitated the development of a method permitting simultaneous immunostaining of large numbers of tissue sections.

Utilizing a mouse monoclonal antibody (19-9) directed at a colorectal carcinoma, we stained tissue sections from 15 surgical cases. All sections were placed in a multislide staining holder and immersed in appropriate dilutions of monoclonal antibody in PBS followed by immersion in solutions of biotinylated antimouse IgG, and Avidin DH-Biotin Perixidase Complex. The chromagen used was 3-amino-9-ethylcarbazole. Three washes in PBS were performed between each step. The results were compared with the staining of duplicate slides incubated flat and developed individually with identical solutions in a moist incubation chamber. Intensity and distribution of tissue staining of paired slides from individual cases was identical by the two procedures.

The use of this batch technique facilitates mass immunohistochemical staining of tissue sections using monoclonal antibodies and is consequently more time effective.

24. MONOCLONAL ANTIBODIES TO CHICKEN HEMOPOIETIC-LYMPHOID CELLULAR DIFFERENTIATION ANTIGENS.

Kimberly Kline, James P. Allison, Jo Anne Lund, and Bob G. Sanders. Dept. of Zoology, University of Texas, Austin, Texas 78712 and the University of Texas System Cancer Center, Science Park, Smithville, TX. 78957.

Characterization of chicken fetal antigen (CFA) using polyclonal antibodies reveal a minimum of 13 CFA determinants localized to hemopoietic-lymphoid cells. CFA determinants are differentially expressed during erythroid cellular differentiation. Splenic lymphocytes from Balb/c mice immunized with SC strain chicken hatch-one (Hl) RBC were hybridized to P3 myeloma cells. The specificity of the antibodies secreted by the hybridomas for CFA was determined by hemagglutination assays. Supernatants from hybrid cells that did not agglutinate RBC from adult chickens but did agglutinate RBC from Hl chickens revealed 84 hybrid cells secreting antibodies reactive to Hl RBC. Five hybridoma clones have been established that secrete monoclonal antibodies (McAb) specific for Hl RBC. Although none of the monoclonal antibodies have given reaction patterns that correlate with any of the 13 CFA determinants detected by polyclonal antisera, differing patterns of antigen expression on peripheral RBC, bursa and thymus cells from different aged embryonic and developing chickens have been observed. Antigens reactive with the McAb were isolated from extracts of ^{125}I-labeled erythroid cell populations and AEV transformed erythroid cells (before and after chemically induced differentiation) and were analyzed by SDS-PAGE. (Supported by NIH Grants CA 12851, BGS; and CA26321 and CA 26981, JPA.)

AUTHOR INDEX

SUBJECT INDEX